Colloid and Interface Science

VOL. II

Aerosols, Emulsions, and Surfactants

Academic Press Rapid Manuscript Reproduction

Proceedings of the International Conference
on Colloids and Surfaces—50th Colloid and Surface Science Symposium,
held in San Juan, Puerto Rico
on June 21–25, 1976

Colloid and Interface Science

Science

VOL. II
Aerosols, Emulsions, and Surfactants

EDITED BY

MILTON KERKER

Clarkson College of Technology
Potsdam, New York

Academic Press Inc.

New York   San Francisco   London   1976
A Subsidiary of Harcourt Brace Jovanovich, Publishers

ACADEMIC PRESS, INC.
111 Fifth Avenue, New York, New York 10003

United Kingdom Edition published by
ACADEMIC PRESS, INC. (LONDON) LTD.
24/28 Oval Road, London NW1

Library of Congress Cataloging in Publication Data

International Conference on Colloids and Surfaces,
    50th, San Juan, P.R., 1976
    Colloid and interface science.

    CONTENTS: v. 2.   Aerosols,
emulsions, and surfactants.–v. 3.   Adsorption,
catalysis, solid surfaces, wetting, surface tension,
and water.–v. 4.   Hydrosols and rheology. [etc.]
    1.   Colloids–Congresses.   2.   Surface chemistry
–Congresses.   I.   Kerker, Milton.   II.   Title.
QD549.I6   1976   541'.345   76-47668
ISBN 0–12–404502–2   (v. 2)

Contents

CONTENTS

List of Contributors

R. Agarwal, Department of Chemistry, Drexel University, Philadelphia, Pennsylvania 19104

J. Allen, Air Pollution Research Laboratory, Department of Mechanical Engineering, Washington University, St. Louis, Missouri 63130

Hiroyuki Akusu, Science University of Tokyo, Department of Chemistry, Faculty of Science, Kagurazaka, Shinjuku-ku, Tokyo, Japan

R.J. Anderson, Graduate Center for Cloud Physics Research, University of Missouri-Rolla, Rolla, Missouri 65401

David L. Bartley, Space Sciences Research Center-Rolla, Graduate Center for Cloud Physics Research, University of Missouri-Rolla, Rolla, Missouri 65401

S.H. Bauer, Department of Chemistry, Cornell University, Ithaca, New York 14853

David Z. Becher, Department of Chemistry, Lehigh University, Bethlehem, Pennsylvania 18015

Raghupathy Bollini, Engineering and Technology, Southern Illinois University at Edwardsville, Edwardsville, Illinois 62026

C. Boned, Laboratoire de Thermodynamique, Institut Universitaire de Recherche Scientifique, Université de Pau et des Pays de l'Adour, B.P. 523 Pau-Université, 64010 Pau, France

R. Buscall, Pharmacy Department, University of Nottingham, University Park, Nottingham NG7 2RD, England

C.J. Cante, General Foods Corporation, Research Department, 1551 East Willow Street, Kankakee, Illinois 60901

A.W. Castleman, Jr., Cooperative Institute for Research in Environmental Sciences, University of Colorado/NOAA and Department of Chemistry, University of Colorado, Boulder, Colorado 80309

K.S. Chan, Department of Chemical Engineering and Anesthesiology, University of Florida, Gainesville, Florida 32611

M. Chiang, Department of Chemical Engineering and Anesthesiology, University of Florida, Gainesville, Florida 32611

K. Choudhury, Pharmacy Department, University of Nottingham, University Park, Nottingham NG7 2RD, England

T.S. Chow, J. C. Wilson Center of Technology, Xerox Corporation, 114, Webster, New York 14580

M. Clausse, Laboratoire de Thermodynamique, Institut Universitaire de Recherche Scientifique, Université de Pau et des Pays del'Adour, B.P. 523 Pau-Université, 64010 Pau, France

Robert W. Coughlin, Department of Chemical Engineering, Lehigh University, Bethlehem, Pennsylvania 18015

E. James Davis, Department of Chemical Engineering, Clarkson College, Potsdam, New York 13676

S.S. Davis, Pharmacy Department, University of Nottingham, University Park, Nottingham NG7 2RD, England

H.T. DelliColli, Chemical Laboratory, Westvaco Corporation, Polychemicals Department, P.O. Box 5207, North Charleston, South Carolina 29406

Edward A. Dennis, Department of Chemistry (M-016), University of California at San Diego, LaJolla, California 92093

Lj. A. Despotović, Laboratory of Colloid Chemistry, "Ruder Boskovic" Institute, 41001 Zagreb, P.O. Box 1016, Croatia, Yugoslavia

R. Despotović, Laboratory of Colloid Chemistry, "Ruder Boskovic" Institute, 41001 Zagreb, P.O. Box 1016, Croatia, Yugoslavia

Pavel Ditl, Department of Chemical Engineering, Lehigh University, Bethlehem, Pennsylvania 18015

Ralph Dlugi, Institut für Meteorologie, Johannes Gutenberg-Universität Basel, Klingelbergstrasse 80, 4056 Basel, Switzerland

Shosuke Ebina, Tokyo University, Chemical Research Institute of Non-aqueous Solutions, 2-1-1 Katahira, Sendai 980, Japan

Hans-Friedrich Eicke, Physikal.-Chem. Institut der Universität Basel, Klingelbergstrasse 80, 4056 Basel, Switzerland

C.G. Essex, Dielectrics Group, Physics Department, Queen Elizabeth College, University of London, Campden Hill Road, Kensington, London W8 7AH, England

K.C. Fan, Department of Chemical Engineering, University of Maryland, College Park, Maryland 20742

N. Filipović-Vinceković, Laboratory of Colloid Chemistry, "Ruder Boskovic" Institute, 41001 Zagreb, P.O. Box 1016, Croatia, Yugoslavia

Fredrick M. Fowkes, Department of Chemistry, Lehigh University, Bethlehem, Pennsylvania 18015

D.L. Fox, The School of Public Health, Department of Environmental Sciences and Engineering, The University of North Carolina at Chapel Hill, Chapel Hill, North Carolina 27514

Stig Friberg, Department of Chemistry, University of Missouri, Rolla, Missouri 65401

David J. Frurip, Department of Chemistry, Cornell University, Ithaca, New York 14853

Felícita Garcia, Catholic University of Puerto Rico, Ponce, Puerto Rico 00731

J.W. Gentry, Department of Chemical Engineering, University of Maryland, College Park, Maryland 20742

W. Gerbacia, Chevron Oil Field Research Company, Box 446, LaHabra, California 90631

James R. Griffith, Organic Chemistry Branch, Chemistry Division, Naval Research Laboratory, Washington, D.C. 20375

Hans-Heinrich Grünhagen, Institut Pasteur, Neurobiologie Moleculaire, 25, Rue du Docteur Roux, F - 75015 Paris, France

Hans Gustavsson, Division of Physical Chemistry 2, The Lund Institute of Technology, Chemical Center, Lund, Sweden

Gottfried Hänel, Institut für Meteorologie, Johannes Gutenberg-Universität, Mainz, Germany

Marju-Ritta Hakala, Department of Physical Chemistry, Abo Akademi, Porthansgatan 305, 20500 Abo (Turku) 50, Finland

B.N. Hale, Department of Physics and Cloud Physics Research Center, University of Missouri-Rolla, Rolla, Missouri 65401

C. Hermansky, Department of Chemistry, Drexel University, Philadelphia, Pennsylvania 19104

Kameo Hirai, Tokyo University, Chemical Research Institute of Non-aqueous Solutions, 2-1-1 Katahira, Sendai 980, Japan

V. Horvat, Laboratory of Colloid Chemistry, "Ruder Boskovic" Institute, 41001 Zagreb, P.O. Box 1016, Croatia, Yugoslavia

W.C. Hsieh, Departments of Chemical Engineering and Anesthesiology, University of Florida, Gainesville, Florida 32611

R.B. Husar, Air Pollution Research Laboratory, Department of Mechanical Engineering, Washington University, St. Louis, Missouri 63130

Gabriel A. Infante, Catholic University of Puerto Rico, Ponce, Puerto Rico 00731

Bernice Irizarry, Catholic University of Puerto Rico, Ponce, Puerto Rico 00731

Gloria Jové, Catholic University of Puerto Rico, Ponce, Puerto Rico 00731

J.L. Kassner, Jr., Graduate Center for Cloud Physics Research, University of Missouri-Rolla, Rolla, Missouri 65401

Milton Kerker, Clarkson College of Technology, Potsdam, New York 13676

J. Kiefer, Department of Physics and Cloud Physics Research Center, University of Missouri-Rolla, Rolla, Missouri 65401

Koji Kihara, Department of Engineering Chemistry, Nagoya Institute of Technology, Gokiso, Nagoya 466, Japan

Oh-Kil Kim, Organic Chemistry Branch, Chemistry Division, Naval Research Laboratory, Washington, D.C. 20375

Charles A. Knight, National Center for Atmospheric Research, P.O. Box 1470, Boulder, Colorado 80302

M.R. Kuhlman, The School of Public Health, Department of Environmental Sciences and Engineering, The University of North Carolina at Chapel Hill, Chapel Hill, North Carolina 27514

J. Lee, Department of Chemical Engineering, University of Maryland, College Park, Maryland 20742

L.A. Liljedahl, Beltsville Research Center, Department of Agriculture, University of Maryland, College Park, Maryland 20742

Israel J. Lin, Department of Mineral Engineering, Technion-Israel Institute of Technology, Haifa, Israel

Björn Lindman, Division of Physical Chemistry 2, The Lund Institute of Technology, Chemical Center, Lund, Sweden

Julián López, Catholic University of Puerto Rico, Ponce, Puerto Rico 00731

Jer Ru Maa, Department of Chemical Engineering, University of California at Berkeley, Berkeley, California 94720 (*Present Address*); Chemical Engineering Department, National Cheng Kung University, Tainan, Taiwna, The Republic of China

R.A. Mackay, Department of Chemistry, Drexel University, Philadelphia, Pennsylvania 19104

M. MacNeil, Memorial University of Newfoundland, St. John's, Newfoundland, Canada

W.H. Marlow, Department of Applied Science, Brookhaven National Laboratory, Associated Universities, Inc., Upton, Long Island, New York 11973

Michael J. Matteson, School of Chemical Engineering, Georgia Institute of Technology, Atlanta, Georgia 30332

D. Mayer, Laboratory of Colloid Chemistry, "Ruder Boskovic" Institute, 41001 Zagreb, P.O. Box 1016, Croatia, Yugoslavia

Kenjiro Meguro, Science University of Tokyo, Department of Chemistry, Faculty of Science, Kagurazaka, Shinjuku-ku, Tokyo, Japan

R.C. Miller, Graduate Center for Cloud Physics Research, University of Missouri-Rolla, Rolla, Missouri 65401

H.R. Munkelwitz, Department of Applied Science, Brookhaven National Laboratory, Upton, New York 11973

Tetsutaro Nakamura, Science University of Tokyo, Department of Chemistry, Faculty of Science, Kagurazaka, Shinjuku-ku, Tokyo, Japan

Juan Negrón, Catholic University of Puerto Rico, Ponce, Puerto Rico 00731

Hirofumi Okabayshi, Department of Engineering Chemistry, Nagoya Institute of Technology, Gokiso, Nagoya 466, Japan

Masataka Okuyama, Department of Engineering Chemistry, Nagoya Institute of Technology, Gokiso, Nagoya 466, Japan

J.Th.G. Overbeek, Van't Hoff Laboratory, University of Utrecht, Padualaan 8, Utrecht, The Netherlands

P.L.M. Plummer, Department of Physics and Cloud Physics Research Center, University of Missouri-Rolla, Rolla, Missouri 65401

T.S. Purewal, Pharmacy Department, University of Nottingham, University Park, Nottingham NG7 2RD, England

E.K. Ralph, Memorial University of Newfoundland, St. John's, Newfoundland, Canada

P.C. Reist, The School of Public Health, Department of Environmental Sciences and Engineering, The University of North Carolina at Chapel Hill, Chapel Hill, North Carolina 27514

Anthony A. Ribeiro, Department of Chemistry (M-016), University of California at San Diego, LaJolla, California 92093

H.L. Rosano, The City College of the City University of New York, New York, New York 10031

Jarl B. Rosenholm, Department of Physical Chemistry, Abo Akademi, Porthansgatan 305, 20500 Abo (Turku) 50, Finland

R. Royer, Department de Mathematiques, Institut Universitaire de Recherche Scientifique, Universite de Pau et des Pays de l'Adour, B.P. 523 Pau-Universite, 64010 Pau, France

Donn N. Rubingh, The Procter & Gamble Company, Miami Valley Laboratories, P.O. Box 39175, Cincinnati, Ohio 45247

Fouad Z. Saleeb, General Foods Corporation, Technical Center, Tarrytown, New York 10591

Steven B. Sample, University of Nebraska, Lincoln, Nebraska

J.M. Sangster, Department of Chemical Engineering, Ecole Polytechnique, Campus University of Montreal, Montreal, H3C 3A7, Quebec, Canada

Tomoo Satahe, Science University of Tokyo, Department of Chemistry, Faculty of Science, Kagurazaka, Shinjuku-ku, Tokyo, Japan

Timothy W. Schenz, General Foods Corporation, Technical Center, Tarrytown, New York 10591

H.P. Schreiber, Department of Chemical Engineering, Ecole Polytechnique, Campus University of Montreal, Montreal, H3C 3A7, Quebec, Canada

John H. Seinfeld, Department of Chemical Engineering, California Institute of Technology, Pasadena, California 91125

R.E. Shaffer, Edgewood Arsenal, Maryland 21010

D.O. Shah, Department of Chemical Engineering and Anesthesiology, University of Florida, Gainesville, Florida 32611

Steven A. Shaya, Research and Development Department, The Procter and Gamble Company, Miami Valley Laboratories, P.O. Box 39175, Cincinnati, Ohio 45247

R.J. Sheppard, Dielectrics Group, Physics Department, Queen Elizabeth College, University of London, Campden Hill Road, Kensington, London W8 7AH, England

Keiichi Shioya, Science University of Tokyo, Department of Chemistry, Faculty of Science, Kagurazaka, Shinjuku-ku, Tokyo, Japan

Marek Sitarski, Department of Chemical Engineering, California Institute of Technology, Pasadena, California 91125

A. Smith, Pharmacy Department, University of Nottingham, University Park, Nottingham NG7 2RD, England

E.M. Stein, Department of Physics and Cloud Physics Research Center, University of Missouri-Rolla, Rolla, Missouri 65401

Per J. Stenius, Department of Physical Chemistry, Abo Akademi, Porthansgatan 305, 20500 Abo (Turku) 50, Finland

B. Subotić, Laboratory of Colloid Chemistry, "Ruder Boskovic" Institute, 41001 Zagreb, P.O. Box 1016, Croatia, Yugoslavia

Yasukatsu Tamai, Chemical Research Institute of Non-aqueous Solutions, Tohoku University, Sendai 980, Japan

I.N. Tang, Department of Applied Science, Brookhaven National Laboratory, Upton, New York 11973

R.J.M. Tausk, Van't Hoff Laboratory, University of Utrecht, Padualaan 8, Utrecht, The Netherlands

O. Theimer, Research Professor of Physics, New Mexico State University, A&S Research Center, Box RC, Las Cruces, New Mexico 88003

Minoru Ueno, Science University of Tokyo, Department of Chemistry, Faculty of Science, Kagurazaka, Shinjuku-ku, Tokyo, Japan

Jack Wagman, Emissions Measurement & Characterization Division, Environmental Sciences Research Laboratory, United States Environmental Protection Agency, Research Triangle Park, North Carolina 27711

Paul E. Wagner, Clarkson College of Technology, Potsdam, New York 13676 (*Present address*) I. Physikalisches Institut der Universität Wien, Strudelhofgasse 4, A-1090 Vienna, Austria

J.H. Whittam, Gillette Company, Gillette Park, Boston, Massachusetts 02106

C.T. Wilder, Department of Chemical Engineering, University of Maryland, College Park, Maryland 20742

Thomas L. Wills, School of Chemical Engineering, Georgia Institute of Technology, Atlanta, Georgia 30332

Wen Hai Wu, Chemical Engineering Department, National Cheng Kung University, Tainan, Taiwna, The Republic of China

Brian Yates, Unilever Research, Port Sunlight, Wirral Merseyside, England

Preface

This is the second volume of papers presented at the International Conference on Colloids and Surfaces, which was held in San Juan, Puerto Rico, June 21–25, 1976.

The morning sessions consisted of ten plenary lectures and thirty-four invited lectures on the following topics: rheology of disperse systems, surface thermodynamics, catalysis, aerosols, water at interfaces, stability and instability, solid surfaces, membranes, liquid crystals, and forces at interfaces. These papers appear in the first volume of the proceedings along with a general overview by A. M. Schwartz.

The afternoon sessions were devoted to 221 contributed papers. This volume includes contributed papers on the subjects of aerosols, emulsions, and surfactants. Three additional volumes include contributed papers on adsorption, catalysis, solid surfaces, wetting, surface tension, water, hydrosols, rheology, biocolloids, polymers, monolayers, membranes, and general subjects.

The conference was sponsored jointly by the Division of Colloid and Surface Chemistry of the American Chemical Society and the International Union of Pure and Applied Chemistry in celebration of the 50th Anniversary of the Division and the 50th Colloid and Surface Science Symposium.

The National Colloid Symposium originated at the University of Wisconsin in 1923 on the occasion of the presence there of The Svedberg as a Visiting Professor (see the interesting remarks of J. H. Mathews at the opening of the 40th National Colloid Symposium and also those of Lloyd H. Ryerson in the Journal of Colloid and Interface Science 22, 409, 412 (1966)). It was during his stay at Wisconsin that Svedberg developed the ultracentrifuge, and he also made progress on moving boundary electrophoresis, which his student Tiselius brought to fruition.

The National Colloid Symposium is the oldest such divisional symposium within the American Chemical Society. There were no meetings in 1933 and during the war years 1943–1945, and this lapse accounts for the 50th National Colloid Symposium occurring on the 53rd anniversary.

However, these circumstances brought the numerical rank of the Symposium into phase with the age of the Division of Colloid and Surface Chemistry. The Division was established in 1926, partly as an outcome of the Symposium. Professor Mathews gives an amusing account of this in the article cited above.

The 50th anniversary meeting is also the first one bearing the new name

Colloid and Surface Science Symposium to reflect the breadth of interest and participation.

There were 476 participants including many from abroad.

This program could not have been organized without the assistance of a large number of persons and I do hope that they will not be offended if all of their names are not acknowledged. Still, the Organizing Committee should be mentioned: Milton Kerker, Chairman, Paul Becher, Tomlinson Fort, Jr., Howard Klevens, Henry Leidheiser, Jr., Egon Matijevic, Robert A. Pierotti, Robert L. Rowell, Anthony M. Schwartz, Gabor A. Somorjai, William A. Steele, Hendrick Van Olphen, and Albert C. Zettlemoyer.

Special appreciation is due to Robert L. Rowell and Albert C. Zettlemoyer. They served with me as an executive committee which made many of the difficult decisions. In addition Dr. Rowell handled publicity and announcements while Dr. Zettlemoyer worked zealously to raise funds among corporate donors to provide travel grants for some of the participants.

Teresa Adelmann worked hard and most effectively both prior to the meeting and at the meeting as secretary, executive directress, editress, and general overseer. She made the meeting and these Proceedings possible. We are indebted to her.

Milton Kerker

EVALUATION OF THE CLASSICAL THEORY OF NUCLEATION
USING EXPANSION CHAMBER MEASUREMENTS OF THE
HOMOGENEOUS NUCLEATION RATE OF WATER FROM THE
VAPOR*

R. C. Miller, R. J. Anderson and J. L. Kassner, Jr.
University of Missouri-Rolla

ABSTRACT

Extensive expansion cloud chamber measurements of the homogeneous nucleation rate of liquid water from the vapor are used in a comprehensive test of the classical liquid drop theory. Homogeneous nucleation was measured over a wide range of rates (50 to $3 \cdot 10^5$ drops/$cm^3 \cdot sec$) and temperatures (-44 to $+17^{o}C$). This data base allows the temperature dependence of the condensation coefficient and surface tension to be determined by fitting the theory to the data.

Although the surface tension so determined differs from the extrapolated table values for bulk water by a maximum of about 12%, the divergence is greatest at the highest temperatures where larger values of the critical cluster size are predicted and where one would expect that bulk values would be approached. The condensation coefficient varies from $\sim 10^{-3}$ at $-40^{o}C$ to $\sim 10^{-9}$ near room temperature. The precipituous decrease to extremely low values at the higher temperatures appears to indicate that the classical theory suffers from a deficiency in its basic temperature dependency. Suggestion is made that further work on the classical theory might involve a revision of the reaction kinetics by allowing the nucleation current to flow through cluster configurations other than that of the minimum free energy using input data from molecular models of the clustering process.

I. INTRODUCTION

Nucleation from the vapor phase is perhaps the easiest phase transition to study and homogeneous nucleation is the simplest form of nucleation to investigate; this is true both for experimental and theoretical studies. Nucleation enjoys a close relationship to critical point phenomena.

*Research supported by the Atmospheric Sciences Section, National Science Foundation, NSF ATM75-16900.

The theory of small vapor phase clusters is directly related to the theory of virial coefficients. Moreover, the theory of the structure of clusters must in the large size limit reduce to the theory of liquids. All theories of nucleation borrow heavily on concepts evolved for homogeneous vapor phase nucleation, giving it a position of central importance.

The central problem in nucleation theory involves the development of a model which adequately represents the dynamic properties of equilibrium surfaces. The latter depends upon the individual and collective properties of the molecules constituting the surface. Nucleation theory has undergone precious little advancement in fundamental understanding from its inception in the late 1920's through the mid 1970's. In the liquid drop model very small clusters of molecules were treated via the capillarity approximation wherein one assumes that their surface properties could be represented by measurements made on bulk condensed systems. The capillary approximation was necessary since very little was known about the actual thermodynamic properties of embryonic clusters. However, the advent of large high-speed computers obviates the necessity for adherence to the capillary approximation. Work on molecular models of very simple cluster systems by a number of authors (1,2,3,4,5) has advanced far enough at this time so that progress in the field of nucleation can benefit markedly from extensive and accurate experimental measurements of homogeneous nucleation rates.

Over a period of years, Kassner and his associates (6,7) have developed the expansion cloud chamber technique so that precision measurements of homogeneous nucleation rates can be made over a wide range of temperatures and nucleation rates. Recently, Miller, Anderson and Kassner (to be published) have completed a unique set of measurements for water. This data is sufficiently extensive to yield a more definitive test of theory than has ever been possible previously. Careful comparison of the data with the classical liquid drop theory reveals several very interesting characteristics which will be discussed in considerable detail.

II. THEORETICAL CONCEPTS

The capillarity approximation offers the most tractable approach to the theory of homogeneous nucleation from the standpoint of mathematical formulation. It is also the version of the theory which has been most widely applied to different nucleation processes. Since most of our primary

2

deductions can be presented in terms of the features of the liquid drop model, we shall discuss its underlying assumptions in some detail.

Strictly speaking, thermodynamics deals with the course of phase transitions under equilibrium conditions where the velocity of progress of the phase transition is zero. It is assumed that these principles can be extended to conditions where the phase transition velocity is small but finite. Questions arise when we try to define the limits of validity of this assumption. Homogeneous nucleation concerns itself with the appearance of the first microscopic fragments of the new phase in the absence of chemically reactive molecular species, ions, or particulates which might catalyze the process. Since large deviations from equilibrium are required to induce an observable nucleation rate, it is not clear *a priori* to what extent equilibrium concepts can be relied upon to describe the process with quantitative accuracy. Moreover, the growth of these microscopic fragments of the new phase can only proceed provided the two phases are not exactly in equilibrium with one another. In fact, all the experimental procedures used to make quantitative measurements of nucleation rates rely upon the growth of these tiny embryos of the new phase to macroscopic proportions where they can be detected by visual observation, light scattering or photography.* In this sense, we do not observe the nucleation event itself but rather a by-product of it. As long as we have reason to believe that we can guarantee a one-to-one correspondence between the actual nucleation events and the observable by-products, we have a bonafide measurement of the nucleation rate.

In nucleation theory we must keep in mind that thermo-dynamics represents only average conditions prevailing throughout the system and that fluctuation theory must be used to describe stochastic processes which give rise to local transient states we refer to as clusters (8). These fluctuations of locally higher chemical potential are characterized by smaller probabilities of occurrence. The primary premise of the classical liquid drop theory of homogeneous nucleation is that fluctuations resulting in clusters consisting of several tens of molecules can be considered to possess the basic classical properties attributed to macroscopic liquid drops. The classical liquid drop model is basically a continuum representation of clusters while in actuality they are made up of a very finite number of discrete molecules. Such a representation would be expected to work well so long as the

---

* (We do not regard nozzle and molecular beam experiments as providing a quantitative measurement of nucleation rates.

number of molecules in the cluster remains large. Since most experimental measurements of nucleation rates involve con- sideration of clusters containing 40 to 80 molecules, the liquid drop model should be considered to be on rather tenuous ground.

Derivation of the classical homogeneous nucleation rate equation may be divided into two parts: (1) the determina- tion of the free energy of formation of the cluster, from which the equilibrium concentration of each cluster species comprising the distribution may be found, and (2) the formulation of the unimolecular reaction kinetics which attempts to represent the steady state flow through the cluster distribution. This current through the cluster distribution is identified as being identical to the nuclea- tion rate. Reaction kinetics is assumed to be sufficiently fast such that a steady state condition is reached very quickly (less than 10^{-6} sec) compared to the duration of an experimental sampling of the nucleation rate (10^{-2} sec in the case of our experimental technique).

Neglecting the translation and rotation factor to which Lothe and Pound (5) focused attention in their 1962 paper, the free energy of formation of a cluster of radius r is given by:

$$\Delta G = -\frac{4}{3}\pi r^3 n_\ell kT \, \ln S + 4\pi r^2 \sigma \qquad (1)$$

where n_ℓ is the number of molecules per cm^3 in the bulk liquid, r is the radius of the cluster, k is Boltzmann's constant, T is absolute temperature, S is the supersaturation ratio and σ is surface tension. For all S greater than 1, ΔG passes through a maximum at

$$r* = \frac{2\sigma}{n_\ell kT \, \ln S} , \qquad (2)$$

where $r*$ is defined as the critical radius. The value of the free energy corresponding to the critical radius is given by:

$$\Delta G* = \frac{16\pi\sigma^3}{3(n_\ell kT \, \ln S)^2} , \qquad (3)$$

The concentration of clusters of size r is given by:

$$N_r = N_1 \exp{-[-\frac{4}{3}\pi r^3 n_\ell kT \, \ln S + 4\pi r^2 \sigma]/kT}, \qquad (4)$$

and finally, the concentration of critical clusters is

$$N_{r*} = N_1 \exp{-[\frac{16\pi\sigma^3}{3kT(n_\ell kT \, \ln S)^2}]}, \qquad (5)$$

where N_1 is the monomer concentration.

The properties of ΔG and N_r as a function of r are shown in Fig. 1. It can be clearly seen that the free energy

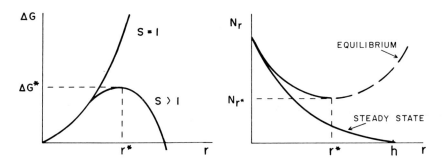

Fig. 1. Free energy of formation ΔG and concentration of cluster N_r for clusters of radius r.

decreases monotonically beyond the critical cluster size so that $\Delta G*$ represents an energy barrier to nucleation. Stochastic fluctuations give rise to the distribution of subcritical clusters, re-supplying those lost through nucleation events. The occurrence of clusters becomes less and less probable as their corresponding ΔG increases. However, at suitably high supersaturations, the concentration of critical clusters, N_{r*}, becomes large enough to allow an observable number of clusters to trickle past the barrier whereupon they grow unrestricted to become macroscopic droplets. The kinetic calculation attempts to estimate this leakage rate.

Volmer (10) estimated the equilibrium nucleation rate by assuming that every critical sized cluster which condensed an additional molecule became free growing and could be counted as a nucleation event. The resulting equilibrium nucleation rate is given by:

$$J_e = s_{r*}\alpha \; \Gamma \; N_{r*},$$

where s_{r*} is the surface area of a critical size cluster, Γ is the rate of impingement of monomers on unit surface area, and α is the condensation coefficient. The impingement rate of vapor molecules on the surface of the cluster is given by the kinetic theory relation for effusive flow, $p/\sqrt{2\pi mkT}$, where p is the partial pressure of the vapor and m is the mass of a vapor molecule. The equilibrium nucleation rate, J_e, becomes:

$$J_e = \alpha 4\pi r*^2 \; \frac{p}{\sqrt{2\pi mkT}} \; N_1 \, exp - [\frac{16\pi\sigma^3}{3kT \, (n_\ell kT \, \ln S)^2}]. \qquad (6)$$

Various corrections have been made for the fact that nucleation disturbs the equilibrium cluster distribution by deplet-
ing the cluster concentrations at larger sizes (11,12,13,14).

An additional correction term known as the Zeldovitch
factor, Z, is required to modify Eq. 6 so that it describes
the steady state situation. This factor accounts for the
fact that the steady state concentration of critical size
clusters is smaller than the equilibrium value and that the
critical and slightly supercritical clusters may re-evaporate.
The Zeldovitch factor is given by:

$$Z = (\frac{\Delta G^*}{3\pi kT g^{*2}})^{1/2} , \qquad (7)$$

where g^* is the number of molecules in the critical cluster.
Inserting the Zeldovitch factor and the appropriate
expressions for r^* and noting that $N_1 = SN_1^0$ where N_1^0 is
the equilibrium concentration of monomers, we may write the
steady state nucleation rate, J_s, as follows:

$$J_s = (\frac{\alpha}{d}) (\frac{2MN_A^3\sigma}{\pi R^4})^{1/2} (\frac{P_e}{T})^2 S^2 \exp - [\frac{16\pi\sigma^3}{3kT (n_\ell kT \ln S)^2}] , (8)$$

in which N_A is Avogadro's number, R is referred to
as the gas constant, and d, M, and P_e are the density,
molecular weight, and saturation vapor pressure of water,
respectively.

The formalism used by Becker and Döring (11) in their
evaluation of the kinetic coefficient is particularly suited
to our discussion of possible deficiencies of the classical
theory. The reduction of the difference equations to a
differential equation, as is done in the treatment preferred
by Frenkel (8) for instance, becomes questionable when the
cluster size is so small that the reaction from a g to a $g+1$
size cluster in no stretch of the imagination represents a
continuum process. Likewise, when the critical cluster
size is small, all of the reactions leading to nucleation
events fall into this category. The method of calculating
the nucleation rate used by Becker and Döring retains
recognition of the fact that the growth of the embryos is a
succession of discrete reaction events until late in the
calculation where they convert a series to an integral.
The sequence of reaction leads to results which are conven-
tionally analogous to a series electrical circuit driven by a
potential difference.

Consider the growth of embryos as a succession of
unimolecular reactions with monomers.

$$A_1 + A_1 \rightleftharpoons A_2$$
$$A_2 + A_1 \rightleftharpoons A_3$$

$$\cdot \qquad \cdot \qquad \cdot$$
$$\cdot \qquad \cdot \qquad \cdot$$

$$A_{g-1} + A_1 \rightleftharpoons A_g \tag{9}$$

$$\cdot \qquad \cdot \qquad \cdot$$
$$\cdot \qquad \cdot \qquad \cdot$$

$$A_{h-1} + A_1 \rightleftharpoons A_h,$$

where A_g represent clusters containing g molecules, where $g = (4/3)\pi r^3 n_\ell$. The set of equations is arbitrarily terminated at some value h > g*. The selection of h is made so that it does not affect the steady state concentration of critical clusters noticeably. Reactions between clusters and the spontaneous fissioning of clusters is neglected because these events are deemed to be too rare to play a significant role in determining the nucleation rate. Note that only one cluster of each size is considered, namely the lowest free energy species. It is through this species that all the nucleation current is assumed to pass.

If the mean current through the cluster distribution in the forward direction is everywhere the same as is assumed in the case of the steady state nucleation rate, we may write

$$J = a_{c,g-1} s_{g-1} N_{g-1} - a_{e,g} s_g N_g . \tag{10}$$

where $a_{c,g}$ and $a_{e,g}$ are the rates of condensation and evaporation, respectively, for a cluster of size g whose surface area is s_g. Defining $\beta_g = a_{c,g-1}/a_{e,g}$ and $N_g' = s_g N_g$, becomes

$$N_g' = \beta_g N_{g-1}' - J\frac{\beta_g}{a_{c,g-1}}. \tag{11}$$

In general, β_g increases monotonically with increasing g until at $g*$, $\beta_g = 1$. Molecular models, especially in the case of water, tend to question the validity of the assumption that all clusters can be characterized by the same value of $a_{c,g}$ and $a_{e,g}$ (15). Nevertheless, Becker and Döring's system of difference equations can be written as

$$N_2' = \beta_2 N_1' - J\,\frac{\beta_2}{a_{c,1}}$$

$$N_3' = \beta_3 N_2' - J\,\frac{\beta_3}{a_{c,2}}$$

$$\cdot \qquad \cdot \qquad \cdot$$
$$\cdot \qquad \cdot \qquad \cdot \qquad\qquad (12)$$

$$N_{g+1}' = \beta_{g+1} N_g' - J\,\frac{\beta_{g+1}}{a_{c,g}}$$

$$\cdot \qquad \cdot \qquad \cdot$$
$$\cdot \qquad \cdot \qquad \cdot$$

$$N_h' = \beta_h N_{h-1}' - J\,\frac{\beta_h}{a_{c,h-1}}.$$

In order to solve this set of equations, Becker and Döring divided the N_{g+1}' equation by $\beta_2 \beta_3 \cdots \beta_{g+1}$ yielding:

$$\frac{N_2'}{\beta_2} = N_1' - J\,\frac{1}{a_{c,1}}$$

$$\frac{N_3'}{\beta_2 \beta_3} = \frac{N_2'}{\beta_2} - J\,\frac{1}{a_{c,2}\beta_2}$$

$$\cdot \qquad \cdot \qquad \cdot$$
$$\cdot \qquad \cdot \qquad \cdot \qquad\qquad (13)$$

$$\frac{N_g'}{\beta_2 \beta_3 \cdots \beta_g} = \frac{N_{g-1}'}{\beta_2 \beta_3 \cdots \beta_{g-1}} - J\,\frac{1}{a_{c,g-1}\beta_2 \beta_3 \cdots \beta_{g-1}}$$

$$\cdot \qquad \cdot \qquad \cdot$$
$$\cdot \qquad \cdot \qquad \cdot$$

$$\frac{N_h'}{\beta_2 \beta_3 \cdots \beta_h} = \frac{N_{h-1}'}{\beta_2 \beta_3 \cdots \beta_{h-1}} - J\,\frac{1}{a_{c,h}\beta_2 \beta_3 \cdots \beta_{h-1}}.$$

When these equations are added, the N_g' terms cancel out in pairs, leaving only N_1' and N_h':

$$\frac{N_h{}'}{\beta_2\beta_3\cdots\beta_h} = N_1{}' - J\left[\frac{1}{a_{c,1}} + \frac{1}{a_{c,2}\beta_2} + \cdots\right.$$

$$\frac{1}{a_{c,g}\beta_2\beta_3\cdots\beta_g} + \cdots + \left.\frac{1}{a_{c,h-1}\beta_2\beta_3\cdots\beta_{h-1}}\right]. \qquad (14)$$

The only recognition of the properties of the intermediate clusters is contained in the $a_{c,g}$ and β_g.

This procedure does not work if one assumes several cluster structures having the same g participate in a reaction scheme where branching is allowed. The latter situation can be readily dealt with by employing a purely numerical solution of the problem. In the usual formation of the classical theory, all the $a_{c,g}$ and $a_{e,g}$ are taken to be the same for all cluster sizes.

Becker and Döring recognized that Eqs. 12 and 13 are analogous to Ohms law for a series circuit composed of h resistors. Thomson's equation is used to interrelate the various $a_{c,g}$. Thus, one can write

$$\frac{N_h{}'}{\beta_2\beta_3\cdots\beta_h} = N_1{}' - \frac{J}{a_c}\left[1 + \frac{1}{\beta_2} + \frac{1}{\beta_2\beta_3} + \cdots \frac{1}{\beta_2\beta_3\cdots\beta_{h-1}}\right],$$

and

$$\Phi_1 - \Phi_h = J\left[R_1 + R_2 + R_3 + \ldots R_{h-1}\right] \qquad (15)$$

In the latter part of the Becker-Döring calculation the sum is converted into an integral which facilitates the determination of the nucleation rate J. By way of contrast Frenkel (8) prefers to convert the original difference equations into a differential equation, making the transition from the discrete to the continuum earlier in the calculation.

The primary assumptions we wish to emphasize is that the theoretical nucleation rate is determined by the current through the lowest free energy configuration of each cluster size in the cluster distribution. Virtually all of the individual properties of the clusters are swept away into smoothly varying functions (for instance via Thomson's equation) so that we have little recognition left of the intermediate discrete steps involved in the actual clustering reactions. As we shall discuss later, the assumption of spherically symmetric Van der Waals forces seems to be inherent in the theory and is essential to the justification of these smoothly varying functions.

III.   COMPARISON OF THEORY WITH EXPERIMENT

Miller, Anderson and Kassner have reported an empirical expression for the nucleation rate of water vapor as a function of the supersaturation ratio and temperature (to be published).

$$J = S^2 \exp[43.36 - .25716T - .01087T^2 - 5.032E{-}5T^3 - (104.26 -$$

$$2.456T + 3.0107E{-}3T^2) \ln^{-2}S], \qquad (16)$$

where T is in $^\circ$C. This equation was obtained from a fit of expansion chamber measurements of rates from 50 to 3.10^5 drops/ cm^3-sec covering a temperature range from -44°C to $+17^\circ$C. The breadth of these measurements makes them especially amenable to a detailed comparison with theory.

Comparison of the data with theory is facilitated if Eq. 8 is rewritten in the slope-intercept form suggested by Hale and Plummer (16):

$$\ln(J/S^2) = \ln A(T) + B(T)[-\ln^{-2}S],$$

where

$$\ln A(T) = \ln(\frac{\alpha}{d})(\frac{2MN_A^3\sigma}{\pi R^4})^{1/2}(\frac{P_e}{T})^2 \qquad (17)$$

and

$$B(T) = 16\pi\sigma^3/3kT(n_\ell kT)^2.$$

Here both σ and α are assumed to be functions of temperature only, making $\ln A(T)$ and $B(T)$ temperature dependent constants. When the nucleation rate data is plotted using the scales $\ln(J/S^2)$ vs. $-\ln^{-2}S$, data for constant final temperature should turn out to be a straight line with slope $B(T)$ and intercept $\ln A(T)$. Not only may a simple comparison between theory and experiment be made, but, assuming the basic correctness of the functional dependencies of the theory, independent experimental values of both $\ln A(T)$ and $B(T)$ can be determined directly from the graph. These values may then be employed to evaluate the temperature dependence of both the condensation coefficient and surface tension respectively. Their behavior can then be compared to what one expects from physical considerations.

The experimental data fit was plotted in the above-mentioned manner for several temperatures in Fig. 2. The fact that the result is a set of straight lines at least indicates that the basic $\ln(J/S^2)$ vs. $-\ln^{-2}S$ functionality is sufficiently correct that deviations from linearity are not apparent. Also included are dashed lines representing the classical theory, where the values used in the theory for P_e (17), d (18), and σ (19) are from fits (or extrapolation of fits in the case of temperatures below 0°C) of reported measurements on bulk liquid water. Obviously extrapolation is necessary since measured values of these quantities do not exist for the low temperatures which were achieved in the nucleation studies. For the theoretical curves the condensation coefficient was given the value of 1.0 for a lack of better information. A realistic temperature dependent value for α would alter the present comparison between theory and experiment.

Several observations can be made from Fig. 2. In the temperature and nucleation rate domain corresponding to our measurements, the theoretical nucleation rate is always greater than the experimentally determined rate. The closest agreement exists at the lowest nucleation rates and for the lowest temperatures but it is felt that this is probably accidental. Thus, the relatively close agreement reported by Heist and Reiss (20) between diffusion chamber measurements and the classical theory for a rate of about 1 drop/(cm^3-sec) near room temperature and above are verified. However, the divergence between theory and experiment as the rate increases could not be observed by Heist and Reiss because the diffusion chamber is able to measure only very low nucleation rates.

Our data agrees well with theory near -40°C. However, a discrepancy in both the absolute nucleation rate and the slope of the nucleation rate curves grows steadily with increasing temperature. All parameters in the classical theory are well known except for the surface tension, σ, which is the only adjustable parameter appearing in the slope, $B(T)$, and the condensation coefficient, α, which appears in the intercept, $\ln A(T)$. Thus, the slope-intercept format facilitates the evaluation of a $\sigma(T)$ and $\alpha(T)$ which reproduces the data. Temperature dependent values for both can be determined by using experimental values for $B(T)$ and $\ln A(T)$ (as determined from Fig. 2) with the aide of Eq. 17. Although this procedure removes any disagreement between the theory and the experiment data, it

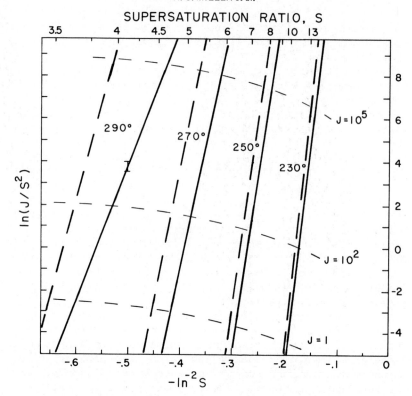

Fig. 2. *Comparison of experimental measurements* (——) *with the classical theory* (- -) *for constant temperature.*

does force $\sigma(T)$ and $\alpha(T)$ to accommodate for any faults which might exist in the classical theory.

The limitation encountered in the reliability with which $\sigma(T)$ and $\alpha(T)$ can be determined is evident from Fig. 3. The data lies in the low range of the ordinant while the nucleation rate lines (evaluated for constant temperature) possess a steep slope and intersect the ordinant at relatively large values. Thus, the intercepts are strong functions of the slopes of the curves, and a relatively small scatter in the data has the effect of introducing a rather large uncertainty in the values inferred for the condensation coefficient.

The results of the above-mentioned analysis are displayed in Figs. 4, 5 and 6. The uncertainties in the estimated quantities are indicated by the shaded areas of the curves. The authors wish to indicate in this manner that in spite of the uncertainties, reasonable interpretation of the data does not allow for significant changes in these parameters.

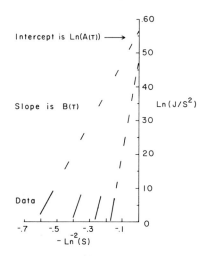

Fig. 3. *Representation of sensitivity of physical parameters upon A(T) and B(T). Small change in slope B(T) produces a large change in the intercept lnA(T).*

It is disconcerting to note that the extrapolation of the table values of the surface tension for bulk water (19), denoted by the broken line in Fig. 4, deviate markedly from the values appropriate to our experimental nucleation rate data, especially at the higher temperatures. Everywhere the surface tensions inferred from fitting the classical theory to our experimental data fall below the bulk measured values. One would expect that since the critical cluster size is larger at the higher temperatures, the use of bulk surface tension values in the theory would yield better results at the higher temperatures, but this does not appear to be the case. The conclusion to be reached is that the surface tension appropriate to the theory, in order to achieve reasonable agreement with the experimentally determined slopes of the nucleation rate curves representing constant temperature, needs to be smaller than the bulk measured values.

The condensation coefficient determined from the nucleation rate measurements range from about 10^{-3} near $-40°C$ to about 10^{-9} near room temperature. One expects the condensation coefficient to increase with decreasing temperature, possibly approaching unity for suitably low temperatures (21,22). The function of α determined from our nucleation rate data possesses this general rudamentary qualitative property as shown in Fig. 5; however, it is difficult to reconcile the extremely small values near room temperature ($\sim10^{-9}$) with values of α determined from droplet growth experiments which range from .01 to 1 (23,24). While it would seem that values of α of the order of 10^{-9} would make the

13

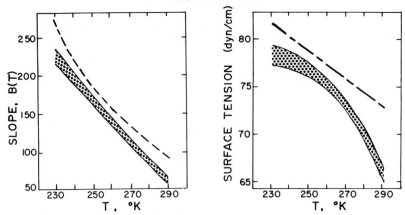

Fig. 4. *Values of the slope B(T) and corresponding surface tension σ used in classical theory (--) and those found experimentally (shaded area).*

establishment of the equilibrium cluster distribution extremely sluggish, our observations indicate that there is no change in the nucleation rate whether more time is allowed for the establishment of the cluster distribution or not (25). One could conclude that some element is missing in the theory which would make the values of α appear to be more reasonable.

Lastly, the critical cluster size may be inferred from the experimental data through the relation

$$g^* = 2B(T)\ln^{-3}S. \qquad (18)$$

In Fig. 6 we chose to show the variation of the critical cluster size with temperature for a nucleation rate of 1 drop/cm^3/sec as obtained from the theory using $\alpha=1$ for all temperatures and σ taken from bulk measured values (the broken line) and from Eq. (18) using our nucleation rate measurements (the shaded area). The increased disagreement as temperature increases is expected as the slopes differed the most at the higher temperatures. It seems important to note that the experimental data suggests a much smaller increase in g^* with increasing temperature than is given by theory.

IV. IMPLICATION TO THEORY

When one makes an attempt to compare an extensive body of nucleation rate data with the classical liquid drop model as we have done in the previous section, any

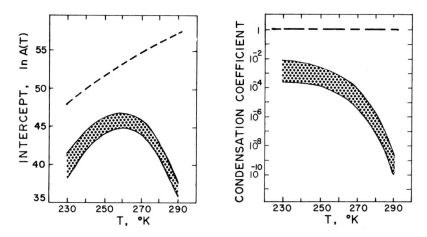

Fig. 5. Intercept and condensation coefficient used in classical theory (--) and those obtained from data.

inadequacies in the theory are swept into the functions $\sigma(T)$ and $\alpha(T)$. Thus, if $\sigma(T)$ and $\alpha(T)$ appear to depart dramatically from bulk measured values, this can be construed to reflect a deficiency in the theory. It would appear that in order to achieve agreement between theory and experiment either un-realistic temperature dependencies must be assigned to $\sigma(T)$ and $\alpha(T)$ or perhaps these functions do not enjoy such a close relationship to their bulk measured counterparts as has been previously assumed. In this section we shall attempt to explore these possibilities.

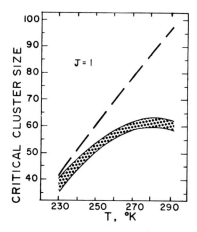

Fig. 6. Critical cluster size for J=1. Classical theory (--) and experimental (shaded area).

Frenkel (8) points out that the molecular complexes are assumed to be held together under the influence of Van der Waals' type cohesive forces. The implications of this assumption on various facets of the theory are far reaching. For instance, there is a notable absence of such effects as saturation in the intermolecular force fields as occurs in the case of chemical binding and the effects of chemical isomerism are absent. It implies that configurational permutations go easily so that clusters encounter a minimum of difficulty in settling down in something approximating the lowest free energy state, corresponding to nearly spherical geometry. All of these features help insure that the energy of association of the cluster increases smoothly with increasing number of molecules because the average conditions of binding of molecules in the surface remains very much the same, except for curvature effects. The compact spherical geometry of the cluster provides conditions under which a surface of tension can be appropriately defined. The average distance between molecules is fairly uniform so that the density of the various cluster sizes remains essentially constant. Accordingly, the surface area of the cluster of a cluster is proportional to $g^{2/3}$. Thus, one attributes any deficit in the association energy, below what one expects from the bulk latent heat, to surface free energy.

Water does not fit these conditions very well due to the association of the molecules via hydrogen bonding. The bonds are tetrahedrially coordinated and much stronger than the background Van der Waals field. Near room temperature the hydrogen bond energy is approximately 10 kT so that permutations are greatly restricted and cluster configurations are possible which deviate markedly from spherical geometry, see Fig. 7, (2-5). Deviations from the close packed spherical geometry, would be expected to cause perceptable deviations from bulk measured values of surface tension because the surface of tension becomes poorly defined. The statistical mechanical treatment of clusters (2-5) does not suffer from this handicap since only molecular properties are used as input in the theory and no mention is made of bulk properties.

Because the hydrogen bond energy is so large compared to kT throughout the range of temperatures of interest to us in the experimental data presented, the individual cluster configurations are stable for times which are long compared to the period of the normal mode oscillations. This is true at least for the smaller clusters. For this reason this particular molecular model is often referred to as a microcrystalline cluster model. Spontaneous decay of such clusters must proceed through the constructive combination of

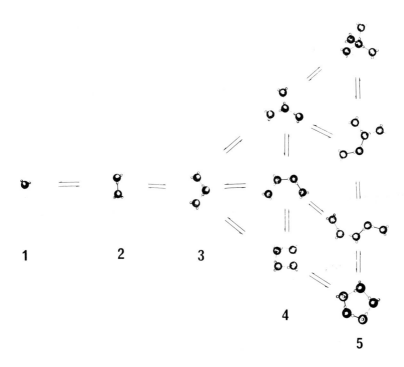

Fig. 7. Schematic of possible paths for current through the small end of the cluster distribution.

normal mode vibrations (15) whereupon a given molecule in the cluster receives an exceptionally large excursion in its vibrational amplitude. If the excursion falls within a narrow energy range, bond breakage may occur without the molecule acquiring sufficient energy to be completely dissociated from the cluster. In this case a configurational permutation may occur. This type of permutation is probably relatively rare, a condensation, evaporation or fissioning reaction being more likely to occur.

Certain cluster species exhibiting a high degree of bonding are predicted to be exceedingly stable, (15,26). Recent molecular beam experiments tend to confirm these theoretical predictions. For both charged and uncharged clusters respectively Searcy and Finn (27) and Lin (28) have evidence indicating the pronounced stability of the 20 molecule clathrate cluster structure where an additional molecule is trapped in the center of the cage. In fact the curve presented by Lin might be interpreted as an indication that this 21 molecule species is less reactive than adjacent cluster sizes and that it might act as a bottleneck to the flow of current through the cluster distribution.

The conclusion to be drawn from the above discussion is that condensation and evaporation coefficients must be determined which are appropriate for particular cluster structures. The condensation coefficient should depend upon the relative availability of bond sites on the cluster and the evaporation coefficient probably should depend primarily upon the number of singly bonded molecules, its temperature dependence following some theoretical dependency of the sort embodied in the treatment by Hagen and Kassner (15).

In the molecular treatments eluded to earlier, work has tended to concentrate on the most stable configurations. One of the weakest facets of the work relates to the methods of handling the configurational entropy. Some of the less stable cluster configurations possess groups of molecules which can undergo hindered rotation about a bond, thereby tending to overcome part of the effect on the free energy arising from the reduction in the state of bonding. Such considerations being the case, it seems prudent that future work should allow the nucleation current to pass through several of the lower free energy cluster configurations as depicted in Fig. 7 for the small end of the cluster distribution. In such a case the free energy appropriate to the classical theory would be a weighted average of the free energies of those cluster configurations through which the majority of the current passes.

The classical liquid drop theory employs a very rudamentary approach in dealing with the collision dynamics of molecules with the cluster. No account is taken of the velocity distribution of the bombarding molecules, the intermolecular interaction potential, or the impact parameter through which the *hardness* of the collision might be ascertained. An approach utilizing the Chapman-Enskog transport theory (29) would comprise a significant improvement.

It can be seen that there is plenty of room for increasing the level of sophistication which can be introduced into nucleation theory. The results of molecular beam experiments and ion mobility experiments may well provide important clues concerning those features which are most essential to nucleation theory. Work is proceeding along these lines in this laboratory.

V. CONCLUSIONS

The classical liquid drop theory has been confronted with an extensive collection of homogeneous nucleation rate data, extending in rates from 50 to $3 \cdot 10^5$ per $cm^3 \cdot sec$ and in temperature from $-44^\circ C$ to $+17^\circ C$. Comparing with theory, where $\alpha = 1$ and bulk measured values of σ are used, there is surprisingly close agreement between theory and experiment near $-40^\circ C$ while a discrepancy in both the absolute magnitude and the slope of the nucleation rate curves grows increasingly pronounced with increasing temperature, reaching differences of the order of 10^3 near room temperature. It is the rate of growth of the discrepancy between theory and experiment which signals a difficulty in the formulation of the theory.

If one fits the experimental data to the theory by allowing the surface tension and the condensation coefficient to be temperature dependent functions, results are obtained which are somewhat surprising. The surface tension everywhere falls below the bulk measured values, the departure from bulk measured values becoming more pronounced as the temperature increases. On the other hand, the condensation coefficient is found to decrease dramatically with increasing temperature, giving values of approximately 10^{-9} near room temperature. It is difficult to concede that there could be such a tremendous difference between values of α determined from droplet growth measurements ($\alpha \sim 10^{-2}$) and the α determined by us from experimental measurements of nucleation rates. However, it must be kept in mind that the $\sigma(T)$ and $\alpha(T)$ determined in this paper are model dependent.

The classical liquid drop theory is exceedingly elementary in its formulation of the kinetic problem. It restricts the current through the cluster distribution to only the lowest free energy clusters. Molecular modeling of physical clusters indicates that the most stable clusters are also the least reactive. It is suggested that, particularly in the case where the molecular species associate through hydrogen bonding, cluster permutations do not proceed easily and reactions contributing to the current through the cluster distribution extend into somewhat higher free energy states as indicated in Fig. 7. Moreover, condensation and evaporation coefficients need to be tailored to take into account the number of available bond sites and the number of singly bonded molecules in the cluster structure.

The value of a comprehensive set of nucleation rate data subjected to the type analysis presented here is quite evident. It offers a unique opportunity to evaluate theory in considerable detail. As a result new horizons are opened up for the further development and refinement of nucleation rate theory.

ACKNOWLEDGEMENTS

The authors wish to thank B. N. Hale, P. L. M. Plummer and D. E. Hagen for many helpful discussions.

REFERENCES

1.	Abraham, F.F., "Homogeneous Nucleation Theory," p. 109. Academic Press, New York, 1974.

2.	Bolander, R.W., Kassner, J.L., Jr., and Zung, J.T., J. Chem. Phys. 50, 4402 (1969).

3.	Daee, M., Lund, L.H., Plummer, P.L.M., Kassner, J.L., Jr., and Hale, B.N., J. Colloid and Interface Sci. 39, 65 (1972).

4.	Hale, B.N., and Plummer, P.L.M., J. Chem. Phys. 61, 4012 (1974).

5.	Rahman, A. and Stillinger, F.H., J. Chem. Phys. 55, 3336 (1971).

6.	Allen, L.B., and Kassner, J.L., Jr., J. Colloid and Interface Sci. 30, 81 (1969).

7.	Allard, E.F., and Kassner, J.L., Jr., J. Chem. Phys. 42, 1401 (1965).

8.	Frenkel, J., "Kinetic Theory of Liquids," Ch. 7. Dover Publications, New York, 1955.

9.	Lothe, G.J., and Pound, G.M., J. Chem. Phys. 36, 2080 (1962).

10.	Volmer, M., Z. Phys. Chem. 25, 555 (1929).

11.	Becker, R. and Döring, W., Ann. Physik. 24, 719 (1935).

12.	Farkas, L., Z. Physik. Chem. (Leipzig) 125, 236 (1927).

13.	Zeldovitch, J.B., J. Exp. Theoret. Phys. 12, 525 (1942).

14.	Zeldovitch, J.B., Acta. Phys. Chem. USSR 18, 1 (1943).

15.	Hagen, D.E., and Kassner, J.L., Jr., J. Chem. Phys. 61 4285 (1974).

16. Hale, B.N., and Plummer, P.L.M., J. Atm. Sci. 31, 1615 (1974).

17. List, R.J., "Smithsonian Meteorological Tables," p. 350. Smithsonian Press, Washington, 1971.

18. Tilton, L.W., and Taylor, J.K., J. Res. Nat. Bur. Std. 18, 205 (1937).

19. "International Critical Tables," 4, p. 447. McGraw-Hill, New York, 1933.

20. Heist, R.H., and Reiss, H., J. Chem. Phys. 59, 665 (1973).

21. Bryson, C.E., Cazcarra, V., Chouarain, M., and Levenson, L.L., J. Vacuum Science and Technology 9, 557 (1971).

22. Mills, A.F., and Seban, R.A., Int. Jour. of Heat and Mass Trans. 10, 1815 (1967).

23. Carstens, J.C., and Carter, J.M., Int. Colloquium on Drops and Bubbles, Calif. Inst. of Technology, August 28-30, 1974.

24. Alty, T., and Mackay, C.A., Proc. Roy. Soc. (London) A199, 104 (1935).

25. Miller, R.C., Ph.D. Dissertation, University of Missouri-Rolla (1976).

26. Kassner, J.L., Jr., and Hagen, D.E., J. Chem. Phys. 64, 1860 (1976).

27. Searcy, J.Q., and Fenn, J.B., J. Chem. Phys. 61, 5282 (1974).

28. Lin, S., Rev. Sci. Instrum. 44, 516 (1973).

29. Chapman, S. and Cowling, T.G., "The Mathematical Theory of Non-Uniform Gases," 2nd ed. Cambridge University Press, New York, 1952.

30. Goodman, F.O. and Wachman, H.Y., "Dynamics of Gas Surface Scattering," chapter 10, Academic Press, New York (1976).

HOMOGENEOUS NUCLEATION OF ICE IN WATER AS A FUNCTION OF
PRESSURE: A TEST OF CLASSICAL NUCLEATION THEORY

Charles A. Knight
National Center for Atmospheric Research[*]

ABSTRACT

Recent data on the homogeneous nucleation of ice Ih in supercooled water at pressures to 2,000 bars (1) allows an interesting test of the classical thermodynamic nucleation theory. Calculations of homogeneous nucleation temperature as a function of pressure are presented and discussed. The thermodynamic theory, within the limits of uncertainty of its own input data, fits the nucleation data very well. The use of high pressure to perform similar measurements in other systems would constitute an interesting test of the applicability of the theory for the nucleation of crystals from supercooled liquids.

I. INTRODUCTION

Nucleation has great importance in many scientific and engineering subjects. It has two very different theoretical approaches. The molecular approach is the more fundamental and shows great promise (2) but appears still to be rather far from practical application. The well-known thermodynamic approach (3) is appealingly simple but has always displeased molecular theorists because of its treatment of small particles as if they were homogeneous, with sharp interfaces. For many materials, particles of the size of critical embryos can not be homogeneous, and the question always has been, how important are the uncertainties introduced by treating them as

---
[*] The National Center for Atmospheric Research is sponsored by the National Science Foundation. The work reported herein was performed as a part of, and with some support from, the National Hail Research Experiment, managed by the National Center for Atmospheric Research and sponsored by the Weather Modification Program, Research Applications Directorate, National Science Foundation.

23

if they were? The classical, thermodynamic theory is directly useful in many practical applications, and it is important
to discover what the limits of its applicability are.

Very painstaking tests of the thermodynamic theory have
been performed using the homogeneous nucleation of liquids
from the vapor, because vapor-liquid interfacial energies can
be measured accurately. Some amazingly good agreement has
been found between measurements and theoretical predictions
for many liquids (4). On the other hand, the theory works
poorly for some liquids near critical points (5), where very
thick interfaces are anticipated.

Tests of classical nucleation theory are rare in the nucleation of crystals, because nucleation rate is extremely
sensitive to small variations of surface energy, and the surface energies of crystals are usually not measurable with the
needed accuracy. In effect, surface energy is an adjustable
parameter in the theory. The recent measurements of the homogeneous nucleation temperature of ice in water as a function of pressure (1) do allow a test of nucleation theory for
crystals, since an ice-liquid interfacial energy can be fixed
by applying the theory to the measured nucleation temperature
at one pressure, and the rest of the nucleation temperatures
can then be predicted by the theory, using measured thermodynamic data on bulk ice and water, and a set of assumptions
that may be argued to be reasonable.

II. THEORY

As explained in numerous texts and reviews (6), the homogeneous nucleation rate J in events per cm^3 per second, for
ice formation in supercooled liquid water, is given by

$$J = Kn(1) \exp(-\Delta F^*/kT),\qquad 1)$$

with
$$\Delta F^* = \frac{16}{3} \Pi \frac{\gamma^3}{(\Delta S - .004 \times 10^7 \Delta T)^2} \Delta T^2 \qquad 2)$$

for the nucleation of ice from pure, supercooled water. In
the right-hand side of eq. 1), $n(1)$ is the number of single
molecules of the liquid per unit volume, and it, multiplied
by the exponential, gives the "steady state" concentration of
critical embryos. ΔF^* is the free energy of formation of a
critical embryo, k the Boltzmann constant, and T the absolute
temperature. K is a kinetic factor expressing the rate at
which a critical embryo gains one molecule and hence grows
without limit. J is very sensitive of ΔF^*, but not very sensitive of variations of K, and therefore $Kn(1)$, the "pre-

exponential factor," can be treated as a constant, which has a value of 10^{25} in the water-ice system (7). In equation 2), γ is the ice-water interfacial energy, ΔT the supercooling, and ΔS the entropy of fusion per unit volume of ice at the external pressure and the stable equilibrium temperature. This equation assumes spherical shape for the embryo, which is probably a very good assumption in this system (7).

The (ΔS - .004 x $10^7 \Delta T$) term, which is in cgs units, is derivable from the entropy change of the crystal and of the liquid in cooling from the equilibrium temperature to the nucleation temperature. This derivation assumes that C_p of ice and of supercooled water are linear functions of temperature. This is very nearly true for ice, but the few measurements of C_p for slightly supercooled water do not define or justify such a relation very well, and it remains an assumption. The equation used for supercooled water is

$$\rho_I \, C_p = - .0033 \text{ x } 10^7 T + 5.113 \text{ x } 10^7, \qquad 3)$$

in cgs units, where ρ_I is the density of ice and C_p is in ergs/cc of ice/K°. Using this formula, we disregard the recent data of Rasmussen and MacKenzie (8). This is done for a reason given in the last section.

In applying eqs. 1) and 2) to the homogeneous nucleation of ice in water at pressures to 2,000 bars, it is assumed that: 1) K is independent of pressure; 2) γ is independent of pressure and temperature; 3) C_p of ice and of supercooled water are independent of pressure; and 4) that supercooling of the liquid does not have a drastic effect upon its properties as applicable to nucleation theory. Pressure effects on ρ_I, the melting point, and ΔS are included, using experimental data (9). Of the four assumptions above, the first is well justified from the behavior of the viscosity of water in this pressure range (9) and the fact that nucleation rate is not very sensitive to K anyhow. The portion of the second assumption involving pressure also appears well justified. In a one-component system, interfacial volume is zero by definition, making the term

$$\left. \frac{\partial \gamma}{\partial P} \right)_V = \left. \frac{\partial V}{\partial A} \right)_P , \qquad 4)$$

where A is interfacial area (10), strictly zero. However, volume is not held constant in the case at interest here. Since ΔV for freezing does increase with increasing pressure in this system (9) the density gradient at the ice-water interface must increase somewhat with increasing pressure also, and one might expect a small increase of γ with increas-

25

ing pressure from this cause alone. With regard to the assumed independence of γ from temperature, it should be noted that the deduced γ is already at -38°C and 1 atmosphere. Rasmussen and MacKenzie (8) deduce an appreciable decrease in γ with decreasing temperature, but using an involved and (admittedly) incomplete treatment. Fletcher (11) also deduces a γ that decreases with decreasing temperature, from an entirely different standpoint. In the present work, we assume constant γ along the T_H line of Fig. 1, and see how the theory fits the results. But it should be remembered that there are reasons why γ might decrease with decreasing temperature and increase with increasing pressure. Assumption 3) is difficult to assess, particularly as regards very highly supercooled water, for which the data are not available. It reduces to assuming that the difference between the pressure effect on C_p of ice and that of water is small. This remains an intuitively reasonable assumption, but it is not objectively justified. Assumption 4) implies ignoring some recent data, and is discussed more fully below.

III. RESULTS AND DISCUSSION

Fig. 1 shows the experimental curve from (1) with points calculated from eqs. 1) and 2), using data sources given above. The procedure here was to find the γ for which J = 10^{10} at -38°C, and then use this value of γ for the calculations at higher pressure. The result is almost completely insensitive to the chosen value of J. We note that the theory reproduces the main feature of the experimental data, the increase in ΔT for homogeneous nucleation with increasing pressure, and the agreement is very good, considering the assumptions on γ and C_p. Within these limits, this constitutes a successful test of the classical, thermodynamic theory of homogeneous nucleation. The critical embryo radii vary from about 6 to 11 Å over the pressure range, and the embryos thus include about 150 to 650 molecules according to the theory. Good agreement is expected only if such small particles do behave substantially like bulk material at the pressure given by the Laplace equation. Since only one substance is involved in the test, however, a fortuitous agreement is possible, and this kind of experiment should be carried out with other materials.
It is notable that no arguments about water structure itself are needed to explain the nucleation behavior. Is is also noteworthy that most of the experiments on the effects of pressure on the heterogeneous nucleation of ice (12) show that higher pressure decreases the supercooling required for

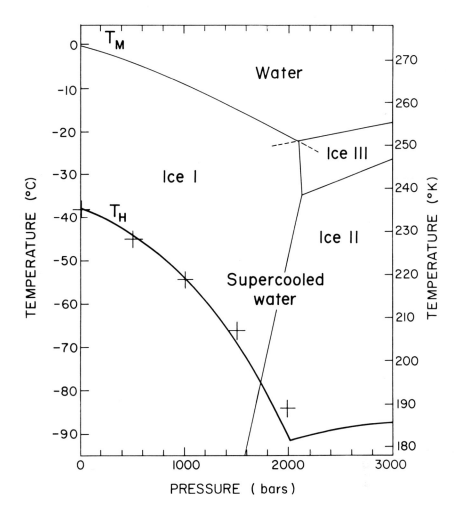

Fig. 1.   The equilibrium melting point of ice Ih, T_M, and the homogeneous nucleation temperature, T_H, are shown by the labelled lines, from data in (1).   Theoretical homogeneous nucleation temperatures, calculated as described in the text, are shown by crosses.   The agreement provides some justification for treating the critical embryos with thermodynamic methods.

nucleation. This may represent a substantial effect of pressure upon the substrate-liquid and/or the substrate-ice interfacial energy, though the situation in heterogeneous nucleation is complex enough to allow a number of possible explanations (13).

IV. THERMODYNAMIC DATA FOR THE METASTABLE PHASE

We have ignored data on the density and the specific heat of supercooled water to about -35°C at 1 atmosphere, recently measured by Rasmussen and MacKenzie (8). These properties were determined from the bulk properties of liquid water emulsions in n-heptane with an emulsifier; 1-10 μ diameter water droplets comprising about 50% of the final emulsion. The measurements show very substantial, "anomalous" changes in density and specific heat for liquid water below -15 to -20°C.

But consider the nucleation theory. It describes an idealization of the energy of formation of a single critical ice embryo from "the liquid." "The liquid" in the theory is that immediately in contact with the embryo, with which the embryo interacts, while the liquid that is available for bulk measurements, at high supercoolings, already contains substantial, steady state concentrations of sub-critical embryos. Indeed, Rasmussen and MacKenzie interpret their results in terms of the increasing contribution of the sub-critical embryos and their interfaces to the bulk properties as the supercooling increases. But the entire nucleation theory assumes that embryos do not grow from other sub-critical embryos. Therefore it seems clear that nucleation theory must use data extrapolated to the metastable condition, rather than data measured there, in cases of this kind.

V. ACKNOWLEDGEMENT

The writer thanks Ms. I. Paluch for programming the calculations.

VI. REFERENCES

1. Kanno, J., Speedy, R. J., and Angell, C. A., Science 189, 880 (1975).
2. Briant, C. L., and Burton, J. J., J. Chem. Phys. 60, 2849 (1974).
3. Gibbs, J. W., "Collected Works," Vol. 1, Longmans Green and Co., New York, 1928; Volmer, M., "Kinetik der Phasenbildung," Steinkopff, Dresden and Leipzig, 1939.

4. Katz, J. L., and Ostermier, B. J., *J. Chem. Phys.* 47, 478 (1976); Katz, J. L., *ibid.* 52, 4733 (1970); Katz, J. L., Scoppa II, C. J., Kumar, N. G., and Mirabel, P., *ibid.* 62, 448 (1975).

5. Heady, R. B., and Cahn, J. W., *ibid.* 58, 896 (1973).

6. Fletcher, N. H., "The Chemical Physics of Ice," Cambridge University Press, Cambridge, 1970.

7. Fletcher, N. H., *J. Chem. Phys.* 29, 572 (1958).

8. Rasmussen, D. H., and MacKenzie, A. P., *J. Chem. Phys.* 59, 5003 (1973).

9. Bridgman, P. W., "The Physics of High Pressure," Bell, London, 1931; Gmelins Handb. der anorganischen Chemie, 8 Aufl., Sauerstoff, Leif. 5 (Verlag Chemie, GMBH, Weinheim/Bergstr., 1963); Hobbs, P. V., "Ice Physics," Clarendon Press, Oxford, 1974.

10. Lewis, G. N., and Randall, M., "Thermodynamics," McGraw Hill, New York, 1961.

11. Fletcher, N. H., *J. Cryst. Growth* 28, 375 (1975).

12. Evans, L. F., *Trans. Faraday Soc.* 63, 1 (1967).

13. Knight, C. A., *J. Atmos. Sci.* 30, 324 (1973); Evans, L. F., and Lane, J. E., *ibid.* 30, 326 (1973).

ANISOTROPIC CRYSTAL GROWTH BY NUCLEATION OF CRYSTALLINE EMBRYOS AT ICE-VAPOR INTERFACES

David L. Bartley*
University of Missouri-Rolla

ABSTRACT

Simple techniques are presented for investigating mono-layer ice-like clusters at ice-vapor interfaces. Under several simplifying assumptions, it is found that the basal surfaces prefer triangular embryos with an orientation which reverses from layer to layer, whereas the most stable clusters on the prism surfaces are rectangular in configuration. The preferred prism clusters are determined to have a significantly lower critical energy of formation than the basal clusters due to differences in both corner free energy and configurational entropy. This phenomenon provides a mechanism for strongly anisotropic crystal growth at high saturations. Monolayer ledge configurations are also treated by analysis of edge jumps, and mean cluster configurations are obtained by investigating restricted ledges.

I. INTRODUCTION

In the present paper we examine an ice crystal growth process which should be significant at high saturations -- namely, we consider the nucleation of small monolayer ice-like clusters at the ice-vapor interfaces. Under the basic assumptions that the ice surfaces are smooth and sparcely covered with monomers, dimers, etc., in near equilibrium with the vapor, and that bond energies dominate the energy of formation G of a cluster of particular configuration, it can be shown (1) that G is given by

$$\Delta G = -nkT \ln S + (2n - n_b) |V_b|. \qquad (1)$$

Here n is the number of molecules in the cluster, n_b is the number of cluster bonds, T is the temperature, S is the supersaturation ratio, and $|V_b| \sim 12kT$ (near freezing) is the H-bond energy. (Edge relaxation and some vibrational effects (2) are neglected.)

---

*Supported by the National Aeronautics and Space Administration under Grant No. NAS8-31150.

Accounting for configurational fluctuations, it is found that the basal surfaces prefer triangular embryos, whereas the most stable clusters on the prism surfaces are rectangular in configuration. At any given saturation ratio, the preferred prism clusters are found to have a critical energy of formation significantly lower than that of the basal clusters, basically because of differences in the cluster corner free energies.

Although refined calculation of temperature dependent growth rates awaits knowledge of the monomer surface densities and diffusion rates, as well as details of vibrational and other contributions to the cluster energies of formation, it appears that these results may have significant implications. The energy gap between prism and basal clusters provides a mechanism for strongly anisotropic crystal growth at high saturations. Moreover, this gap should afford a rich variety of growth forms as the pressure is increased from its equilibrium value: leaving a regime where only step propagation is observed (analogous to the supercooled vapor regime), entering a region where lateral nucleation growth is significant, and finally one where nucleation growth can occur both in lateral and axial directions. Furthermore, there may be related aspects of dislocation growth mechanisms for ice crystals in vapor. Finally, because of their simplicity, the techniques presented for the analysis of cluster shape and configurational entropy may prove fruitful for investigations of heterogeneous ice nucleation and of two-dimensional nucleation phenomena of other materials as well.

In Section II a broad class of ice-like clusters is examined -- namely, we consider all (irregular 12-sided) clusters formable by straight-line monolayer cleavages which leave closed rings. Minimization of the energy of formation at roughly fixed n then leads to the minimal structures (i.e., triangular and rectangular) described above. In Section III, we vastly enlarge the class of structures considered and thereby obtain estimates of the configurational entropy in a number theoretic framework. Configurational entropy is shown to dominate bond energy in the small S limit. Section IV contains an analysis of basal clusters by a study of the distribution of "kinks" along restricted monolayer ledge segments.

II.   DUODECAGONAL CLUSTERS

In order to ascertain the minimal energy cluster configurations for given n, we consider the class of structures which can be formed by straight-line monolayer cleavages

which leave closed rings. Details of the ice surfaces lead
to the bonding arrangements as in the typical clusters shown
in Figs. 1 and 2.

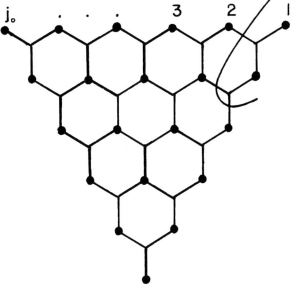

Fig. 1. *Typical basal cluster.*

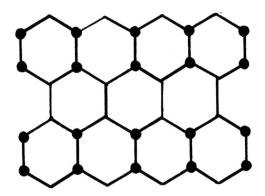

Fig. 2. *Typical prism cluster.*

Here vertices represent water molecules, lines represent
hydrogen bonds, and the solid dots show bonds between
cluster and the ice surface.

We now parameterize 12-sided closed-ring structures as
shown in Fig. 3 (here the six-membered ring is used for
the natural units of dimension).

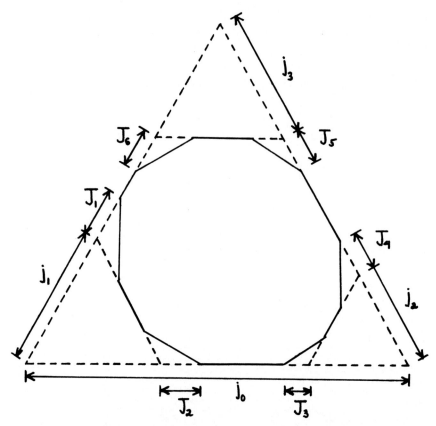

Fig. 3. Parameterization of duodecagonal clusters.

It is possible to show that the edge parameter ε, defined by

$$\varepsilon = 2n - n_b \qquad (2)$$

(see Eq. 1), is given by

$$\varepsilon_{basal} = j_0 \qquad (3)$$

for the basal clusters. Similarly, for the prism clusters we have

$$\varepsilon_{prism} = 3/2\ j_0 - 1/2(j_1 + j_2 + j_3) - 1/2(J_1 + J_4). \qquad (4)$$

The total number of particles in the cluster (on either basal or prism surface) can be expressed in the form

$$n = j_0^2 - \vec{j} \cdot \vec{j} - \vec{J} \cdot \vec{J} \qquad (5)$$

neglecting $O(j_o)$, $O(\vec{j})$ and $O(\vec{J})$ corrections. Interestingly, the parameterization adopted for the cluster leads to a 10-dimensional Lorentz invarient for n, as seen in Eq. 5.

We can now easily find the minimal energy clusters. Eqs. 3 and 5 immediately show that the preferred basal clusters are triangles with

$$\vec{j} = 0 \qquad (6a)$$

$$\vec{J} = 0 \qquad (6b)$$

($\vec{j}$ and $\vec{J}$ are like the momenta of a relativistic particle of rest mass n with the kinetic energy ε_{basal} which is minimum when the momentum vanishes). Similarly, minimizing ε_{prism} holding n roughly fixed gives

$$J_i = 0, \quad i \neq 1,4 \qquad (7a)$$

$$j_1 = j_2 = j_3 = J_1 = J_4 = 1/3 \; j_o, \qquad (7b)$$

which represents a rectangular structure with dimensions $\frac{1}{2}n^{1/2}$ [a, c/2], where a and c are the ice lattice parameters.

We can now eliminate j_o, $\vec{j}$, $\vec{J}$ for the minimal energy clusters in favor of n. The result is that the edge parameter ε for the preferred (triangular) basal clusters is assymptotically equal to that of the (rectangular) prism clusters and is given simply by

$$\varepsilon = n^{1/2}. \qquad (8)$$

This expression, of course, neglects the $O(n^o)$ corner energy contributions, which will be dealt with in Section III. Eqs. 1 and 8 give critical cluster sizes n* of the order of

$$n* = (V_b/2 \; kT\ln S)^2, \qquad (9)$$

corresponding to a critical energy of formation $\Delta G*$ given by

$$\Delta G* = V_b^2/4kT\ln S. \qquad (10)$$

III. CONFIGURATIONAL ENTROPY

In the present section, shortcomings of the calculation of Section II will be cleared up by considering the perimeter

free energy exactly (including corner contributions).
Furthermore, the class of structures considered is broadened
so that configurational entropy can be reasonably discussed.
Finally, we no longer make dubious continuous variations
over discrete variables. We begin discussion with the basal
clusters.

Consider a low energy basal cluster parameterized by j_o
as shown in Fig. 1. The number of molecules n and the
edge parameter ε are found to be given by the simple, exact
expressions:

$$n = j_o^2 \tag{11}$$

$$\varepsilon = j_o. \tag{12}$$

We now consider all structures which can be made by
"decimating" such completed "parent" structures by removing j
molecules breaking only two bonds at a time. This procedure
yields all of the 12-sided structures considered in Section II
as well as a multitude of others. The rationale here is that
removal of a molecule, breaking only two bonds, leaves the
edge parameter ε exactly invarient and changes the free
energy only slightly (as compared, for example, with
breaking three bonds) for $kT\ln S << |V_b|$, as is always the
case in situations of physical interest. If $\zeta(j)$ is the
number of ways of so decimating a parent triangular structure,
then the configurational entropy S_{con} is given by

$$S_{con} = k\ln \zeta(j), \tag{13}$$

Subtracting the configurational free energy from the energy
of formation (Eq. 1) for a particular cluster, we obtain the
energy of formation ΔG_n of an n-cluster in the form

$$\Delta G_n = -nkT\ln S + j_o |V_b| - kT\ln\zeta(j) \tag{14a}$$

$$n = j_o^2 - j. \tag{14b}$$

The difficulty now lies in calculating $\zeta(j)$. Focus
attention on the upper right corner of the triangle in Fig. 1,
and think of rows of molecules running diagonally upward and
to the right. We can remove an arbitrary number of molecules
from the row labelled 1 (breaking two bonds at a time). If
an even number of molecules is removed from row 1 (say, four,
as indicated in Fig. 1), then up to an equal number can be

removed from row 2. However, if an odd number is taken from
row 1, then only up to the odd number minus one can be taken
from row 2. The other rows can be considered similarly.
The result is that the number $\eta(j)$ of ways of removing j
molecules from a single corner, breaking two bonds at a time,
is equal to the number of ways the integer j can be
represented as a sum of other integers where odd integers
appear only once. The number $\zeta(j)$ for the entire triangle
can similarly be represented in a simply stated though
difficult number theoretic framework. In Figure 4 is
plotted $\ln\zeta(j)$, obtained by direct counting (valid,
strictly, for $j<2j_o-1$). Also plotted in Fig. 4 is an
assymptotic expression, $\ln \zeta_{ass}(j)$, where

$$\zeta_{ass}(j) = \frac{3}{16j^{3/2}} \, exp(\frac{3\pi^2}{2} \, j)^{1/2}. \tag{15}$$

The expression $\zeta_{ass}(j)$ is obtained by methods analogous to
those employed by Hardy and Ramanujan (3) in their treatment
of the theory of the partition of integers.

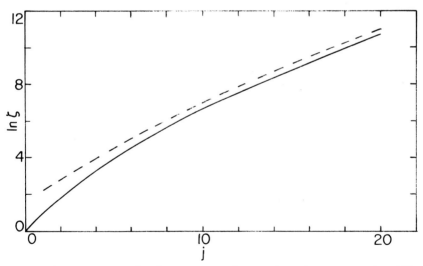

*Fig. 4. Basal cluster configurational entropy $\ln\zeta(j)$
vs. j (solid curve). Assymptotic approximation to $\ln\zeta(j)$
(dashed curve).*

Knowing $\zeta(j)$, we can now find ΔG_n by means of Eq. 14.
In Figs. 5 and 6 are plotted ΔG_n vs. n at two saturation
ratios (setting $V_b = -12kT$ and $T \sim T_o$) for n in the
vicinity of the critical size n* (Eq. 9).

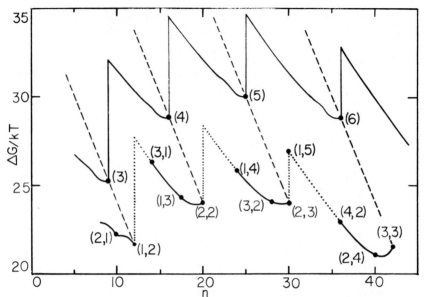

Fig. 5. ΔG_n vs. n at S = 3.3 for basal (upper curve) and prism (lower curve) clusters. Dashed lines represent ΔG neglecting configurational entropy. Parent triangles (j_Q^n) and (prism) rectangles (J_1, J_2) are indicated by solid dots.

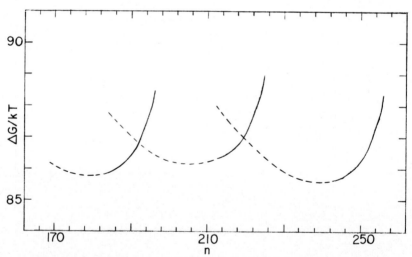

Fig. 6. ΔG_n vs. n at S = 1.5 showing onset of overlapping among basal clusters. (Dashed segments calculated via Eq. 15).

Also shown in Fig. 5 is ΔG_n for prism clusters, for which the configurational entropy was obtained by tedious counting. Lowering by $\underset{\sim}{\sim} |V_b|/2$ below the basal clusters is clear.

As is easily seen, the microscopic aspects of this problem yield discontinuous curves for ΔG_n. Each continuous segment of these curves represents decimations of a particular parent triangular cluster (as in Fig. 1), which itself is represented by the point corresponding to the largest number of molecules (in a continuous segment). The jaggedness of ΔG_n may be simply interpreted physically as representing activation barriers to (one-dimensional) nucleation of new cluster edges. Note that the configurational entropy, which, if neglected gives the straight lines in the figures, is clearly significant. In fact, at lower saturations, configurational entropy dominates bond energies and the critical structures "melt" into a more nearly circular shape as overlapping between the decimations of different parent triangles occurs (see Fig. 6).

IV. MONOLAYER LEDGES

We now briefly consider monolayer ledge configurations on the basal surfaces of an ice crystal. The subject is of importance to the study of crystal growth both by nucleation and by step migration. The latter process has been observed experimentally (4,5) for ice growing on covellite, and is probably dominant at low supersaturation for other ice crystal forms as well, steps being extant in accord with the Frank (6) spiral dislocation hypothesis. In the present paper we describe ledges on the basal ice surfaces in terms of a distribution of "kinks", in close analogy to the work of Burton and Cabrera (7,8) on the simple cubic structure. Relevant kinks are determined in accordance with the ideas of Section III, and the formalism is applied to ledges bounding the clusters considered in the above sections.

In Fig. 7 is shown a representative ledge segment showing the types of kinks considered. The edge is completely specified in terms of its position at a single point (e.g., at $x = 0$) and by the jump type at each value of (integral) x. The jump type may be labelled by (r, α) where $r = 0, \pm 1, \pm 2, \ldots$, and $\alpha = \pm$. For example, in Fig. 7 there is a $(2,+)$ jump at $x = 1$, $(2,-)$ at $x = 2$, $(-4,+)$ at $x = 3$, $(0,+)$ at $x = 4$, etc. This is somewhat different from the case with the simple cubic structure, where a jump may be specified simply in terms of its height r. Here the kink type and orientation of the (x,y)-axes

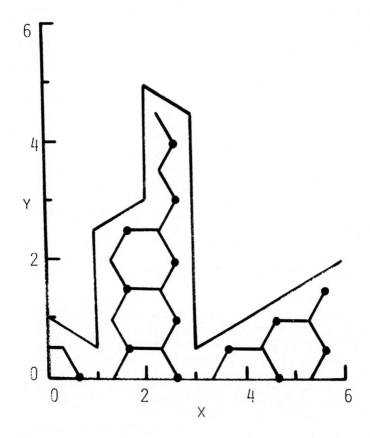

Fig. 7.   Illustrative basal ledge.

have been carefully chosen in order to account for all the decimations considered in Section III for edges with slope decreasing from $1/\sqrt{3}$ to $-\infty$ with increasing x.

We can now apply this ledge description to estimate the corner rounding of clusters caught in any one of the local "potential wells" as seen, for example, in Fig. 5. Fix a point of the edge at x = 0, y = 0, and consider all edges which extend out to a fixed value x_{max} of x, and no farther. Let $n_r^+(x)$ be the probability of an $(r,+)$-jump at x. $\frac{n_r^+}{r}$ will be well defined at S>1, if x_{max} is small enough.

Define $g_r^+(x)$ by

$$g_r^+ \equiv n_r^+/n_o^+.$$ (16)

It is possible to show (9) that for $x < x_{max}$,

$$g_r^+ = g_r^- \equiv g_r,$$ (17a)

$$g_{\pm|r|}(x) = [g_{\pm 1}(x)]^r,$$ (17b)

$$g_1(x) g_{-1}(x) = e^{-|V_b|/kT},$$ (17c)

$$g_1(x) = S^{-2x} g_1(o),$$ (17d)

$$n_o^-(x) g_1(x) = n_o^+(x) S e^{-|V_b|/kT}.$$ (17e)

Certainty of some sort of jump at each x implies

$$\sum_{r,\alpha} n_r^\alpha(x) = 1.$$ (18)

The remaining independent variable $g_1(o)$ is determined by a boundary condition at $x = x_{max} - 1$ to be

$$g_1(o) = S^{2x_{max}} e^{-|V_b|/kT}.$$ (19)

Now the mean height $\langle y \rangle_x$ of the edge is difficult to obtain; however, the mean slope, $\tan\theta \equiv \langle y \rangle_x - \langle y \rangle_{x-1}$, is given simply by

$$\tan\theta = 1/2 \sum_{r=0}^{\infty} (2r+1) [n_r^- - n_{-r}^+]$$

$$+ 1/2 \sum_{r=1}^{\infty} (2r-1) [n_r^+ - n_{-r}^+].$$ (20)

Eq. 20 is easily summed, and $\langle y \rangle_x$ may then be approximated by integrating $\tan\theta$. The explicit result for $\langle y \rangle_x$ at

$$x_{max} = \frac{|V_b|}{4kT \ln S} ,$$ giving a critical cluster mean edge segment,

is

$$\langle y \rangle_x = \frac{1}{2\ln S} \ln \left| \frac{(e^w + S)(1 - e^{-w} S^{-2x})(1 - e^{-w} S^{2x})}{(1 - e^{-w})^2 (e^w S^{-x} + S^{x+1})} \right| ,$$ (21)

41

where w is defined by

$$w = |V_b|/2kT. \tag{22}$$

Eq. 21 is plotted in Fig. 8 for $S = 1.5$. The radius of curvature ρ at the corner is given (in units of length along x-axis of Fig. 7) by

$$\rho = \frac{1}{3\ln S} [1-S+S(1+S)^{1/2}]. \tag{23}$$

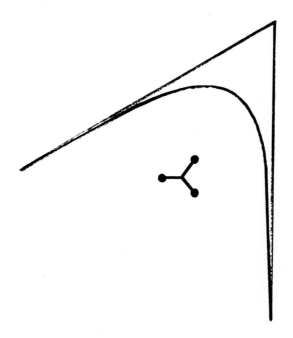

Fig. 8. Mean edge segment at S = 1.5. Tetramer fixes orientation and scale.

As can be seen from Eqs. 21 and 23, critical clusters over a range of S appear to have roughly similar shapes. It should be noted, however, that for S small enough that the overlapping described in Section III becomes important, the kink analysis of restricted edges becomes unreliable. In actuality, the radius of curvature ρ may increase much faster than as given by Eq. 23 as $S \to 1$. Furthermore, it is not obvious that the large clusters for small S remain monolayered. These comments are pertinent also to large (e.g., spiral) growth ledges and pose an intriguing problem as to their description.

V. REFERENCES

1. For details regarding Secs. I, II, and III, see Bartley, D.L., J. Chem. Phys. (1976, to appear).

2. For a treatment of oscillations of three-dimensional clusters, see Plummer, P.L.M. and B.N. Hale, J. Chem. Phys. 56, 4329 (1972).

3. Hardy, G.H., and S. Ramanujan, Proc. London Math. Soc. (2) 16, 131 (1917).

4. Mason, B.J., G.W. Bryant and A.P. Van der Heuvel, Phil. Mag. 8, 505 (1963).

5. Bryant, G.W., J. Hallett and B.J. Mason, J. Phys. Chem. Solids 12, 189 (1960).

6. Frank, F.C., Disc. Faraday Soc. no. 5, 48 (1949).

7. Burton, W.K., and N. Cabrera, Disc. Faraday Soc. no. 5, 33 (1949).

8. Burton, W.K., N. Cabrera and F.C. Frank, Phil. Trans. Roy. Soc. 243, 299 (1951).

9. Bartley, D.L., to be published.

MODELS FOR WATER AND ICE CLUSTERS ON SIMPLE SURFACES*

P. L. M. Plummer, B. N. Hale, J. Kiefer and E. M. Stein
University of Missouri-Rolla

ABSTRACT

In order to formulate a microscopic theory for water and ice clusters on simple surfaces we have developed and combined several theoretical models. A partition function formulation for the vapor-cluster-surface system and the molecular model previously applied to clusters in the vapor are used to study the surface stability of cluster structures. Energies of formation for specific structures indicate that the most stable ice clusters on the basal plane of ice I_h have two-dimensional, triangular configurations. Two complementary calculations have been made to obtain more detailed information about the monomer-surface interaction: a CNDO-INDO quantum mechanical treatment of a water molecule on a 7 to 20 molecule ice surface; and an empirical potential calculation of the interaction of a water monomer on a larger (50 to 486 molecule) ice crystal. The latter uses a central field effective pair potential and predicts that several bonding configurations are possible for the monomer per site. These models are also being used to study monomer diffusion rates. Future work will include the relaxation of the ice surface molecules and the effect of charge centers and discontinuities in the surface. Preliminary results for the interaction of the water monomer with simple surfaces such as graphite and silver iodide are briefly discussed.

I. INTRODUCTION

For the past several years we have been developing a molecular model (1) for the nucleation of water and ice, both from a supersaturated vapor and on surfaces. In general, the approach has been to assume a specific low energy molecular structure for the water aggregate and to calculate the free energy of formation from an approximate cluster canonical partition function. This partition function was assumed factorable into Z_{CM}, Z_B, Z_{VIB}, Z_{PROT} and Z_{CONF} corresponding

---

*Supported in part by NASA grant NAS8-31150 and the Atmospheric Science Section, National Science Foundation, GA-32386.

to contributions from center of mass motion, cluster binding
energy, the vibrational free energy and proton and
configurational entropy. This model has yielded results which
are in reasonable agreement with experiment for the process
of homogeneous nucleation (2) from a supersaturated vapor.

The nucleation of ice and water on surfaces has many
additional complications. A formalism (3) has been developed
which parallels that for vapor phase clusters but includes
the vapor, clusters on the surface, the surface and the bulk
solid. In our initial study (4) of water clusters on an
idealized ice surface, the surface clusters were modelled as
consisting of complete six-membered rings. Two classes of
these cluster structures were considered: those having only
one layer of rings perpendicular to the surface and those
having two or more layers perpendicular to the surface.
These studies suggested that on the basal face of ice I_h, the
preferred structure (5) for the clusters was a single layer
of rings arranged in a roughly triangular pattern. However,
problems were encountered in formulating structure dependence
on n for minimal energy clusters and in approximating the
contribution from configurational entropy for intermediate
sized clusters. Recent work by Bartley (6) has derived an
analytic form for the shape and n dependence of minimal energy
monolayered crystalline clusters for the basal and prism
planes of ice I_h. These results also predict triangular
clusters on the basal plane but with some incomplete rings.
These extra molecules increases the number of surface-
cluster bonds and maximize the total number of bonds. These
minimal energy clusters have small configurational entropy
and have j_o^2 molecules (where $j_o = 1, 2, 3 \ldots$) (see Fig. 1).
For such clusters the total number of cluster bonds, cluster-
surface bonds and the number of two, three and four-bonded
molecules can be written explicitly in terms of the total
number of molecules.

For these minimal energy clusters, at a supersaturation
ratio of 1.13 at $260^O K$, the critical size cluster contains
about 2000 molecules. The corresponding energy of formation
is of the order of 270 kT. These results would indicate
that not only is the formation of ice-like clusters from
the vapor unlikely (7) but also the "epitaxial" nucleation of
ice clusters on the surface is very improbable (see Fig. 2).
These results do not, however, include the formation of less
rigid clusters or cluster formation on a "relaxed" surface.
In order to understand these predictions and under what
conditions they are valid, we have attempted a more complete
analysis of the interaction of the water monomer with a surface.

The first surface chosen for study was ice I_h. Even
though no exact intermolecular potential for a system of
water molecules is known, recently a number of model

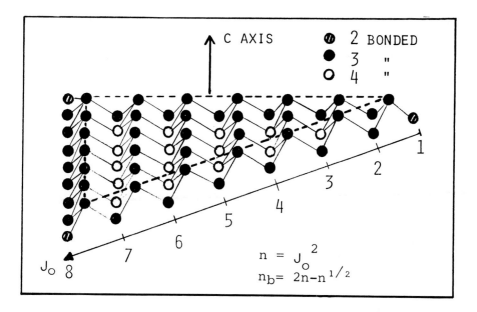

Fig. 1. *Minimal Energy Basal Plane Cluster on Unrelaxed ice I_h.*

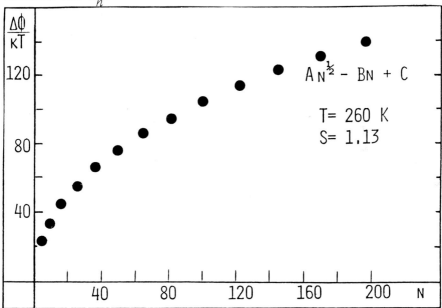

Fig. 2. *Cluster energy of formation, $\Delta\phi/kT$, as a function of the number of water molecules in the cluster. Clusters are monolayered and monocrystalline as shown in Fig. 1.*

potentials (8) have been proposed. For small sections of the
ice surface, it is also possible to perform quantum mechanical
calculations using the CNDO/INDO technique. In the next
three sections, we will describe and compare the potentials
used and the results obtained for a water *monomer* interacting
with an "unrelaxed" surface of ice I_h. Interaction with both
the basal face and a prism face are considered. In part II
we discuss the quantum mechanical model and its results.
In part III the central force potential model is presented
with a discussion of bonding sites, binding energies, and
normal mode frequencies. In part IV we compare the two
approaches quantitatively and discuss implications for future
work. This is essentially a summary of two rather extensive
studies and more details will be published separately. (P. L.
M. Plummer and E. M. Stein, J. Chem. Phys. and B. N. Hale and
J. Kiefer, J. Chem. Phys.)

Concurrently with the investigation of the ice surface,
we are examining the surfaces of A_gI and graphite and the
interactions of a water monomer with these surfaces. These
calculations will be described briefly in part IV. Some
preliminary results are available but quantitative conclusions
would be premature.

II. QUANTUM MECHANICAL STUDIES OF THE MONOMER ICE I_h SURFACE INTERACTION

For the quantum mechanical studies the ice surface,
either basal or prism face, was represented by water molecules
arranged in the lattice positions of ice I_h. Figure 3
illustrates the relative planes of the oxygen atoms and the
specific hydrogen orientations assumed for the basal surface.
Figure 4 shows the arrangement of molecules used to represent
the prism face.

For a system of this size, 42 nuclei and 420 electrons,
exact *ab initio* calculations are not practical or even
feasible. Therefore we have used an approximate molecular
orbital theory which includes all valence electrons for both
the monomer and the molecules in the surface. In addition,
multicenter integrals were neglected and some two electron
integrals were parameterized using atomic spectral data.
The specific technique employed was a semiempirical self-
consistent field approach known as CNDO/INDO (9). These
acronyms stand for *complete* or *intermediate neglect of
differential overlap*. In Dirac notation this means

$$(\mu\nu|\lambda\sigma) = \delta_{\mu\nu}\delta_{\lambda\sigma}(\mu\mu|\lambda\lambda) \tag{1}$$

where ϕ_μ, etc., represent Slater type orbitals centered on
the nuclei. The technique also includes the following

approximations:

$$(\mu|V_B|\nu) = \delta_{\mu\nu}V_{AB} \tag{2}$$

where $-V_B$ is the potential due to the nucleus of charge Z_B and the inner shell of atom B and $V_{AB} = Z_B\int s_A{}^2(1)(r_{1B})^{-1}d\tau_1$. Also

$$(\mu\mu|\lambda\lambda) = \gamma_{AB} \tag{3}$$

ϕ_μ on atom A

ϕ_ν on atom B

$$H_{\mu\nu} = \beta_{AB}^O S_{\mu\nu} \tag{4}$$

where the electron repulsion integral γ_{AB} is calculated as $\gamma_{AB} = \int\int s_A{}^2(1)(r_{12})^{-1}s_B{}^2(2)\ d\tau_1 d\tau_2$. The off-diagonal core matrix elements, $H_{\mu\nu}$, are set proportional to the overlap integral, $S_{\mu\nu}$, where β_{AB}^O is determined from atomic spectral data for atoms A and B. The basis set used for these calculations is a *valence* basis consisting of Slater orbitals of the type 1s for each hydrogen and 2s, $2p_x$, $2p_y$ and $2p_z$ for each oxygen. The results reported here are CNDO/2 calculations. (The /2 identifies the specific parameterization (9) used for the integrals.)

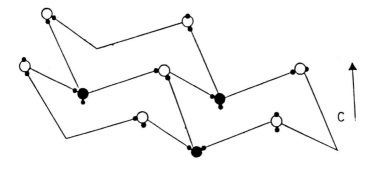

Fig. 3. Perspective drawing of basal face illustrating the relative planes of the oxygen atoms (large circles) and the specific hydrogen positions.

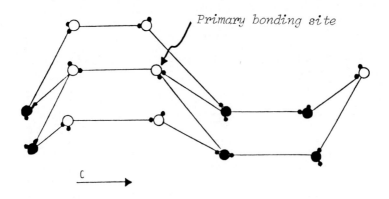

Primary bonding site

Fig. 4. Perspective view of prism plane showing specific hydrogen configurations.

Four possible bonding sites were examined for the basal plane and one for the prism plane (10). Only one orientation of the approaching monomer were considered for each site. However, the effects of monomer rotation were examined for several of the bonding sites. This effect was found to be minimal, of the order of 0.5 kcal/mole, indicating that almost free rotation is allowed. Three of the orientations for the monomer approaching the basal face had one oxygen-hydrogen bond pointing directly toward the surface. The maximum interaction was -8.0 kcal/mole while the other sites for this monomer orientation exhibited binding energies in the neighborhood of -4 kcal/mole (see Table 2, section IV). The remaining site on the basal face was located between two surface molecules with the monomer approaching the surface with both hydrogens down. This position would be exhibited by a monomer in the process of surface diffusion. In order to mimic a "bi-pedal walk" of a monomer on the surface, we calculated the barrier for rotation from a bonding position with a single hydrogen bond to the surface into this 'bifurcated' bonding position. The barrier was found to be about 2.5 kcal/mole to rotate into the bifurcated position and about 1.5 kcal/mole to rotate back to a single bonded configuration. These results together with the frequencies calculated for bond stretching (see Table 2) are used to estimate a residence time and a mean path length for the monomer on the surface (see section IV).

For the prism face instead of calculating the interaction energy for a variety of surface sites, we generated
an electron density map of the surface to better visualize
the charge distribution for the entire face. This map is
shown in Fig. 5. The density is plotted for a plane parallel
to the *c-axis* and through the upper layer of molecules as
indicated by the open circles in Fig. 4. The maximum density
peaks correspond to the oxygens. The distance between grid
points is 0.25Å and the viewer is oriented 45° clockwise from
the *c-axis* and 30° above the plane.

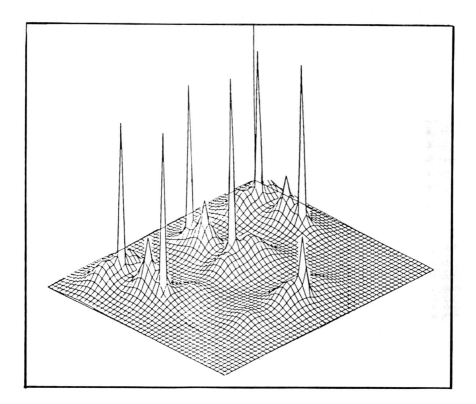

*Fig. 5. Electron density map of the prism face of Ice
I_h (from quantum mechanical calculations).*

III. THE CENTRAL FORCE MODEL FOR THE MONOMER-ICE I_h SURFACE
INTERACTION

The central force effective pair potential assumes a
three point charge model for the water molecules as proposed
by Lemberg and Stillinger (11). The specific model of this
type was that used by Rahman, Lemberg and Stillinger (12) in a
molecular dynamics calculation. The effective charge
(.32983e) was determined from the measured dipole moment of
the monomer (13) at the equilibrium positions of point masses.
The O-O, O-H, and H-H potentials are rather complicated and
non-unique in functional form (11). However, they reproduce
approximate bond stretch and bend frequencies. The mean
O-H bond length and HOH angle is preserved and hydrogen
bonding between water molecules is allowed. This potential
does permit dissociation of the molecule but the latter is
improbable at ordinary temperatures. Fig. 6 indicates the
form of potential used.

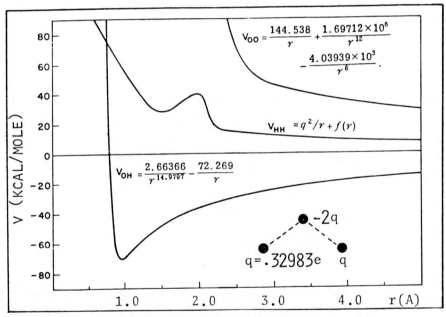

*Fig. 6. The Central Force model effective pair
potential model as described in reference 11.*

In this study the bulk-surface oxygen atoms were fixed
in an ice I_h lattice and their associated hydrogens were
placed such as to reproduce the tetrahedral bonding configuration of ice I_h (HOH bond angles $\sim 109°$). The monomer was

allowed to move freely above the surface so as to minimize
its binding energy. Several basal plane surface sites
(differing by orientation of surface molecule hydrogens) were
studied: (1) a site with all protons down (Fig. 7); (2) a
site with all protons up (Fig. 8); and (3) two sites with
hydrogens on alternating positions. For these sites we found
from five to eight bonding positions for the monomer - depending on the surface hydrogen orientations in and surrounding
the particular site. The strength of the monomer surface bond
ranged from about -3 to -9 kcal/mole. The nine by nine monomer
potential energy matrix was diagonalized and normal modes

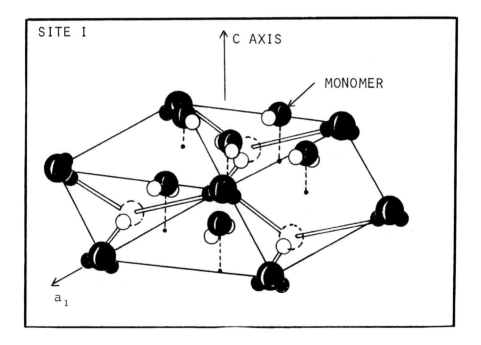

*Fig. 7. The six binding configurations found for
the water monomer on a basal plane site with all surface
protons down. The projection of the monomer center
of mass onto the basal plane is indicated by vertical
dotted lines.*

53

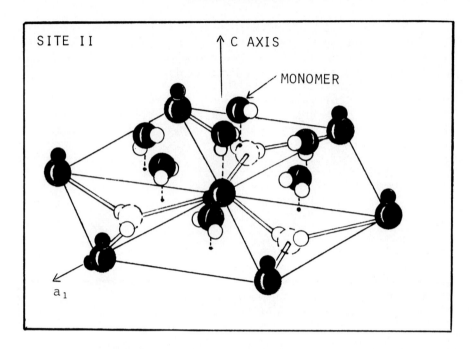

Fig. 8. The seven binding configurations found for the water monomer on a basal plane site with six surface protons up and the center surface protons down. Dotted lines indicate the projection onto the basal plane of the center of mass of the adsorbed monomer.

were calculated. Table 1 shows the average binding energy, monomer distance above the surface and the average intra-molecular, librational and intermolecular frequencies found from the sites studied.

TABLE 1
Central Field Results - Average from 31 Sites on the Basal Plane

$E_b (k_{cal/mole})$	Z(A)	$\nu(cm^{-1})$		
		INTRA.	LIB.	INTER.
-7.0	2.7 2.0	4100 3470 1870	1130 650 420	300 150 74

As a means of studying the monomer diffusion from
site to site we fixed the projection of a rigid water monomer
oxygen over the basal (xy) plane and minimized the monomer-
surface interaction potential with respect to the oxygen
height above the surface (z) and the water monomer Euler
angles. The monomer H-O-H angles was fixed at 105°. In this
manner a minimal energy path is generated by which such a
monomer can traverse bonding sites. The procedure
yielded barrier energies of about 2.0-2.5 kcal/mole. Using
this, together with the normal mode frequencies and an average
distance between monomer bonding sites of about 2.3A, the
monomer mean time and mean path length on the basal plane is
estimated (see section IV).

After studying the monomer minimal energy paths between
sites it appeared more enlightening to generate a monomer
minimal energy surface. This was done by minimizing the
monomer surface potential energy over a grid of xy points at
0.25A intervals. For each grid point (x,y) the oxygen-
surface distance, Z, and monomer Euler angles were varied to
find a minimum energy monomer configuration. The results
are shown in Fig. 9 for a basal plane site with alternating
surface hydrogens. The central surface molecule has both
hydrogens directed away from the surface. The potential
energy is plotted in the vertical direction with the
horizontal plane parallel to the basal plane surface.
For all points shown in the figure the monomer binding
energy was less than -1.9 kcal/mole. About 95% of the
binding energy surface is less than -5.0 kcal/mole with the
central region between -5.0 and -6.5 kcal/mole. The flat area
along the upper right hand edge is between -6.5 and -8.0
kcal/mole as is the flat region in the lower left corner. The
two troughs pointing toward the upper corner have regions
where the binding energy is between -8.0 and -9.5 kcal/mole.
In general, the areas near the *surface* molecules with hydrogens
up have higher binding energies varying from -6.0 kcal/mole
to -1.9 kcal/mole. The latter high binding energies indicate
an energy barrier for a monomer with hydrogens pointing toward
the surface. The energy surface is four-dimensional and
additional calculations are necessary to re-examine these
regions.

IV. RESULTS AND DISCUSSION

Since a complete and general intermolecular potential
does not exist for systems of water molecules, we have
performed calculations which model the water-ice surface
using two widely different approaches. It was hoped that the
detailed examination of these results can provide a much
more realistic picture of this interaction than either technique

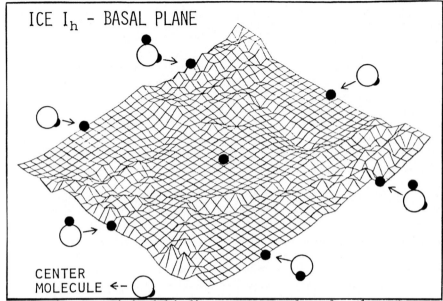

Fig. 9. Minimal binding energy surface for the water monomer above a basal plane site of ice I_h.

used alone. In addition, since the size of surface studied here is near the practical limit for the quantum mechanical approach, it is valuable to determine whether the computationally simpler central field model reproduces the primary results of the CNDO/2 calculations. A second objective was to determine how well a small section of the surface can mimic the bulk surface. (In previous calculations (1,2) on vapor phase water clusters we had assumed that the majority of the local environment was determined by nearest neighbor interactions). In the initial central field calculations, a 486 molecule, 9x9x6 molecular dimensional array, was used. It was found that >98% of the binding energy could be obtained using a 50 molecule section to represent the surface. Similarly, for the quantum mechanical studies (10), 94% of the interaction energy could be attributed to nearest neighbor interactions with 5.4% from second nearest neighbors. Thus, for both models the nearest neighbor interaction determined the magnitude of the binding. Fig. 10 illustrates the comparable basal surfaces used in these calculations.

In Table 2 we have compared the central force model and the quantum mechanical model results for four monomer bonding sites on the basal plane of ice I_h. The surface site on the basal plane has a center molecule with both hydrogens down and six surrounding molecules arranged on a hexagon with the surface molecules alternating with respect to hydrogens up and down. In Table 2 we see that the binding energies range from -4.0 to -8.4 kcal/mole and the two models are in rough agreement. The quantum mechanical model (QM) in general has oxygen-surface equilibrium distances, R_{os}, smaller than the central force model, (CF). However the ranges of R_{os} (QM: 1.7Å - 2.7 Å and CF: 1.9 - 2.8 Å) are still in reasonable agreement and are in accord with the central force model's relatively higher 0-0 separation for the dimer: (11) 2.86Å rather than 2.76Å; and the lower 0-0 separation in the CNDO/2 dimer (2.55Å). The R_{os} stretching frequencies are

TABLE 2
Comparison of results obtained with the quantum mechanical and central field models for similar sites on the basal and prism faces of ice I_h

MODEL	E KCAL/MOLE	Z (Å)	ν (CM^{-1})
BASAL FACE			
QM	-8.0	2.6	280
CF	-6.3	2.7	180
QM	-4.0	1.8	180
CF	-8.4	1.9	170
QM	-6.8	1.7	-
CF	-8.4	1.9	150
QM	-4.2	0.5	-
CF	-7.7	2.0	160
PRISM FACE			
QM	-7.8	2.4	266
CF	-3.8	2.6	—
BASAL FACE-DIFFUSION BARRIER			
QM	1.5 -- 2.5 KCAL/MOLE		
CF	2.0 -- 2.5 KCAL/MOLE		

also encouragingly comparable and probably the greatest disparity between the models is in the dipole orientations of the minimal energy adsorbed monomer. This may well be expected since the central force model cannot be expected to reproduce all the orientational effects of the electrons distributed as they are about the H_2O molecule.

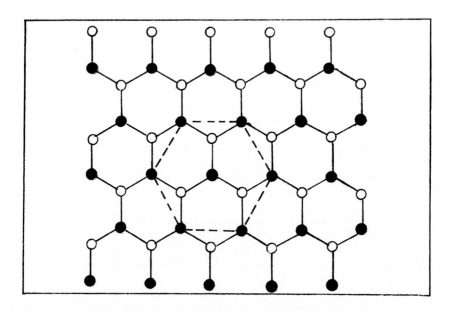

Fig. 10. The 50 molecule model used for the central field calculations on the basal surface. The closed circles indicate (3-bonded) surface water molecules and the open circles the "bulk" molecules in the second layer of the model ice I_h surface. The dashed lines enclose the area of surface examined for bonding sites. The dashed lines also delineates the surface used in the quantum mechanical studies.

From this somewhat limited study we estimate a value for X_s, the mean path length of the adsorbed monomer on the basal plane of ice I_h and τ_s, the mean time it spends on the surface. In making this estimate we assume the following expressions for X_s and τ_s:

$$X_s = a \frac{\nu'}{\nu} \exp[(E_A - E_D)/2kT];$$ (5)

and

$$\tau_s = \nu^{-1} \exp(E_A/kT).$$ (6)

For E_A an average binding energy equal to ~7.0 kcal/mole was used. The diffusion energy, $E_D \simeq 2.5$ kcal/mole, was approximated from average barrier heights. The frequency, ν', associated with the diffusion, was assumed to be one of the average librational frequencies of the monomer (700 cm^{-1}) and $\nu = 200$ cm^{-1} was estimated from the bond stretching frequencies. These values at T=260K predicted an X_s equal to about 330A and a τ_s equal to about 10^{-7} seconds. A similar analysis is underway for the prism face of ice I$_h$.

In conclusion, it is felt that the complimentary studies have been quite illuminating and extensive work along these lines is planned and partly in progress. We plan to examine other basal surface sites for adsorbed monomer minimal energy surfaces and to treat the prism plane in the same detail. An essential modification will be to relax the surface molecules of ice I$_h$ on both planes and observe the effects of relaxation on the surface and on the adsorbed monomer. Plans are also being made to add several monomers to the surfaces and study the effects of monomer-monomer interaction. This could generate a water-like layer on the ice surface and we plan to study this carefully. Future work will also include a study of defects, vacancies, ledges, kinks, impurities and charge centers on the ice surface.

It is also hoped to model AgI and graphite surfaces in a similar fashion. For unrelaxed basal plane AgI we have already generated the electro-static potential 1.5A above the surface. This was done using fourier analysis and lattice sums (14), (see Fig. 11). Quantum mechanical calculations for a graphite surface are also underway. Preliminary results show that edge effects will play a much larger role than in ice. Thus a more extensive section of surface will be required. Using surfaces of 6 to 10 carbon atoms yield surface-water monomer interaction energies of about 10 kcal/mole. This is certainly an over-estimate of the strength of the interaction due to the use of an insufficient number of atoms to represent the bulk surface. Additional studies are planned.

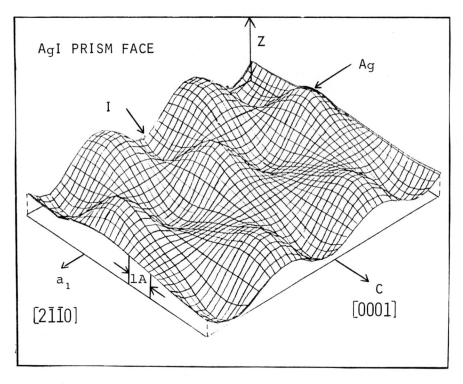

Fig. 11. The electrostatic potential 1.5 A above the prism plane of AgI (unrelaxed) $q_{eff.}$ = *.6e.*

ACKNOWLEDGEMENTS

We would like to thank Kathryn Berkbigler and Kerry Whisnant for computational assistance, Dr. Gerald Alldredge for his help with the AgI surface calculations, the computer center at University of Missouri for providing much of the computer time for these studies, and Professor James Kassner for his continued support and enthusiasm for this project.

REFERENCES

1. Plummer, P.L.M. and B.N. Hale, J. Chem. Phys. 56, 4329 (1972).

2. Hale, B.N. and Plummer, P.L.M., J. Atmos. Sci, 31, 1615 (1974).

3. Hale, B.N. and Kiefer, J., J. Stat. Phys., 12, 437 (1975);
   Hale, B.N. and Kiefer, *International Symposium on Snow and Ice*, IUGG, Grenoble, France, Aug. 1975.

4. Plummer, P.L.M. and Hale, B.N., *Proceedings of the VIII International Conference on Nucleation* (Galvoronsky, Editor), pp. 158-165, Moscow 1975.

5. Plummer, P.L.M., *International Symposium on Snow and Ice*, IUGG, Grenoble, France, Aug. 1975.

6. Bartley, D. L., J. Chem. Phys., (in press).

7. Hale, B.N. and Plummer, P.L.M., J. Chem. Phys. 61, 4012 (1974).

8. Lie, G.C. and Clementi, E., J. Chem. Phys., 62, 2195 (1975). Shipman, L.L. and Scheraga, H.A., J. Phys. Chem. 78, 909 (1974). Weres, O. and Rice, S.A., J. Am. Chem. Soc. 94, 8984 (1972). Ben-Naim, B. and Stillinger, F.H., *Water and Aqueous Solutions*, (R.A. Horne, Editor), Wiley-Interscience, New York, 1972, Chap. 8.

9. Pople, J.A. and Beveridge, D.L., *Approximate Molecular Orbital Theory*, McGraw-Hill, New York, 1970.

10. Plummer, P.L.M. and Stein, E.M., *Proceedings of the International Conference on Cloud Physics*, Boulder, Colorado, July, 1976.

11. Lemberg, H.L. and Stillinger, F.H., J. Chem. Phys., 62, 1677 (1975).

12. Rahman, A., Stillinger, F.H. and Lemberg, H.L., J. Chem. Phys., 63, 5223 (1975).

13. Kern, C.W., and Karplus, M. in "Water, a Comprehensive Treatise" (F. Franks, Ed.) Vol. 1, p. 37, Plenum, New York, 1972.

14. Gevirzman, R., and Kozirovski, Y., J. of the Chemical Society, Faraday Transactions II, 68, 1699, 1972.

THE EFFECT OF FRICTION REDUCING SUBSTANCES ON THE DROPLET SIZE DISTRIBUTION FROM SPRAY NOZZLES

C.T. Wilder and J.W. Gentry
Department of Chemical Engineering
University of Maryland

L.A. Liljedahl
Agricultural Research Service
U.S. Department of Agriculture

ABSTRACT

The droplet size distribution emitting from a Whirljet and a Fulljet nozzle were measured for a 25 ppm polyacrylamide solution as a function of radial distance from the center of the cone and the fluid pressure of 30-75 psi. Corresponding values were measured for water.

The measurements employed on optical array particle probe to measure the droplets in the size ranges from 7.5 to 2300 microns.

Results for both nozzles showed that the polymer solution produced a significant increase in mean droplet size as compared to water with the largest effect observed at low pressures toward the center of the spray cone. Similarly droplet mass was concentrated toward the center of the spray cone but mass distribution approached that of water as the pressure was increased.

Measurements were made of the surface tension and viscosity of the polymer solutions. Comparison of the experimental measurements were made with existing literature correlations.

I. INTRODUCTION

Spray nozzles are used for many purposes but very few studies of nozzle performance are available. Nearly all of these studies are confined to Newtonian fluids and measure the average drop size throughout the spray cone.

The purpose of this study was:

1) to measure the drop size distribution with respect to pressure and spatial position.
2) to determine the effect of a friction reducing polymer on nozzle performance.

Two common pressure nozzles, a Fulljet and a Whirljet were selected for study at pressures from 30 to 75 psi.

II. LITERATURE SURVEY

A. Particle Probes

Kollenberg[1] has shown that by using a coherent collimated light beam produced by a laser it is possible to determine the size of aerosol particles by the size of the shadow it produces. The number of photodetectors triggered by the shadow is directly proportional to the diameter of the particle.

B. Polymer Solution

Polyacrylamide is a long chain polymer which in solution reduces the friction of flow at turbulent shear rates.

Little[2] has reviewed the studies made with a number of polymers and evaluated several models for friction reduction. None of these models is completely successful but the most plausable all postulate a reduction in the intensity of turbulence.

Hoyt[3] photographed the breakup of a jet in air. It was found that the addition of polyethyleneoxide, a polymer with friction reducing properties, hindered disturbances on the surface of the fluid jet.

C. Fulljet Nozzle

Fulljet is the brand name of the Spraying Systems Co. solid cone nozzle. The body of this nozzle is tubular with a sharp contraction at the orifice. A propeller shaped insert ahead of the orifice number imparts a swirling motion to the fluid.

Perry[4] and the manufacture claim this nozzle produces a uniform spray distribution but the atomization efficiency is lower than a swirl atomizer.

D. Whirljet Nozzle

The Whirljet nozzle is one form of swirl atomizer manufactured by Spraying Systems Co. This nozzle consists of a conical chamber with fluid entering at a tangent to the base of the nozzle and leaving through an orifice at the apex of the cone. Watson[5] gives a graphical correlation for Sauter mean diameter as a function of pressure and discharge coefficient only. This was intended for use with fuels.

Doumas[6] studied the flow of the fluid within the nozzle. An air core surrounded by the swirling fluid was measured passing through the length of the nozzle while in operation. Particle volume was found to be concentrated in a cone with little volume falling on the nozzle axis.

Turner[7] gives a correlation for geometric volume mean drop diameter for Whirljet nozzles having greater flow rates than used here. Measurements were obtained by solidifing hydrocarbon material emitted from the nozzle.

Taylor[8] has derived equations of motion for the fluid film within a swirl nozzle. These calculations indicate that the majority of fluid leaves the nozzle through the flow boundary layer between the cone wall and the freely circulating fluid. Wang[9] gives similar equations extended to include non-Newtonian fluid.

III. EQUIPMENT AND PROCEDURE

Particle counts were recorded in a counting unit having 15 size categories. Input was from either of two remote optical array probes. The small range probe had 15 micron wide channels centered on from 15 to 225 micron diameter particles. The larger probe had channels centered on 150 to 2250 microns with 150 micron width channels.

These two probes were mounted in the bottom of a cabinet with the nozzle mounted 16-1/2 in. above the probes and aimed directly downward. The mounting allowed horizontal movement to position the nozzle either directly over the probes or up to 10 in off axis.

Particle counts were made for 10 sec. periods alternating between the two probes. Measurements were made at 2 in. intervals at from 0 to 10 in. off-set between the nozzle and probe. Pressures used were 30, 40, 50 and 75 psi. This set of measurements was repeated for water and a poly-acrylamide solution mixed in a concentration of 0.1 g per 1 gal. of water. Nozzles tested were a 1.0 Whirljet and a 1.5 Fulljet.

Raw data from the probes were reduced to a usable form by computer. Probe sensitivity was a function of particle diameter so all counts were first converted to a common basis. An additional correction was next applied to equalize the counts from each probe based on the overlaping range of particle diameters. Mean diameters were calculated from the corrected counts by the following formulas.

$$\bar{D}_{10} = \frac{\Sigma D(n)F(n)}{\Sigma F(n)} \quad \text{count mean diameter}$$

$$\bar{D}_{32} = \frac{\Sigma D(n)^3 F(n)}{\Sigma D(n)^2 F(n)} \quad \text{Sauter mean diameter}$$

$$\log \bar{D}_{gvm} = \frac{\Sigma D^3 \log(D(n))f(n)}{\Sigma D^3 F(n)} \quad \text{geometric volume mean diameter}$$

Where $D(n)$ is the mean diameter of channel n, $F(n)$ is the particle count in n and summation is over all size categories.

IV. RESULTS AND DISCUSSION

A. Whirljet Nozzle

Recorded particle counts were assumed to represent the average particle population over an annulus 2 in. wide around the nozzle axis. Therefore particles from all positions were summed by a weighting factor based upon the area of this ring. The resulting mean diameters were calculated versus pressure to give the results shown in Table 1. Mean diameters for the Whirljet nozzle predicted by several sources are also given in this table. Wang's prediction was based on the calculated boundary layer thickness which was found for a grooved core pressure nozzle to be approximately equal to the Sauter mean diameter.

The experimental mean diameters were larger than the predicted diameters but the decrease in diameter as pressure increased was also larger. This is probably due to the relatively low pressures used where the nozzle flow pattern was not fully developed.

TABLE 1

CORRELATION OF MEAN DIAMETER WHIRLJET NOZZLE

Water

$$\bar{D}_{10} = 442 \ P^{-.50} \qquad\qquad R^2 = .85$$

$$\bar{D}_{gvm} = 743 \ P^{-.25} \qquad\qquad R^2 = .98$$

$$\bar{D}_{32} = 926 \ P^{-.35} \qquad\qquad R^2 = .97$$

PAM

$$\bar{D}_{10} = 1388 \ P^{-.68} \qquad\qquad R^2 = .93$$

$$\bar{D}_{gvm} = 8754 \ P^{-.73} \qquad\qquad R^2 = .97$$

$$\bar{D}_{32} = 8120 \ P^{-.74} \qquad\qquad R^2 = .97$$

Predicted

Turner   (water)

$$\bar{D}_{gvm} = 353 \ P^{-.12} \qquad\qquad R^2 = .71$$

Wang   (water)

$$D_{32} = 348 \ P^{-.0879} \qquad\qquad R^2 = .998$$

Wang   (PAM)

$$D_{32} = 160 \ P^{-.191} \qquad\qquad R^2 = .997$$

Figures 1 and 2 compare the Sauter mean diameters with respect to position for water and the
polymer solution respectively.

Water showed essentially the same size distribution at all pressures. That is the smallest
particles occured on the nozzle axis. Drop size
increased reached a peak and then decreased slightly
toward the edge of the spray cone. At low pressures (less than 50 psi) the polymer solution produced the largest particles on the nozzle axis with
drop size declining as off-set increased. At 50
psi the polymer produced a size pattern similar to
water at only 30 psi with the smallest drops on axis.

Figures 3 and 4 show the particle volume dis-

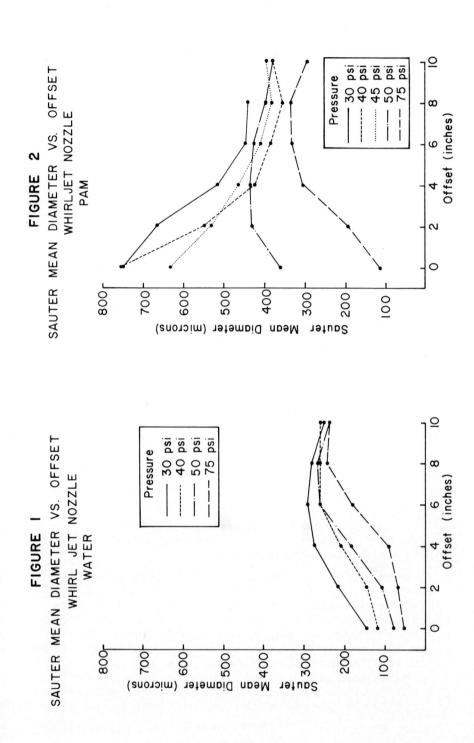

FIGURE 2

SAUTER MEAN DIAMETER VS. OFFSET
WHIRLJET NOZZLE
PAM

FIGURE 1

SAUTER MEAN DIAMETER VS. OFFSET
WHIRL JET NOZZLE
WATER

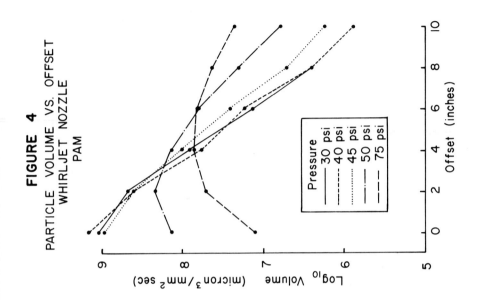

FIGURE 4

PARTICLE VOLUME VS. OFFSET
WHIRLJET NOZZLE
PAM

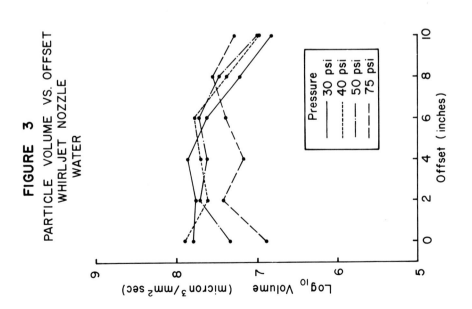

FIGURE 3

PARTICLE VOLUME VS. OFFSET
WHIRLJET NOZZLE
WATER

tribution for water and polymer as a function of position and pressure. Water showed a nearly flat profile in the inner region of the spray cone which decreased toward the edge. At low pressures the polymer showed an increase in volume at the nozzle axis as compared to water. Volume rapidly fell below water as off-set increased.

At 50 and 75 psi polymer volume showed a sharp drop on axis. Volume increased and then delined toward the edge of the cone.

B. Fulljet Nozzle

Data similar to the Whirljet nozzle was taken and processed for a Fulljet nozzle. Due to space limitations only correlations in Table 2 are given although figures for diameter and volume versus position are described.

TABLE 2

Correlation of Mean Diameter Fulljet Nozzle

Water

$$\bar{D}_{10} = 101 \ P^{-.11} \qquad\qquad R^2 = .16$$

$$\bar{D}_{gvm} = 820.2 \ P^{-.25} \qquad\qquad R^2 = .43$$

$$\bar{D}_{32} = 1034 \ P^{-.35} \qquad\qquad R^2 = .58$$

PAM

$$\bar{D}_{10} = 28.0 \ P^{.25} \qquad\qquad R^2 = .80$$

$$\bar{D}_{gvm} = 1436 \ P^{-.33} \qquad\qquad R^2 = .98$$

$$\bar{D}_{32} = 1248 \ P^{-.33} \qquad\qquad R^2 = .93$$

The polymer solution produced a significant increase in the Sauter mean drop size with the constant term in the correlation increasing by 21% over water while the exponent on pressure changed only from -0.33 for water to -0.35 for polymer. The Sauter mean diameter with water showed no consistent variation with position. The polymer solution produced the largest drops at the center of the spray cone for all pressures. That is a pattern similar to the Whirljet nozzle with polymer for pressures less than 50 psi.

Particle volume showed a increased spread away from the nozzle axis as the pressure increased with both water and polymer. The polymer solution produced a concentration of volume toward the nozzle axis as compared to water. Again behavior was similar to the Whirljet nozzle at low pressures using polymer.

C. Explanation of Polymer Effects

Two explanations of the observed changes in nozzle behavior when using the polymer solution are plausable.

1) The change in viscosity results in an increase in the boundary layer thickness resulting in increased drop size. Wang's equations indicate this occures for the Whirljet nozzle but fluid behavior in the Fulljet nozzle is not known.

2) The polymer hinders the breakup of the fluid into droplets. Hoyt's photographs show this does happen but with a considerably different geometry for the fluid.

Data available does not indicate which mechanism is actually responsible for observed results.

V. REFERENCES

1. Knollenberg, R.G., J. Appl. Meteorol. Vol. 9.
2. Little, R.C. et al., Ind. Eng. Fundam. Vol. 14, No. 4, 1975.
3. Hoyt, J.W., Taylor, J.J., & Runge C.D., J. Fluid Mech., Vol. 63 Part 4.
4. Perry, "Chemical Engineers' Handbook", 5th Ed., McGraw-Hill, 1973.
5. Watson, E.A., Proc. Inst. Mech. Eng. (London) Vol. 158, No. 2, 1948.
6. Doumas, M. & Laster, R., Chem. Eng. Progr., Vol. 49, No. 4, April 1953.
7. Turner, G.M. & Mouton, R.W., Chem. Eng. Prog., Vol. 49, No. 4, April 1953.
8. Taylor, G.I., Quart. J. Mech. Appl. Math., Vol. III, Part 2 (1950).
9. Wang, K.H., Nakano, & Chi Tien, "Atomization and Drop Size of Non-Newtonian Fluid", Syracuse University Research Institute.

ON THE MECHANISM OF DROPWISE CONDENSATION

Jer Ru Maa and Wen Hai Wu
Cheng Kung University

ABSTRACT   The behavior of the condensation rate-super-
saturation relationship for the condensation of water
vapor on chilled immiscible liquids agree with that
predicted by the heterogeneous nucleation theory.
This confirms that the dropwise condensation is a
nucleation phenomenon. It is shown that the heterogeneous nucleation theory can be used to predict the
heat transfer rate of the dropwise condensation process
if the distribution of the strength of the active
nucleation sites is considered. Problems concerning
the calculation of the drop size distribution by the
population balance are also discussed.

I.   INTRODUCTION

Considerable interest and study have been centered on
dropwise condensation because of its high heat transfer
coefficient as compared to filmwise condensation. Dropwise
condensation is a complex process. In the case of condensation of water vapor, the substrate surface has to be
sufficiently hydrophobic in order to maintain the mode of
condensation dropwise. Condensate drops form on active
sites, which are pits, scratches or microscopic specks of
wettable particles on the substrate surface. The site
density, the diffusion of water vapor through the layer of
noncondensable gas accumulated at the vicinity of the
condensate surface, the resistance of the vapor-liquid interface, the increase of vapor pressure caused by the curvature
of the drop surface, the conduction of heat through the
drop body and the condenser wall, and the constriction
resistance caused by the presence of neighboring drops, are
all important factors influencing the heat transfer rate
of the dropwise condensation.

Dropwise condensation is a transport phenomenon and has been treated by many authors with success using this approach. However, it is also an interfacial phenomenon, a nucleation phenomenon as well as a particulate process with stochestical natures. This work is a discussion of the dropwise condensation from these points of views.

II. CONDENSATION ON SURFACE WITHOUT ACTIVE SITE

The mechanism of dropwise condensation is complicated by the fact that drops are formed on the nucleation sites. The problem can be much simplified if one is able to study the condensation of water vapor on substrate surfaces without active nucleation site. This was realized experimentally by using a jet tensimeter (1). A chilled laminar jet of an immiscible liquid was exposed to water vapor in a chamber maintained at a constant pressure. The time of exposure was sufficiently short that the collision and merging of the condensate droplets and the temperature increase caused by the latent heat of condensation are of negligible importance. The condensation rate on this jet surface was determined by measuring the supplying rate of water vapor to the chamber.

It was observed that the rate of condensation depended highly on the interfacial properties. At the same super-saturation, the water vapor condensed much faster on the surface of less hydrophobic immiscible liquid. The experimental results for the condensation of water vapor on a laminar jet of n-dodecane at various temperature are shown as condensation rate versus supersaturation, p_v/p_1 ,

Figure 1. It seems that for each temperature there exists a "critical supersaturation." The nucleation rate is small for values of supersaturation below this characteristic value and extremeley high above it. This is a typical characteristic for nucleation process. The critical super-saturation values for each temperature are estimated by extrapolating the experimental curve to zero condensation rates as shown by the dotted lines in this figure.

The smooth surface of the immiscible liquid can also be considered as a surface with nucleation sites of the same strength existing everywhere. The nucleation rate, I , on such a surface can be computed by the classical hetero-geneous nucleation theories. The derivation was presented

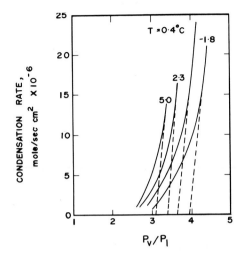

Fig. 1.   Condensation rate of water
vapor on n-dodecane.

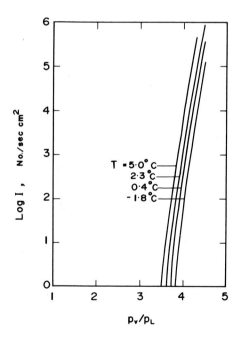

Fig. 2.   Steady-state nucleation rates
as functions of supersaturation.

elsewhere (2). The calculated nucleation rates for various temperatures are shown in Figure 2 as functions of super-saturation, p_v/p_1 . The "critical supersaturation" for each temperature is taken as the p_v/p_1 value at which I equals unity (3,4). The values of the critical super-saturation for several temperatures obtained theoretically and experimentally are listed in Table I for the purpose of comparison. The agreement varifies that the dropwise condensation is indeed a nucleation phenomenon.

TABLE I Critical Supersaturations Calculated Using the Nucleation Theory Compared with the Experimental Values

T_1	°C	5.0	2.3	0.4	−1.8
Critical Supersaturation	Theoretical	3.5	3.6	3.7	3.8
	Experimental	3.1	3.4	3.7	3.95

III. THE VARIATION OF SITE ACTIVITY

In the processes of dropwise condensation on solid surfaces, such a critical supersaturation is not observed; condensation starts at very low supersaturation, its rate increases gradually as the temperature is increased. This is because   condensate drops form on sites. No two sites are identical and the effective supersaturation vary from site to site. On some of the sites, condensate nuclei may form at very low supersaturation. On some of the other sites a higher supersaturation may be necessary for the nucleation to begin. One has to take this variation of site activity into consideration if the nucleation theory is used to compute the rate of dropwise condensation.

If the effective values of contact angle between the condensate and the substrate surface, θ , are all the same, the nucleation rate of the condensate drops, $I(\theta,N_s)$, can be computed using the heterogeneous nucleation theory (5). It is a function of θ and the site density, N_s . The experimentally determined contact angle, θ_E , is a macro-scopic property. Because of the variation of the site activity, the effective contact angle at the vicinity of the nucleation sites, θ , may not equal to the observed value,

θ_E. It may be assumed that the θ value distributes normally, the fraction of surface imperfections having θ value in the range of θ to $\theta + d\theta$ is

$$f(\theta)d\theta = \frac{1}{\sigma\sqrt{2\pi}} \exp\left[-\frac{(\theta-\theta_E)^2}{2\sigma^2}\right]d\theta \qquad (1)$$

where σ is the standard deviation. The actual value of the nucleation rate is therefore:

$$\bar{I} = \int_0^\pi I(\theta)f(\theta)d\theta . \qquad (2)$$

The heat flux of dropwise condensation can be estimated from the $\bar{I}$ value based on a generally accepted experimental observation that in the process of dropwise condensation, major portion of the heat is transfered by small drops which grow mainly by direct condensation; negligibly small fraction of the heat is transfered by large drops which grow mainly by coalescence (6). The boundary between these size ranges may be taken as $r = r_o$, where r_o is half of the distance between the active nucleation sites. $r_o = 1/2\ N_s^{1/2}$ if it is assumed that the sites are arranged in a square array. Since the coalescence of drops with radius smaller than r_o can be neglected and these drops are responsible to most of the heat transmission, the heat flux can be approximated by:

$$Q = \bar{I}\ V_o\rho\lambda . \qquad (3)$$

Where ρ is the condensate density, λ is the latent heat of condensation and V_o is the volume of drop of radius r_o. Heat flux values estimated by this equation are approximately in agreement with the experimental data as shown in Figure 3.

IV. THE STUDY OF DROPWISE CONDENSATION BY POPULATION BALANCE

The correct prediction of heat flux of dropwise condensation requires knowledge of the growth rate of single drop and the distribution of drop size. Since dropwise

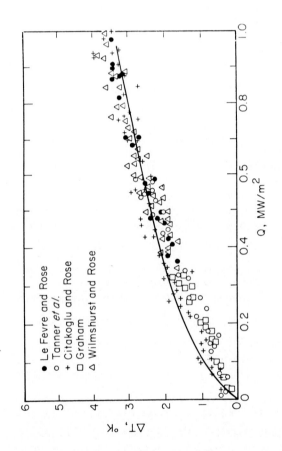

Fig. 3.   The heat flux as function of ΔT for the dropwise condensation of water vapor of 760 Torr on copper surface promoted with dioctadecyl disulfide. The curve is computed using Equation 3. The data points are quoted from Reference 8.

condensation is a particulate process of stochestical nature, it is logical to compute the drop size distribution by the method of population balance.

When a bare hydrophobic substrate surface is exposed to a supersaturated water vapor, condensate nuclei form on the active sites and the growing and merging process begin. The drops grow by direct condensation. They also combine with the near by ones if touched. Since the new drop formed by coalescence occupies less space than the original ones, some bare substrate surface is exposed and is available for the formation of more initial droplets. This process continues until a drop reaches its critical size and slides down. It sweeps all drops on its way and leaves the substrate surface bare again. The population balance equation for the process of dropwise condensation can be written as:

$$\frac{N(r,t)}{\partial t} = B(r,t) - D(r,t) - \frac{\partial}{\partial r}[N(r,t)G(r)]$$
$$+ \frac{\partial}{\partial r}\left[b(t)\delta_{rr_{min}}\right].$$

(4)

In this equation, $N(r,t)$ is the population density of the drops, and $N(r,t)dr$ is the number of drops within the size range $(r, r + dr)$ per unit area at a spot whose exposure time, or the time since this area is swept bare, is t. $B(r,t)$ and $D(r,t)$ are the birth function and death function, $B(r,t)dr$ and $D(r,t)dr$ are the number of drops within the size range $(r, r + dr)$ produced and destroyed by coalescences and merging per unit area per unit time at a spot whose exposure time is t, respectively. $G(r)$ is the growth rate of a single drop of radius r by direct condensation. Since a quantity $N(r,t)G(r)$ represents the flux per unit area per unit time of drops whose radii grow passing the value r, the third term on the right-hand side of Equation 4 denotes the net drop number to be stored at radius r. $b(t)$ is the rate of generation of drops of minimum radius, r_{min}, on the bare area exposed due to the coalescences of other drops per unit condenser surface area at a spot whose exposure time is t. r_{min} is the minimum radius of drop which can grow and $b(r)\delta_{rr_{min}}$ is the Dirac delta function of strength $b(r)$.

Since $N(r,t)$ is the population density of the drops at a spot of exposure time t, one needs to average the $N(r,t)$ value over the exposure period, τ, or the length of time between two consecutive sweeps by the falling drops:

$$\bar{N}(r,\tau) = \frac{1}{\tau} \int_0^\tau N(r,t)dt \ .$$ (5)

Moreover, the exposure period τ varies with the location on the condenser surface. τ value is longer at the top part of the condenser surface and is shorter at the bottom part. In order to compute the average heat flux of the entire condenser surface, one needs to average the $\bar{N}(r,\tau)$ value over all locations:

$$\bar{\bar{N}}(r) = \frac{1}{\ell} \int_0^\ell \bar{N}(r,\tau)dx$$ (6)

where ℓ is the height of the condenser surface and x is the distance from the top.

Note that the exposure period, τ, depends also on the condensate-substrate contact angle, θ. This is another point where interfacial properties come into play.

In Equations 4, 5 and 6, the growth rate, $G(r)$, is a well understood function of drop radius, but $B(r,t)$, $D(r,t)$, $b(t)$ and $\tau(x)$ are all complicated functions. It is obvious that one needs certain degrees of simplification in order to solve for the population density.

In the work of Tanaka (7), the coalescence of a drop of radius r with another one larger than itself was considered as its death. The coalescences with drops smaller than itself was considered as part of its growth. The calculation was further simplified by introducing a concept of "hydraulic diameter" and the transient population densities were obtained for drops of radius larger than 5 microns.

Another approach to solve for the drop size distribution is to assume it is steady and homogeneous, or in other words, the probability of finding a drop of a certain size is the same at any position on the condenser surface at any time.

In the process of dropwise condensation, only the small drops are important to the heat transmission. The coalescence between these small drops is negligible and they grow mainly by direct condensation. Hence, for the purpose of computing the heat flux, only a steady state size distribution of these small drops is necessary. The birth and death functions can be ignored and the mathematics much simplified.

V. DISCUSSION

The process of dropwise has been studied by many authors since the pioneering paper of Schmidt and his coworkers, and much progress has been made over the last fourty years toward the economical industrial application and complete understanding of its mechanism. It seems to these authors that, at this stage, the two major break throughs need to be made are: to find appropriate material which would economically promote long lasting dropwise condensation and to develop a method of predicting the active site density and the distribution of site strength on the promoted substrate surface.

VI. REFERENCES

1. Maa, J. R., and Hickman, K., Desalination 10, 95-111 (1972).

2. Wu, W. H., and Maa, J. R., Condensation and Nucleation on the Surface of Immiscible Liquid, paper accepted by J. of Colloid and Interface Sci. (1976).

3. Zettlemoyer, A. C., ed.,"Nucleation," Dekker, New York, 1969.

4. Abraham, F. F., "Homogeneous Nucleation Theory," Academic Press, New York, 1974.

5. Wu, W. H., and Maa, J. R., A Mechanism for Dropwise Condensation and Nucleation, paper accepted by Chem. Eng. J. (1976).

6. Graham, C., and Griffith, P., Int. J. Heat Mass Transfer 16, 337 (1973).

7. Tanaka, H., <u>J. of Heat Transfer, Trans. ASME, Series C.</u> 97, 72 (1975).

8. LeFevre, E. J., and Rose, J. W., Bicentenary of the James Watt Patent, Proc. Symp. Univ. of Glasgow, p. 165, 1969.

ON THE SIMILARITY BETWEEN BOILING AND CONDENSATION

Jer Ru Maa
Cheng Kung University

ABSTRACT Different modes of boiling and condensation
processes are compared on the Q - ΔT plots to show their
differences and similarities. The phenomena concerning
transitions of modes, the inception superheat and
supersaturation, and the boiling and condensation on
surfaces without nucleation site, and the approaches
of achieving better heat transfer efficiencies are
discussed in terms of interfacial and nucleation theories.

I. INTRODUCTION

Boiling and condensation are efficient fashions of heat
transmission. They have been studied extensively for the
understanding of their basic mechanisms and for the improvement of equipment design. They are both complex processes
with similarities and differences. Boiling processes are
usually complicated by the two phase flow hydrodynamics and
instabilities. The diffusion of the condensing vapor
through the uncondensable gas is usually a major controlling
factor to the rate of condensation.

Boiling and condensation are usually regarded as
transport phenomena. The treatment of film condensation by
Nusselt and that of film boiling by Bromley are typical
examples, and the similarity between their resulting working
equations are remarkable. However, they are also nucleation
phenomena as well as interfacial phenomena with stochestical
natures, especially in the cases of nucleate boiling and
dropwise condensation.

The mode of boiling and condensation depends very much
on the solid-liquid interfacial properties. For water, the
condensation can be made dropwise if the solid surface is
hydrophobic, and nucleate boiling is more likely to happen
on a hydrophilic surface.

II.   THE Q - ΔT PLOTS

The natures of the boiling and condensation processes are often displayed on the heat flux - ΔT plot, where ΔT is defined as the difference between the saturation temperature of the condensing vapor and the substrate surface temperature in the case of condensation and as the difference between the temperature of the heating surface and the saturation temperature of the boiling liquid in the case of boiling. Figure 1 and 2 are the Q - ΔT plots for boiling and condensation respectively. They are constructed not exactly to the scale for the sake of clarity.

In Figure 1, Curve A is the "boiling curve" first prepared by Nukiyama based on his experimental data. This curve can be divided into several regions, they are, from left to right: free convection region, nucleate boiling region, unstable film boiling region, stable film boiling region, with transition regions between the neighboring ones. The effects of many factors on the shape of the boiling curve have been studied extensively. These factors are: finish and other properties of the heating surface, pressure, liquid properties, subcooling, velocity, impurities, body forces, etc. Rougher heating surface or higher system pressure move the boiling curve to the left, as Curve B in Figure 1. Smoother heating surface or lower system pressure has the opposite effect, as shown by Curve C in this figure.

In Figure 2, Curves A and B are the Q - ΔT relationships of dropwise and filmwise condensations respectively. They have also been studied extensively by many authors since the pioneering work of Nusselt on the filmwise condensation and that of Schmidt on the dropwise condensation.

III.   THE TRANSITIONS OF THE MODES OF CONDENSATION

It has been shown experimentally that as ΔT increases, the mode of condensation may change from dropwise to filmwise through a transition region (1,2), as indicated by Curve A' in Figure 2. This is very similar to the transition from the nucleate to film boiling in Figure 1. As ΔT further increases, liquid in contact with the cooling surface may freeze and the conduction through the solid plays an important role at higher ΔT (2). This is equivalent to film boiling at high ΔT where the heat transmission by radiation is increasingly important.

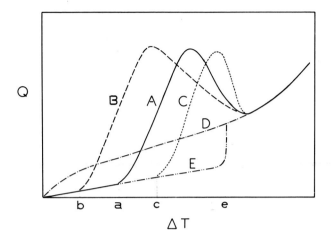

Fig. 1.   The Q - ΔT plot for boiling.

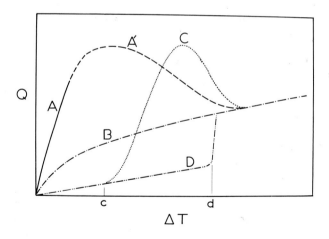

Fig. 2.   The Q - ΔT plot for condensation.

IV. THE INCEPTION ΔT

Under ordinary conditions, it seems that a certain ΔT has to be reached before the starting of the nucleate boiling; such an inception ΔT is not required for the case of dropwise condensation. This is due to the difference in the sizes of their critical nuclei.

In the case of condensation, the critical drop radius is:

$$ r_c^* = \frac{2\sigma v_1}{kT \ \ln \ (p_v/p_s)} . \tag{1} $$

The critical bubble radius in the case of boiling is:

$$ r_b^* = \frac{2\sigma}{p_s - p_1} \tag{2} $$

r_c^* and r_b^* are the minimum drop and bubble sizes which can possibly grow, respectively (3). In both the nucleate boiling and dropwise condensation, the vapor bubbles and liquid drops grow on sites, which are usually pits, scratches or other surface irregularities. If one assumes that the active sites are about the same size as that of the critical drops or bubbles, the necessary site radii for dropwise condensation and nucleate boiling, r , can be estimated using Equations 1 and 2 respectively. The site radii estimated by this method are plotted in Figure 3 as functions of ΔT . It is clear that the required size of the boiling sites is more than one thousand times larger than that of the condensation sites.

In Figure 1, the inception boiling superheat indicated by Point a represents that at certain surface roughness, site larger than the size corresponding to this ΔT determined by the upper curve in Figure 3 is not available. For a rougher surface, Point b in Figure 1 indicates that sites of larger size are available; and Point c indicates that for a smoother surface the maximum size of available site is smaller.

In Figure 2, Curve A indicates that dropwise conden- sation may begin at very small ΔT . This is because the required site size for such small ΔT is about 10^{-5} cm as indicated by the lower curve of Figure 3, and that on

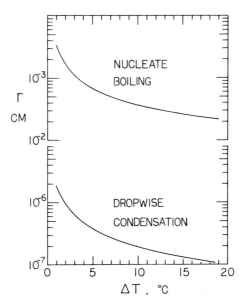

Fig. 3.   Estimated site radii for
nucleate boiling and dropwise condensation
as functions of ΔT.

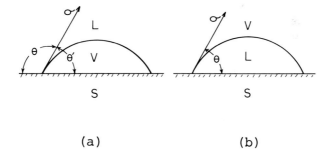

(a)   (b)

Fig. 4.   a) Vapor nucleus on smooth
substrate.   b) Condensate nucleus on smooth
nucleus.

ordinary condenser surfaces, sites of this size are always available. If a hydrophobic surface is carefully polished that the maximum site size on this surface is made much smaller, a "condensation curve" as Curve C in Figure 2 should be possible. Point c on Figure 1 indicates a ΔT value corresponding to the maximum site size on this surface.

V. BOILING AND CONDENSATION ON SURFACES WITHOUT SITE

If the heating surface is perfectly smooth without site, the chance of bubble formation should be the same everywhere on this surface. The nucleation rate of bubbles as shown in Figure 4a is (4):

$$J_b = \left(\frac{\Delta G_b^*}{3\pi k T i_b^{*2}}\right)^{1/2} \cdot \frac{p^*}{(2\pi m k T)^{1/2}} \cdot 2\pi r_b^{*2}(1 - \cos\theta')$$

$$\cdot n_{sl} \exp\left(\frac{-\Delta G_b^*}{kT}\right) \tag{3}$$

where

$$\Delta G_b^* = \frac{16\pi\sigma^3\phi(\theta')}{3(p^* - p_1)^2} \tag{4}$$

$$\phi(\theta') = (2 + \cos\theta')(1 - \cos\theta')^2/4 \tag{5}$$

$$p^* = p_s \exp\frac{v_1(p_1 - p_s)}{kT} \tag{6}$$

$$r_b^* = 2\sigma/(p^* - p_1) \tag{7}$$

$$i_b^* = 4\pi r_b^{*3} p^*\phi(\theta')/3kT . \tag{8}$$

Figure 4b shows a liquid drop on a perfectly smooth surface without site. If one assumes that the growth of such drops is mainly by direct condensation of the vapor molecules on the drop surface, their nucleation rate is (4):

$$J_c = \left(\frac{\Delta G_c^*}{3\pi k T i_c^{*2}} \right)^{1/2} \cdot \frac{P_v}{(2\pi m k T)^{1/2}} \cdot 2\pi r_c^{*2} (1 - \cos \theta)$$

$$\cdot n_{sv} \exp \left(\frac{-\Delta G_c^*}{kT} \right) \tag{9}$$

where

$$\Delta G_c^* = \frac{16\pi \sigma^3 \phi(\theta)}{3\Delta G_v^2} \tag{10}$$

$$\Delta G_v = - \frac{kT}{v_1} \ln \frac{P_v}{P_s} \tag{11}$$

$$r_c^* = -2\sigma/\Delta G_v \tag{12}$$

$$i_c^* = \frac{4\pi r_c^{*3} \phi(\theta)}{3v_1} \cdot \tag{13}$$

Both the boiling nucleation rate, J_b, and the condensation nucleation rate, J_c, are functions of the contact angle, θ, and ΔT. Computation results using Equations (1) to (13) show that in the case of condensation, for each set of contact angle and cooling surface temperature, there exists a critical p_v or ΔT. The nucleation rate is negligibly small when p_v or ΔT is below the critical value, and becomes very high when this critical p_v or ΔT is exceeded. This is a typical characteristic for nucleation processes. The critical p_v and ΔT values for the condensation on a 100°C smooth surface without site are computed and plotted in Figure 5 as functions of contact angle θ. In the region above the curve in this figure, everywhere on the substrate surface is a favorable spot for the formation of liquid drop and the condensation is film-wise. In the region below this curve, there is no condensation if the substrate surface is perfectly smooth, the heat transfer is contributed only by convection as indicated by Curve D in Figure 2. As ΔT reaches the critical value

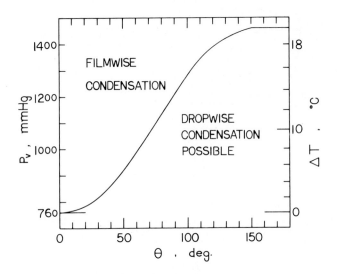

Fig. 5.   Effect of contact angle on the mode of condensation.   T = 100°C

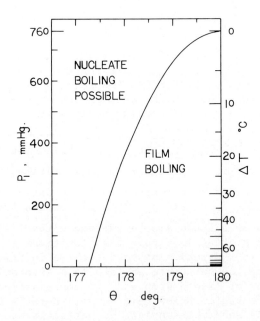

Fig. 6.   Effect of contact angle on the mode of boiling.   T = 100°C

at Point d , filmwise condensation starts suddenly.
Condensation process of the type of Curve D in Figure 2 was
realized experimentally by condensing water vapor on the
surface of chilled laminar jet of various organic liquids
(5). In the region below the curve in Figure 5, if sites
are available on the substrate surface, drops will form
at the positions of the sites, the mode of condensation may
be filmwise or dropwise depending on the site density on the
substrate surface, removing rate of large drops, and other
factors.

Similar computation was made for the case of boiling.
The curve in Figure 6 indicates the critical p_1 and ΔT
for the boiling of water from a surface of 100°C as
functions of contact angle, θ . On the lower right side
of this curve, the J_b value is high, everywhere on the
substrate surface is a favorable spot for the formation of
vapor bubble, the mode of phase transition is filmwise.
It may be stable film boiling or unstable film boiling
depending on the two phase flow hydrodynamics and
instabilities. This type of film boiling, represented by
Curve D in Figure 1, can happen only if the heating surface
is highly hydrophobic. It can be realized only if the
contact angle is larger than 177.5° as indicated in Figure 6.
On the upper left side of the curve in Figure 6, the J_b
value is negligibly low, there is no boiling if the heating
surface is perfectly smooth, the heat transfer is contributed only by convection as indicated by Curve E in Figure 1.
As ΔT reaches the critical value at point e , film
boiling starts suddenly. Boiling process of the type of
Curve E in Figure 1 was realized by heating up water drops
submerging in immersible host liquids (6) or resting on
clean mercury surface (7,8). On the upper left side of the
curve in Figure 6, if sites are available on the heating
surface, the boiling process will follow a curve such as
A, B or C in Figure 1.

VI. SIMILARITIES AND DIFFERENCES BETWEEN BOILING AND
    CONDENSATION

By comparing Figures 1 and 2, one finds that boiling
and condensation processes are indeed qualitatively similar.
For each curve on the Q - ΔT diagram for boiling, it is
likely to find a corresponding one on the q - ΔT diagram
for condensation. However, there are also differences.

Figure 6 indicates that unless the heating surface is highly hydrophobic, the mode of boiling will be nucleate. The transition from nucleate to film boiling is usually caused by the vapor bubbles not being removed fast enough from the heating surface, not by reason due to the liquid-substrate interfacial properties. The type of film boiling of Curve D in Figure 1 is very unlikely to happen. Figure 5 shows that the liquid-substrate properties play a far more important role in determining the mode of condensation. Both dropwise and filmwise condensation are possible within a wide range of super saturation depending on the value of contact angle. At high heat flux, if the large condensate drops are not removed fast enough from the condenser surface, the mode of condensation may transit from dropwise to filmwise. This is similar to the problem of removing vapor bubbles from the heating surface in the case of nucleate boiling, except that it is usually less critical because the condensate drops occupy much less space than the vapor bubbles.

The ultimate goal of studying the boiling and condensation process is to achieve high heat flux at low ΔT. In the case of dropwise condensation, this objective can be approached by finding the right promoters which gives optimum condensate — substrate contact angle and active site density. In the case of nucleate boiling, the major influence of the interfacial properties is their effect on the frequency of bubble detachment. To improve the means of increasing the size of active sites and techniques of two phase flow hydrodynamics seem to be the more effective approach to achieve better heat transfer efficiency.

VII. NOTATIONS

ΔG_b^* Gibbs free energy of formation of a critical vapor bubble, erg

ΔG_c^* Gibbs free energy of formation of a critical condensate drop, erg

ΔG_v Gibbs free energy change per unit volume of condensate, erg cm^{-3}

i_b^* number of molecules in the critical vapor bubble

i_c^* number of molecules in the critical condensate drop

J_b bubble nucleation rate, sec^{-1} cm^{-2}

J_c drop nucleation rate, $\sec^{-1} \text{cm}^{-2}$

k Boltzmann's constant, 1.38×10^{-16} erg $\deg^{-1}$

m molecular mass, gm

n_{sl} number of condensate molecules per unit area, cm^{-2}

n_{sv} number of vapor molecules per unit area, cm^{-2}

p^* pressure of vapor in the critical bubble, dyne cm^{-2}

P_1 pressure of the liquid, dyne cm^{-2}

P_s saturation pressure corresponding to the system temperature, dyne cm^{-2}

P_v pressure of the vapor, dyne cm^{-2}

Q heat flux, erg cm^{-2}

r estimated size of the nucleation sites, cm

r_b^* radius of the critical vapor bubble, cm

r_c^* radius of the critical condensate drop, cm

T system temperature, °K

T_v saturation temperature corresponding to P_v, °K

T_1 saturation temperature corresponding to P_1, °K

ΔT equals $(T_v - T)$ for condensation; equals $(T - T_1)$ for boiling

v_1 volume per liquid molecule, cm^3

σ surface tension of condensate, dyne cm^{-1}

ϕ function defined by Equation (5)

θ equilibrium contact angle, rad.

θ' $\pi - \theta$

VIII. REFERENCES

1. Wilmshurst, R., and Rose, J. W., Paper Cs 2.4, Proc. of the 5th Int. Heat Transfer Conf. Tokyo, 1974.

2. Takeyama, T., and Shimizu, S., Paper Cs 2.5, Proc. of the 5th Int. Heat Transfer Conf. Tokyo, 1974.

3. Keenan, J. H., "Thermodynamics," p. 434-8. John Wiley, New York, 1948.

4. Hirth, J. P., and Pound, G. M., "Condensation and Evaporation," p. 41, 149. Pergamon Press, London, 1963.

5. Maa, J. R., and Hickman, K., Desalination 10, 95-111 (1972).

6. Trefethen, L., J. Appl. Phys. 28, 923 (1957).

7. Novaković, M., and Stefanović, M., Int. J. Heat Mass Transfer 7, 801 (1964).

8. Henry, R. E., Quinn, D. J., and Spleha, E. A., Paper No. B.3.5, Proc. 5th Int. Heat Transfer Conf. Tokyo, 1974.

THE ABSORPTION OF SULFUR DIOXIDE
BY WATER DROPLETS DURING CONDENSATION

Thomas L. Wills and Michael J. Matteson
School of Chemical Engineering
Georgia Institute of Technology

It is known that the rate of absorption of a gas by a liquid can be accelerated by surface renewal, but current theories do not explain this phenomenon adequately. An experimental study was made to measure the rate of absorption of sulfur dioxide by water droplets falling through a gas mixture while simultaneously condensing water vapor on the droplets. The droplets were allowed to fall through tubes of 27, 77, 127, and 177 cm length, and the supersaturation ratios tested were 1.0, 1.5, 2.0, and 2.5. The SO_2 concentration in Nitrogen was varied between 500 and 3,000 ppm. The results of this study were compared with the predictions of current theories, and an alternative theory based on surface effects has been postulated.

I. INTRODUCTION

The non steady-state mass transfer of sulfur dioxide from the gas phase to water droplets growing by water vapor condensation takes place at high altitudes, during cloud formation, and in the growth of condensation nuclei in contaminated urban atmospheres. Upon the release of fossil fuel stack gases, containing water vapor, which are rapidly cooled from about 125°C to ambient temperatures, much of the associated SO_2 is absorbed as the water condenses. When Sulfur Dioxide is inhaled in the presence of aerosol particles, the relatively cool submicron droplets from the atmosphere undergo rapid growth by sudden exposure to warmer humid conditions inside the respiratory tract, thereby trapping the trace gases in the inhaled aerosol and allowing penetration deeper into the alveoli. Once captured, the SO_2 is likely to be complexed by any number of metal salts that may act as condensation nuclei and be converted either to the sulfate or sulfuric acid (1).
    There is evidence to suggest that the transfer of SO_2 to droplets growing by water vapor condensation is many times

greater than transfer in steady-state situations. Bogaevskii
(2) has reported 4.7 to 7.2 times more SO_2 is absorbed during
droplet growth by condensation as compared with the quiescent
state. Other investigators (3,4) have shown that mass transfer rates during drop formation are much higher than at any
other period.

This investigation, then, was made in an effort to determine how water vapor condensation affects the uptake of SO_2 by
droplets and how this absorption compares with situations
where no water vapor is condensing. We decided to use a falling drop technique since SO_2 is very quickly absorbed. By
letting the drop fall various distances, and knowing the drop
size, we were able to vary exposure time. The SO_2 and water
vapor concentration in the gas through which the drop fell
were controlled, so that we were able to measure the amount of
SO_2 absorbed as a function of time, water vapor condensation
and SO_2 concentration. By holding the drop temperature to
around 7°C and the humid gas at about 25°C, we could adjust
the water vapor concentration in the gas such that supersaturation ratios (SSR) of 1.0, 1.5, 2.0 and 2.5 were obtained.
The SSR is the ratio of water vapor pressure in the gas to
that in equilibrium with the drop.

II. EXPERIMENTAL PROCEDURE

In our experiments, deionized, deoxygenated, distilled
water of a known droplet diameter (0.282 cm) was dropped at a
frequency of 2.86 drops per second through controlled atmospheres of N_2, H_2O vapor and SO_2. The capillary was a 24-gauge
hypodermic needle with the tip ground flat and polished. The
test cell (see Figure 1) consisted of tube sections which
could be extended 27, 77, 127, and 177 cm. This allowed exposure times of 0.236, 0.404, 0.525 and 0.627 seconds for a droplet to fall and a drop formation time of 0.349 seconds.

The droplets were collected in a container continuously
flushed with nitrogen to prevent further contact with the SO_2
gas. Five ml of the solution collected were then mixed with
one ml of three percent hydrogen peroxide solution, and the pH
of the resultant mixture was measured.

The gas mixture flowed at a rate of 3.6 liters per minute
from top to bottom in the test cell, and was sampled by continuously removing a portion from the top of the test cell
near the water inlet. A Cambridge Systems Model 880 Thermo-
electric Dew Point Hygrometer was applied to measure water
vapor content, and a Beckman Model 215A Infrared Analyzer monitored the SO_2 concentrations in the 500-3000 ppm range.

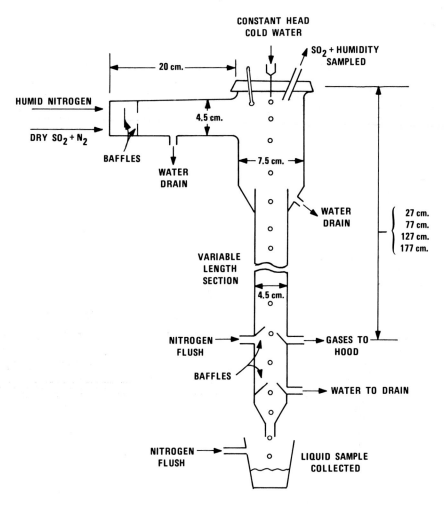

Fig. 1. Absorbtion Chamber

III. RESULTS

The first experiments at a supersaturation ratio (SSR) of 1.0 were undertaken to provide a basis with which to compare SO_2 absorption during water vapor condensation (SSR = 1.5, 2.0, 2.5). The results of these tests are shown in Figures 2, 3, and 4, where the relative average concentration C/C_S is plotted as a function of time. C_S is the equilibrium solubility calculated from Henry's Law for that particular gas phase concentration. All test results expressed in Figures 2 - 4 were normalized to this value. These tests agree quite well with the mass tranfer model proposed by Groothuis and Kramers (5):

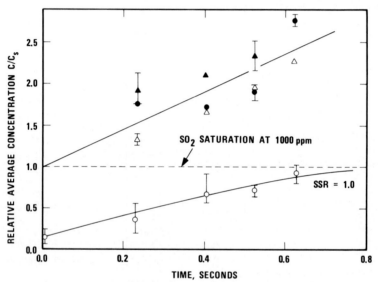

Fig. 2. Rate of absorption of SO_2 by water droplets undergoing condensation in the presence of nitrogen, 1000 ppm SO_2 and water vapor at supersaturation ratios of: ○ = 1.0, ▲ = 1.5, △ = 2.0, ● = 2.5.

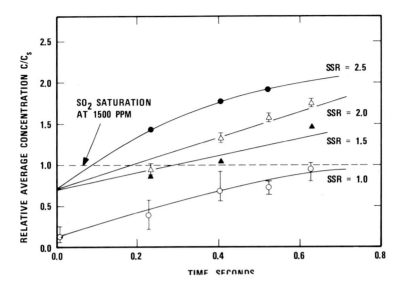

Fig. 3. Rate of absorption of SO_2 by water droplets undergoing condensation in the presence of N_2 and 1500 ppm SO_2.

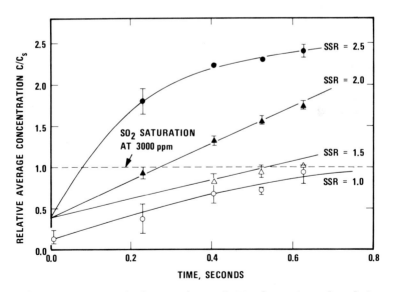

Fig. 4. Rate of absorption of SO_2 by water droplets undergoing condensation in the presence of N_2 and 3000 ppm SO_2.

99

$$N_{SO_2} = (\frac{4\,D_{AB}\,v}{\pi D})^{\frac{1}{2}} (C_s - C) \qquad (1)$$

where

N_{SO_2} = molar flux in $\dfrac{\text{moles}}{\text{cm}^2\,\text{sec}}$

D_{AB} = diffusivity of SO_2 in H_2O = 2×10^{-5} cm^2/sec

v = slip or droplet surface velocity

C_s = equilibrium solubility of SO_2 in water

C = average concentration within the droplet

next let:

$$N_{SO_2} = \frac{W_{SO_2}}{A} = \frac{V}{A}\frac{dC}{dt} \qquad (2)$$

where

V = droplet volume

A = droplet surface area

so that

$$\frac{dC}{dt} = \frac{6}{D} (\frac{4\,D_{AB}\,v}{\pi D})^{\frac{1}{2}} (C_s - C) \qquad (3)$$

For droplets falling from rest v varies from zero to terminal velocity v_t, according to

$$m\frac{dv}{dt} = mg - \tfrac{1}{2}\rho A_p\, v^2 f \qquad (4)$$

assuming the droplet remains spherical with mass m and projected area A_p.

Since the friction factor f varies with velocity, a numerical solution of equation (3) was made and is represented by the SSR = 1 line in Figures 2, 3, and 4. It must be emphasized that the above theory is only a rough approximation. Since we have ignored

a) distortions of the droplet

b) the likelihood that the surface film moves at a somewhat slower velocity than the average relative droplet-air velocity

c) nonuniformity in the concentration of the SO_2 over the surface of the droplet

d) variations between droplet surface and bulk temperatures

e) surface tension instabilities

Nevertheless this approach offers an estimate which seems to agree quite well with experimental data obtained for the case of no water vapor condensation.

In testing at higher water vapor concentrations, however, we noticed that the absorption of SO_2 exceeded the SSR = 1 case by a significant amount (Figures 2, 3, 4: SSR = 1.5, 2.0, 2.5). The dashed line at C/C_s = 1.0 represents the equilibrium solubility of SO_2 in water that one would expect in accordance with Henry's Law for that gas phase concentration. Independent measurements were made to verify the saturation concentration for SO_2 in water at 8°C. These were

Gas Phase Concentration ppm	Saturation Concentration in Water moles/cm^3
1000	2.91×10^{-6}
1500	4.35×10^{-6}
3000	8.65×10^{-6}

The fact that, in most cases, where water vapor condensation took place the average droplet concentrations were much greater than what could be expected, indicates that the mechanism of droplet growth by vapor condensation is an **important step** in concentrating trace gases in the droplet. Also it indicates that this mechanism does not follow accepted mass transfer theories.

In order to test the effect of the carrier gas, duplicate experiments at an SSR = 2.0 and SO_2 = 2000 ppm and at four exposure times were made in the presence of air rather than nitrogen. There was no significant difference in the results of these tests when compared to the results using nitrogen.

An effort was made to correlate SO_2 absorption with the amount of water vapor condensed. This is shown in Figure 5, where the amount of SO_2 absorbed per cm^2 of droplet surface is plotted as a function of the number of moles water vapor condensed per cm^2 of droplet surface. These results indicate that the following relation holds:

$$\frac{d(SO_2)}{dt} = k(Y_{SO_2})^n \frac{d(H_2O)}{dt} \tag{5}$$

such that

$$(SO_2) = k(Y_{SO_2})^n (H_2O) + C_1 \tag{6}$$

Where (SO_2) and (H_2O)[1] represent moles of each species absorbed or condensed in the droplet per cm^2 of drop surface and C_1 is the concentration of SO_2 absorbed by the droplet at the time of release from the capillary tip; Y_{SO_2} is the mole fraction concentration of SO_2 in the gas phase. From slope-intercept

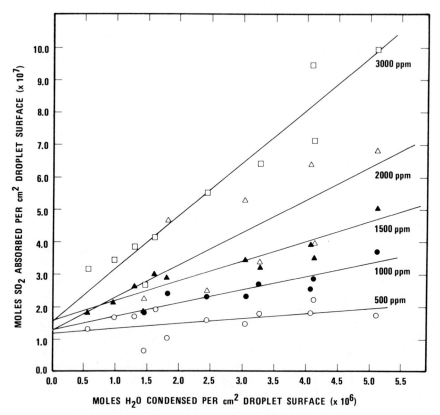

Fig. 5. *SO$_2$ Absorption as a Function of Water Vapor Condensation.*

values of Figure 5 the following constants were obtained

$$(SO_2) = 3.06 \times 10^3 \ (Y_{SO_2})^{1.3} \ (H_2O) + 1.4 \times 10^{-7} \quad (7)$$

This can be compared with Henry's Law for SO_2 at $8°C$:

---

[1]Appendix A contains an outline of the procedure used to calculate (H_2O) and also the temperature change for the droplet.

$$(SO_2) = 5.22 \times 10^{-2} \, Y_{SO_2} \, (H_2O) \tag{8}$$

so that in the condensed water film we are dissolving SO_2 in concentrations about 5 (five) orders of magnitude greater than what is obtained from equilibrium measurements.

Many questions arise from these findings which have prompted continued interest in this research. It is not clear whether this enhanced absorption via water vapor condensation is peculiar to SO_2 or is exhibited by gases in general. Sulfur dioxide will react with OH radicals produced by photochemical decomposition of water vapor (Wood, 6) with the resulting formation of sulfuric acid nuclei. However we did not notice any difference in absorption rates in the absence of light. The transport mechanism of SO_2 to the water droplet may be a result of the highly polar nature of SO_2 causing interaction with the water vapor molecules to form clusters of nuclei, these being driven by the condensing water vapor. One means of testing this would be to measure the absorption rate of a non-nucleating, non-polar gas, oxygen for instance, under similar circumstances to see if equally high levels of absorption concentration enrichment could be achieved.

It must be emphasized that, once water vapor condensation is discontinued, the amount of SO_2 dissolved in the droplet may return to an equilibrium saturation level. However, if certain metal salts are present in the aqueous phase, they could act to complex and react with the dissolved SO_2, thereby preventing or slowing its release. This mechanism, then, would offer an explanation for the low pH - high sulfate content often observed in atmospheric particulates and in rainfall. It would also explain the high conversion rates of SO_2 from the gas to the liquid phase in stack gas plumes.

IV. REFERENCES

(1) Matteson, M. J., W. Stober, and H. Luther, Ind. and Eng. Chem., 8, 677 (1969).
(2) Bogaevskii, O. A., Zh. Fiz Khim, 43, 1298 (1969).
(3) Angelo, J. B., E. M. Lightfoot, and D. W. Howard, J. Am. Inst. Chem. Engr., 12, 751 (1966).
(4) Licht, W. and J. B. Conway, Ind. Eng. Chem., 42, 1151 (1950).
(5) Groothuis, H. and H. Kramers, Chem. Eng. Sci., 4, 17 (1955).
(6) Wood, W. P., A. W. Castleman, Jr., and I. N. Tang, "Mechanisms of Aerosol Formation from SO_2", presented at the 67th Annual Meeting of the Air Pollution Control Association, Denver, Colorado, June 9-13, 1974.

V. APPENDIX

A. Calculation of Heat and Mass Transfer to Falling Droplet

Because of both a temperature gradient and condensation, the droplet is increasing in temperature according to:

$$mC_p \frac{dT}{dt} = \Delta H_{cond} \, dm + hA \, (T_\infty - T)$$

Here we have neglected heats of mixing and radiation. Also we have ignored conductive heat transfer within the drop to assume that the droplet temperature is uniformly at a temperature T at some time t.

m = droplet mass, gms
C_p = heat capacity, cal/gm $^\circ$K
ΔH = Heat of condensation cal/gm
$\dot{dm}$ = increase in droplet mass due to condensation, gms/sec
h = heat transfer coefficient for falling droplet, cal/ cm^2 sec $^\circ$K
$T_\infty - T$ = temperature gradient between droplet and surroundings
A = surface area of droplet (assumed constant and spherical)

The heat transfer coefficient h is obtained from the Nusselt Number expression for a sphere immersed in a moving fluid:

$$Nu = \frac{hD}{k} = 2.0 + Re^{0.50} \, Pr^{0.33}$$

where

Re = Reynold's No. = $Dv\rho$
Pr = Prandtl's No. = $\dfrac{C_p \mu}{k}$

D = droplet diameter, cm
ρ = gas density, gm/cm^3
μ = gas viscosity, gm/cm sec
k = gas thermal conductivity, cal/cm sec $^\circ$C

The incremental mass transferred dm is obtained from the Sherwood Number expression for mass transfer to a sphere immersed in a flowing fluid:

$$Sh = \frac{k_x D_{AB}}{C_f D} = 2.0 + Re^{0.50} \, Sc^{0.33}$$

$$Sc = \text{Schmidt's No.} = \frac{\mu}{\rho D_{AB}}$$

k_x = mass transfer coefficient, $\dfrac{\text{moles}}{\text{cm}^2 \text{ sec mm } Hg}$

$$\frac{dm}{dt} = K_x A \ (P_\infty - P_s) \ \text{moles/sec}$$

However, since the droplet is accelerating the Reynold's Number is continuously increasing until terminal velocity is reached. Therefore the following steps must be taken to solve numerically for ΔT and Δm, the incremental changes in droplet temperature and mass:

    1. For a given initial mass m and temperature T, find the incremental change in velocity and distance for a specified time increment, Δt, cording to:

$$m \frac{\Delta v}{\Delta t} = mg - \tfrac{1}{2}A_p \ \rho v^2 \ C_D$$

$$\Delta v = \frac{1}{C} \ \tanh \ (C \ g \ \Delta t)$$

$$\Delta X = (\frac{1}{c^2 g}) \ \ln \cosh \ (c \ g \ \Delta t)$$

where

$$C = 3/8 \ (\frac{1}{gr}) \ \frac{\rho}{\rho_d} \ C_D$$

    2. Solve for dimensionless groups: Pr, Sc, Nu and Sh
    3. Solve for $k_x h$
    4. Solve for $\Delta m = K_x A \ (P_\infty - P_s) \ \Delta t$
    5. **Solve** for $\Delta T = \Delta H_{cond} \ \Delta m + hA \ (T_\infty - T) \ \frac{t}{C\rho} \ (\frac{1}{m})$
    6. Calculate new T, m, v, C_D, X, P
    7. Repeat 1 - 7 until χ = desired length.

105

BROWNIAN COAGULATION IN THE TRANSITION
AND NEAR FREE MOLECULE REGIME

Paul E. Wagner and Milton Kerker
Clarkson College of Technology

ABSTRACT

The coagulation process in aerosols is well understood when the particle radius is much larger or much smaller than the mean free path of the surrounding gas (very small or very large Knudsen numbers). However, in the so-called transition regime between these limiting cases, no adequate theory is available. This paper reports an experimental investigation on Brownian coagulation in the transition regime. The change of the particle size distribution of an initially well-defined monodisperse aerosol is monitored by means of light scattering measurements. Increasing Knudsen numbers are achieved by reducing the gas pressure rather than by reducing the particle size. Previous measurements are extended towards the free molecule regime. The results are compared with Fuchs' semi-empirical interpolation formula. Estimates of the influence of particle evaporation, diffusional mixing, deposition, settling and gradient coagulation are given.

I. INTRODUCTION

Dynamical processes in colloidal dispersions depend upon the Knudsen number

$$Kn = \lambda/a \tag{1}$$

where λ denotes the mean free path of the molecules in the ambient medium and a the particle radius. Although small Knudsen numbers (continuum regime) prevail in hydrosols, this condition is not generally fulfilled in aerosols.

A theoretical description of Brownian coagulation in the continuum regime was given by Smoluchowski (1). Fuchs (2) applied the kinetic theory of dilute gases to describe the coagulation of aerosols at large Knudsen numbers (free molecule regime). No satisfactory theoretical description is available for the case of intermediate Knudsen numbers (transition regime). Fuchs (2) has proposed a semi-empirical interpolation formula based upon a concentration jump model, but it has not been possible to assess the validity of his result because of the difficulty in specifying an actual physical process.

Experimental investigations of aerosols have been even less satisfactory. Most of the experimental investigations on aerosol coagulation (3-8) have been restricted to monitoring the particle number concentration rather than the evolution of the size distribution. The first attempt to measure changes of the size distribution during coagulation was made in this laboratory (9) at low Knudsen numbers. Later, these measurements were extended to include the initial part of the transition regime (10, 11).

Here, modifications in the experimental procedure are reported, which have led to considerable improvement of the accuracy and have permitted extension of the experiment throughout most of the transition regime. The results are in good agreement with Fuchs' equation.

II. EXPERIMENTAL PROCEDURE

Changes of the aerosol size distribution in a flowing system were monitored by means of light scattering. The polarization ratio was measured as a function of the scattering angle at various stages in the life history of the aerosol. Inversion of data, obtained from the initial, monodisperse aerosol, yields the modal radius a_M and the breadth parameter σ_O (12). From the gravimetrically determined mass concentration the initial number concentration N_O can be calculated.

The initial size distribution is then utilized to calculate the size distribution at later times with the aid of Fuchs' formula, and polarization ratios corresponding to the coagulated system are also calculated. These are then compared with the experimental results, obtained at various positions along the flowing system. Each such position corresponds to a particular experimental time t_E. Best fits between the calculated polarization ratios and the experimental values at each light scattering position determine the theoretical coagulation times t_F.

The actual hold-up times t_E of the aerosol in the flowing system are determined by the flowrate and the volume of the coagulation tube under the assumption that the aerosol is in Poisseuille flow. The ratio of the actual coagulation rate and Fuchs' prediction will be given by t_F/t_E.

The average particle radius of the aerosol, formed by condensation of di(2-ethylhexyl)sebacate (DEHS) upon AgCl nuclei was maintained at approximately 0.2 μm. The Knudsen number was changed by varying the pressure of the helium carrier gas from atmospheric pressure to 67 Torr. Considerable effort was required at each pressure in order to develop the various generator conditions necessary to produce stable aerosols of comparable size distributions and number concentrations.

The experimental arrangement, as shown in Fig. 1, consists of an aerosol generator, a cylindrical coagulation tube, followed by a filter manifold for gravimetric analysis, and a vacuum pump. Pressure and flowrates in the system can be set

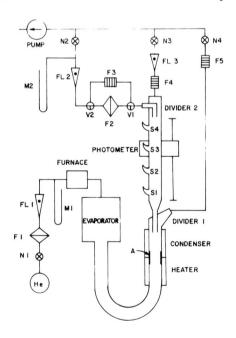

Figure 1. Schematic diagram of the experimental arrangement. He: He-tank; N1 - N4: Needle valves; F1 - F5: Filters; FL1 - FL3: Float flowmeters; M1, M2: Absolute manometers; A: Asbestos ring; S1 - S4: Light scattering cells; V1, V2: Bypass valves.

independently by proper adjustment of the needle valves N1 -
N4. The pressure is monitored by absolute Hg-manometers M1
and M2 and the flowrates are measured by float flowmeters FL1 -
FL3, which were calibrated separately for each of the pressures
used.

The aerosol generation is performed essentially in three
steps (13, 14). Submicroscopic solid particles are formed by
vaporization of AgCl in a furnace (620°C) and subsequent
condensation. Secondly, DEHS-vapor, supplied by a falling
film evaporator (170°C) (13), condenses upon the AgCl-particles.
A number of modifications were necessary in order to adapt the
falling film evaporator to low pressure operation.

Finally, the aerosol is evaporated in a heater (335°) and
recondensed in a condenser (30°C). Because some DEHS condenses at the cool walls of the condenser, an asbestos ring A is
placed on the inside of the tube between the heater and
condenser in order to ensure a steady return of this condensed
DEHS to the heater. This arrangement provides stable aerosol
output over several hours of continuous operation. Otherwise,
intermittent trickling of drops of condensate into the heater
with subsequent sudden vaporization creates instabilities.

The aerosol formed in the condenser is not uniform due to
both axial and radial temperature gradients (14). However, if
only the central fraction of the aerosol is permitted to enter
the coagulation tube by means of flow divider 1, a more
spatially uniform aerosol is obtained.

The cylindrical coagulation tube is equipped with four
light scattering cells S1 - S4 in order to facilitate *in situ*
light scattering measurements at different coagulation times.
The light scattering photometer can slide along the coagulation
tube so that measurements can be made at the various locations
without perturbing the system. It is designed so that only
the central fraction of the coagulating aerosol is observed.
The velocity is constant through this region and equal to twice
the average velocity through the tube.

In order to determine the mass concentration of this
central fraction, samples are extracted by means of flow
divider 2 and collected on a Millipore filter F2. To ensure
constant flow conditions and minimal pressure changes during
change of sampling filters, the aerosol flow can be switched
to a glass wool filter F3 by means of the bypass valves V1 and
V2. The total flow through the coagulation tube is obtained
as the sum of the readings of the flowmeters FL2 and FL3.

III. ERROR ANALYSIS

A number of disturbing influences were investigated in order to estimate the systematic experimental error.

A. Particle Evaporation

In order to check whether evaporation takes place during coagulation and aerosol sampling, samples were collected at a pressure of 94 Torr for sampling times varying between 5 and 90 min. Figure 2 shows the values of the sampled mass as a function of the sampling time.

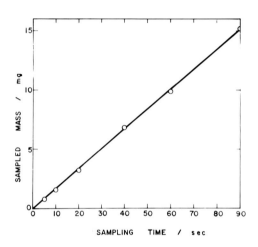

Figure 2. Particulate mass, sampled on a Millipore membrane filter as a function of the sampling time.

The strictly linear dependence proves that evaporation is negligible over a period of at least 90 min. Accordingly, evaporation of particles during coagulation and subsequent sampling can be disregarded.

B. Diffusional Deposition

Davis and Yates[1]   have calculated the simultaneous effects of Brownian coagulation and diffusional deposition in a cylindrical tube with radius R. They show that deposition can be neglected compared with coagulation, if

---

(1) Private communication.

$$C \equiv \tfrac{1}{2}R^2 N_O K/D > 50 \qquad (2)$$

Here, N_O denotes the number concentration, D the diffusion coefficient and K the coagulation function of the initial monodisperse aerosol. In the present experiment, $C > 7400$.

C. Parabolic Velocity Profile

Although only the central fraction of the aerosol is observed, finite velocity gradients will cause non-uniform hold-up times. As estimate of this error including particle diffusion in a radial direction results in a maximum error of 1% at the experimental conditions.

D. Particle Settling

The experimental coagulation time is taken to be the hold-up time of the carrier gas. However, due to gravitational settling, the actual hold-up time will be somewhat greater. A maximum error of less than 0.5% was calculated.

E. Flowmeter Calibration

The accuracy of the calibration of the float flowmeter was checked by comparison between independently calibrated flowmeters. A maximum error of 3% was observed.

F. Filter Efficiency

A Millipore membrane filter (pore width 3 μm) was used for aerosol sampling. A number of test runs were performed with a second Millipore filter (pore width 0.05 μm) set up in series with the sampling filter. The filter efficiency was always found to be better than 98.5%.

IV. RESULTS AND DISCUSSION

The results obtained during a typical single measuring run are illustrated by means of run 133, which was performed at atmospheric pressure. Figure 3 shows the degree of polarization

$$P \equiv \frac{i_2 - i_1}{i_2 + i_1} = \frac{\rho - 1}{\rho + 1} \qquad (3)$$

at cell S1 as a function of the scattering angle θ where i_1 and i_2 are the scattered radiances with electric vectors polarized perpendicular and parallel to the scattering plane, respectively. The open and closed circles refer to the first and final measurements, respectively. It can be seen that the aerosol size distribution did not change significantly during the measuring run. The curve in Fig. 3 corresponds to the best theoretical fit.

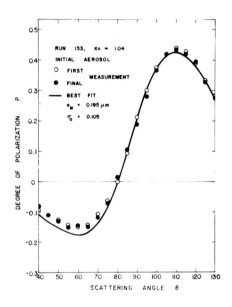

Figure 3. Run 133, initial aerosol: Experimental values of the degree of polarization P versus scattering angle θ at cell S1 and the best theoretical fit.

The degree of polarization at cells S2 - S4 as a function of the scattering angle is shown in Fig. 4 together with the best fits at various theoretical coagulation times t_F. The corresponding theoretical size distributions of the initial aerosol at cell S1 and the coagulated aerosol at cells S2 - S4 are shown in Fig. 5.

Figure 6 illustrates the uniqueness of the fit between the calculated and the experimental values of the polarization data. Each curve corresponds to a particular light scattering cell - S2, S3, or S4. The ordinate is a measure, at each theoretical coagulation time, of the deviation between the calculated (P_{th}) and the experimental (P_{exp}) values of the degree of polarization, here taken to be $\Sigma(P_{th} - P_{exp})^2$. The best fit always occurred at a single and very deep minimum.

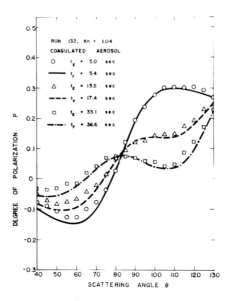

Figure 4. Run 133, coagulated aerosol: Experimental values of the degree of polarization P versus scattering angle θ at cells S2 - S4 and the best theoretical fits.

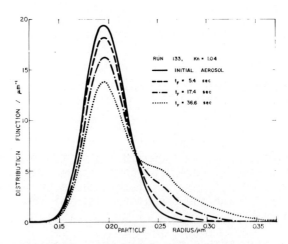

Figure 5. Run 133: Theoretical size distribution corresponding to the best fits of the polarization ratios.

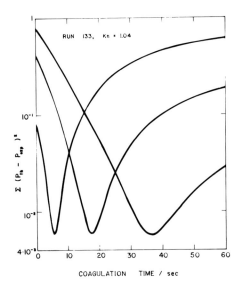

Figure 6. Run 133: Sum of the squares of the differences between the experimental values of the degree of polarization at cells S2 - S4 (deviation measure), as a function of the theoretical coagulation time. The minima correspond to the best fits.

Measurements were taken at five different pressures and accordingly at five different Knudsen numbers. Between 9 and 11 complete measuring runs were performed at each pressure. The experimental results are summarized in Table 1. Columns 1-3 list the values of some experimental parameters, averaged over the runs at each pressure. The percentages in column 3 indicate the relative standard error of the mean and are an estimate of the statistical variation of the results. This scatter is mainly a measure of the variation of the aerosol from run to run, rather than the experimental error. Column 4 lists the dimensionless coagulation function β_F for equal sized spheres, as calculated from $< a_M >$ and $< p >$ by means of Fuchs' formula. The quantity defined by

$$\beta_E \equiv \beta_F \ t_F/t_E \qquad (4)$$

is listed in column 5. This quantity, if used as the coagulation function rather than β_F, will give coagulation times in agreement with experiment. Accordingly it can be considered as the experimental coagulation function.

TABLE 1
Comparison of Experimental Results with Fuchs' Equation

$\langle Kn \rangle$ [a]	$\dfrac{\langle p \rangle}{\text{Torr}}$ [a]	$\dfrac{\langle a_M \rangle}{\mu\text{m}}$ [a]	$\langle \sigma_o \rangle$ [a]	$\dfrac{\langle N_o \rangle}{10^7\ \text{cm}^{-3}}$ [a]	$\langle t_F/t_E \rangle$ [b]	β_F	β_E	Δ_{EF}
1.02 (3.0%)	748.6 (0.6%)	0.198 (2.7%)	0.10 (10.3%)	1.76 (8.5%)	0.977 (1.8%)	2.26	2.20	2.7%
1.91 (3.0%)	376.3 (0.6%)	0.209 (3.0%)	0.08 (7.0%)	1.60 (16.4%)	0.991 (5.9%)	3.45	3.42	0.9%
2.95 (3.5%)	267.4 (0.4%)	0.190 (3.2%)	0.10 (10.0%)	1.63 (7.9%)	1.044 (2.0%)	4.74	4.95	-4.2%
8.52 (4.8%)	94.0 (0%)	0.188 (4.9%)	0.14 (13.5%)	1.37 (16.3%)	1.088 (2.3%)	10.14	11.03	-8.1%
12.2 (2.8%)	67.0 (0%)	0.184 (2.8%)	0.14 (9.2%)	1.37 (14.1%)	1.126 (2.8%)	12.55	14.14	-11.2%

a. Percentages indicate the relative standard deviation of the measuring results.

b. Percentages indicate the relative standard error of the mean.

116

The last column shows the percent difference between the results and is an indication of the agreement between these experiments and Fuchs' interpolation formula.

Figure 7 is a graphical representation of the results. The influence of the coalescence efficiency is illustrated by the lower figure which has been calculated with the Fuchs' equation with a coalescence efficiency of 0.75. It can be seen that Fuchs' interpolation agrees with the experiment with the estimated maximum error of approximately 10% (see the above error analysis assuming a coalescence efficiency of unity). The present experimental evidence demonstrates that Fuchs' flux-matching procedure successfully describes Brownian coagulation in a rarified medium. Of course, a more satisfactory theoretical description must be based upon a rigorous statistical mechanical treatment.

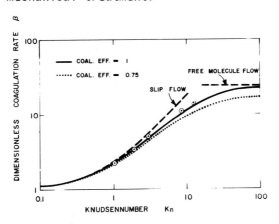

Figure 7. Dimensionless coagulation function β versus Kundsen number. Fuchs' interpolation formula is shown for coalescence efficiencies of unity (upper curve) and 0.75 (lower curve). The Smoluchowski law with Cunningham slip flow correction and the free molecule limit are shown by dashed curves. The experimental results are indicated by the circles.

V. REFERENCES

1. Smoluchowski, M.v., Z. Phys. Chem. Leipzig 94, 129 (1917).

2. Fuchs, N.A., "Mechanics of Aerosols," Pergamon Press, New York, 1964.
3. Whytlaw-Gray, R., and Patterson, H.S., "Smoke: A Study of Aerial Dispersed Systems," E. Arnold, London, 1932.

4. Cawood, W., and Whytlaw-Gray, R., Trans. Faraday Soc. 32, 1059 (1936).

5. Nolan, P.J., and Kerman, E.L., Proc. Roy. Irish Acad. 52A, 171 (1949).

6. Fuchs, N.A., and Sutugin, A.G., J. Colloid Sci. 20, 492 (1965).

7. Devir (Weinstock), S., J. Colloid Interface Sci. 23, 80 (1967).

8. Mercer, T.T., and Tillay, M.J., J. Colloid Interface Sci. 37, 785 (1971).

9. Nicolaon, G.A., and Kerker, M., Faraday Symp. of the Chem. Soc. 7, 133 (1973).

10. Nicolaon, G.A., Kerker, M., Cooke, D.D., and Matijević, E., J. Colloid Interface Sci. 38, 460 (1972).

11. Chatterjee, A., Kerker, M., and Cooke, D.D., J. Colloid Interface Sci. 53, 71 (1975).

12. Kerker, M., "The Scattering of Light and Other Electromagnetic Radiation," Academic Press, New York, 1969.

13. Nicolaon, G.A., Cooke, D.D., Kerker, M., and Matijević, E., J. Colloid Interface Sci. 34, 534 (1970).

14. Nicolaon, G.A., Cooke, D.D., Davis, E.J., Kerker, M., and Matijević, E., J. Colloid Interface Sci. 35, 490 (1971).

SIMULTANEOUS CONVECTIVE DIFFUSION AND
COAGULATION OF AEROSOLS AND HYDROSOLS

E. James Davis and Brian Yates
Unilever Research

I. *ABSTRACT*

This work is a theoretical analysis of simultaneous coagulation and particle diffusion to the wall of a circular tube for laminar flow of an aerosol or hydrosol through the tube. Using Smoluchowski theory to describe the coagulation of an initially (at the tube inlet) monodisperse system of particles, the convective diffusion equation is solved in terms of well-known functions to obtain expressions for the particle number density and the fraction of the initial particles remaining in the flow field at any axial plane. Solutions are obtained by means of a perturbation technique and by a method of successive approximations. The solutions are compared with the theory for pure coagulation of an initially monodisperse sol to establish the effects of deposition.

II. INTRODUCTION

Aerosols and hydrosols consisting of particles sufficiently small to involve significant Brownian motion of
the dispersed phase tend to deposit particles on the walls
of containers and conduits enclosing them. Furthermore, the
coagulation of particles affects the deposition characteristics. When coagulation and deposition occur simultaneously
in a flowing system, the deposition of particles depends upon
the coagulation characteristics, the flow field and the diffusional transport processes. The limiting cases of coagulation without deposition and deposition without coagulation
have been extensively studied, as indicated by the reviews
of Fuchs (1), Davies (2), Hidy and Brock (3) and Fuchs and
Sutugin (4), but the theory of simultaneous coagulation and
convective diffusion has received far less attention because of the apparent intractability of the governing
equations.
The simplest and most convenient flow system used to
study particle deposition is laminar flow in a circular tube

(Poiseuille flow), and Nicolaon and Kerker (5) used a laminar flow system to study Knudsen aerosol coagulation, measuring the size distribution of the aerosol at the inlet and exit of their laminar flow coagulation chamber by light scattering methods. Because of the relatively large ratio of volume to surface area of their coagulation chamber they neglected deposition on the walls of the coagulation chamber.

When the volume to surface area ratio is small, particle deposition and radial diffusion of particles can become significant. The problem of convective diffusion of aerosols and hydrosols in laminar flow in a circular tube with no coagulation has been treated by a number of investigators whose work has been reviewed by Fuchs (1). This problem is mathematically identical to that of heat transfer in laminar flow, and the literature related to that problem is extensive, as indicated by Davis (6). A brief review of that problem is in order because we shall use the results of this earlier work in the theoretical analysis of this paper.

Most of the mathematical solutions available are for the following non-dimensional convective diffusion equation for the case of negligible axial diffusion (Pe→∞).

$$(1-\rho^2) \frac{\partial \phi}{\partial \xi} = \frac{1}{\rho} \frac{\partial}{\partial \rho} \left(\rho \frac{\partial \phi}{\partial \rho} \right) + \frac{1}{Pe^2} \frac{\partial^2 \phi}{\partial \xi^2} , \qquad (1)$$

where $\rho = r/R$, R is the tube radius, $\xi = x/RPe$, the Peclet number is defined by $Pe = 2R\bar{U}/D$, $\bar{U}$ is the average velocity through the tube, and D is the diffusion coefficient of the diffusing particles in the continuous phase. The dimensionless number density (or concentration) of particles is defined by $\phi = n/n^0$, where n^0 is the number density at the tube inlet ($x=0$).

When axial diffusion is negligible (for Pe>100), the boundary conditions that apply are

$\phi(\rho,0) = 1$ (specified inlet concentration)   (2)

$\frac{\partial \phi}{\partial \rho}(0,\xi) = 0$ (symmetry about $\rho=0$)   (3)

$\phi(1,\xi) = 0$ (vanishing concentration at the wall, $\rho=1$)   (4)

The solution of Equations (1)-(4) has the form

$$\phi(\rho,\xi) = \sum_{m=1}^{\infty} A_m e^{-\lambda_m^2 \xi} y_m(\rho) , \qquad (5)$$

where the eigenvalues, λ_m, and the eigenconstants, A_m, have
been reported elsewhere (7-9), the most accurate being those
of Liao (10). The eigenfunctions, $y_m(\rho)$, have usually been
obtained by numerical computation, but Lauwerier (11) and
Davis (6) have shown that solutions of this problem can be
obtained in terms of well-known, tabulated functions.
Recently Hsu (12) and Deavours (13) solved the problem with
the axial diffusion term included, however, the latter paper
was for the related problem of flow between parallel plates.

Equation (5) can be integrated across the cross
sectional area of the tube ($0<\rho\leq1$) to obtain the fraction of
the inlet particles that cross a plane at ξ. In this way
Gormley and Kennedy (14) obtained

$$\bar{\phi}(\xi) = 0.819e^{-3.657\mu} + 0.097e^{-22.3\mu} + 0.032e^{-57\mu}+..., \qquad (6)$$

where $\mu=2\xi$ and the constants are of limited accuracy owing to
the numerical methods used by Gormley and Kennedy.

When coagulation of particles occurs, the particle con-
servation equation, Equation (1), must be modified to take
into account the decrease in the number density due to
coagulation. This requires introduction of a depletion term
(sink term) into the governing equation, and the depletion
term, which describes the coagulation kinetics, significantly
complicates analysis. In general, the coagulation term is a
function of the number density of particles, and for hydrosols the coagulation rate is a function of the electrolyte
and particle properties as well. The laminar flow diffusion/
coagulation system offers a convenient way to study these
phenomena, but to interpret experimental data on inlet and
outlet particle concentrations it is necessary to solve the
problem of simultaneous convective diffusion and coagulation.

It is the purpose of this paper to develop solutions of
the equations governing combined diffusion and coagulation
for laminar flow in a circular tube and to determine the
conditions for which the combined phenomena can be de-
coupled to provide simpler analysis of the individual effects.

III. THE GOVERNING EQUATIONS

Let us consider the aerosol or hydrosol to consist of an
arbitrary number of groups of particles, each group containing particles of the same size and mass, characterized by
length a_i and mass m_i. For spheres a_i is the radius of all
particles in the ith group. Now particles of size a_i can
diffuse to the wall of the tube or can coagulate with other
particles to reduce the number density of the ith group. A

population balance on the ith specie leads to the following partial differential equation

$$2\bar{U}(1-r^2/R^2)\frac{\partial n_i}{\partial x} = D_i \frac{1}{r}\frac{\partial}{\partial r}(r\frac{\partial n_i}{\partial r}) - \psi_i \quad , \qquad (7)$$

where the diffusion coefficient, D_i, is, in general, a function of the particle size, and ψ_i is the net rate of removal of the ith species per unit volume due to coagulation.

For a polydisperse system with a continuum of particle sizes the variation of the number density, $n(m)dm$, of the particles having masses between m and $m+dm$, as given by Müller (15) and discussed by Fuchs (1) and Davies (2) is

$$\frac{\partial n(m)}{\partial t} = \frac{1}{2}\int_O^m K(m_1,m-m_1)n(m_1)n(m-m_1)dm_1$$

$$-n(m)\int_O^\infty K(m,n_1)n(m_1)dm_1 \quad , \qquad (8)$$

where $K(m,m_1)$ is the coagulation constant for particles of mass m and m_1.

The discrete form of Equation (8) is

$$-\psi_i = \frac{\partial n_i}{\partial t} = \frac{1}{2}\sum_{j=1}^{i-1} K_{i-j,j} n_{i-j}n_j - n_i \sum_{j=1}^{\infty} K_{i,j}n_j \quad , \qquad (9)$$

where $K_{i,j}=K_{j,i}$, and n_i and n_j refer to particles of mass m_i and m_j, respectively.

Now Equation (7) with Equation (9) for ψ_i applies to each of the species present, so we have a coupled set of partial differential equations describing the particle concentrations $n_1,n_2......$, and these equations must be solved subject to appropriate boundary conditions to obtain the concentrations as a function of position in the flow field.

In general, the solution of this coupled set of partial differential equations is a formidable task, particularly in view of the fact that the equations are nonlinear because of the form of the coagulation rates. But considerable simplification can be accomplished by considering the case of an initially monodisperse sol with concentration n_1^o at the tube inlet ($x=0$). In this case there are no particles with mass m greater than m_1 at the inlet, that is, $n_2^o=n_3^o = ... = 0$, and the governing equations can be written in non-dimensional form as

$$(1-\rho^2)\frac{\partial\phi_i}{\partial\xi_i} = \frac{1}{\rho}\frac{\partial}{\partial\rho}(\rho\frac{\partial\phi_i}{\partial\rho}) + \frac{1}{2}\sum_{j=1}^{i-1}k_{i-j,j}^{(i)}\phi_{i-j}\phi_j - \phi_i\sum_{j=1}^{\infty}k_{i,j}^{(i)}\phi_j$$

(10)

with boundary conditions

$$\phi_1(0,\rho) = 1 \quad , \quad \phi_2(0,\rho) = \phi_3(0,\rho) = \ldots = 0 \tag{11}$$

$$\frac{\partial\phi_i}{\partial\rho}(\xi_i,0) = 0 \quad , \quad i = 1,2,3, \ldots \tag{12}$$

$$\phi_i(\xi_i,1) = 0 \quad , \quad i = 1,2,3, \ldots \ , \tag{13}$$

where $\rho=r/R$, $\phi_i=n_i/n_1^o$, $\xi_i=D_i x/2R^2\overline{U}$, and the dimensionless coagulation constants $k_{i,j}^{(\ell)}$ are defined by

$$k_{i,j}^{(\ell)} = \frac{R^2 n_1^o}{D_\ell} K_{i,j} \tag{14}$$

Since ϕ_2, $\phi_3,\ldots$ are zero at the tube inlet, we can obtain information about the concentration ϕ_1 in the upstream region by examining the reduced problem,

$$(1-\rho^2)\frac{\partial\phi_1}{\partial\xi_1} = \frac{1}{\rho}\frac{\partial}{\partial\rho}(\rho\frac{\partial\phi_1}{\partial\rho}) - k_o\phi_1^2 \tag{15}$$

$$\phi_1(0,\rho) = 1 \tag{16}$$

$$\frac{\partial\phi_1}{\partial\rho}(\xi_1,0) = 0 \tag{17}$$

$$\phi_1(\xi_1,1) = 0 \ , \tag{18}$$

where $k_o=k_{1,1}^{(1)}$.

We have applied two techniques to solve Equations (15)-(18): 1) a perturbation analysis and 2) the method of successive approximations. The latter method can also be applied to the system of Equations (10).

IV. PERTURBATION THEORY

When the coagulation rate is not too large compared with the rate of deposition, we can treat the effects of coagulation as a perturbation on the convective diffusion. The dimensionless parameter k_o, defined by Equation (14), is a measure of the importance of coagulation compared with diffusional deposition, for it is the ratio of the coagulation rate at the inlet to the diffusive flux there. For $k_o \gg 1$ we can expect coagulation to predominate and for $k_o \ll 1$ the diffusion mechanism predominates to eliminate particles from the flow field. Since the tube radius enters as R^2, increasing the tube radius decreases the importance of diffusion. Now let

$$\phi_1(\xi_1,\rho) = \Phi_o(\xi_1,\rho) + \Phi_1(\xi_1,\rho) , \qquad (19)$$

where $\Phi_o(\xi,\rho)$ is the solution of Equations (15)-(18) for $k_o = 0$, that is, for no coagulation, and $\Phi_1(\xi,\rho)$ is the perturbation due to coagulation. Substituting this assumed solution into Equations (15)-(18), and linearizing the sink term, $k_o \phi_1^2$, we obtain

$$(1-\rho^2) \frac{\partial \phi_o}{\partial \xi} = \frac{1}{\rho} \frac{\partial}{\partial \rho} (\rho \frac{\partial \phi_o}{\partial \rho}) \qquad (20)$$

$$\Phi_o(0,\rho) = 1 \qquad (21)$$

$$\frac{\partial \Phi_o}{\partial \rho}(\xi,0) = 0 \qquad (22)$$

$$\Phi_o(\xi,1) = 1 \qquad (23)$$

and

$$(1-\rho^2) \frac{\partial \Phi_1}{\partial \xi} = \frac{1}{\rho} \frac{\partial}{\partial \rho} (\rho \frac{\partial \Phi_1}{\partial \rho}) - k_o(\Phi_o^2 + 2\Phi_o \Phi_1) \qquad (24)$$

$$\Phi_1(0,\rho) = 0 \qquad (25)$$

$$\frac{\partial \Phi_1}{\partial \rho}(\xi,0) = 0 \qquad (26)$$

$$\Phi_1(\xi,1) = 0 , \qquad (27)$$

where we have dropped the subscript on ξ.

As shown by Davis (6) and Davis and Liao (18), Equations (20)-(23) have the solution given by Equation (5) with

$$y_m(\rho) = e^{-\lambda_m \rho^2/2} M(\frac{2-\lambda_m}{4}, 1, \lambda_m \rho^2) , \qquad (28)$$

where $M(a,b,z)$ is the confluent hypergeometric function and the eigenvalues satisfy the transcendental equation

$$M(\frac{2-\lambda_m}{4}, 1, \lambda_m) = 0 . \qquad (29)$$

The eigenconstants are given by

$$A_m = \frac{\int_0^1 \rho(1-\rho^2) y_m(\rho) d\rho}{||y_m||^2} , \qquad (30)$$

where $||y_m||^2 = \int_0^1 \rho(1-\rho^2) y_m^2(\rho) d\rho$.

The first six values of λ_m, A_m and $||y_m||^2$ are listed in Table 1. Note that because of the rapid increase in λ_m^2 the series in Equation (5) rapidly converges as ξ increases.

TABLE 1
A Partial List of Mathematical Constants

m	λ_m	λ_m^2	A_m	$||y_m||^2$	$\int_0^1 \rho(1-\rho^2) y_m(\rho) d\rho$
1	2.704364	7.313585	1.476435	0.0939338	0.138687
2	6.679031	44.609455	-0.806124	0.0375198	-0.030246
3	10.673379	113.921019	0.588762	0.0234421	0.013802
4	14.671078	215.240530	-0.475850	0.0170471	-0.008112
5	18.669871	348.564083	0.405022	0.0133936	0.005425
6	22.669143	513.890044	-0.355757	0.0110298	-0.003924

Now the solution of Equations (24)-(27) can be obtained by applying the Green's function approach of Davis and Liao (18). The result is

$$\Phi_1(\xi,\rho) = -k_0 \int_0^\xi \int_0^1 \rho' [\Phi_0^2(\xi',\rho') + 2\Phi_0(\xi',\rho')\Phi_1(\xi',\rho')] G(\rho,\xi-\xi';\rho') d\rho' d\xi',$$

$$(31)$$

where the Green's function, $G(\rho,\xi-\xi';\rho')$, is given by

$$G(\rho,\xi-\xi';\rho') = \sum_{m=1}^\infty \frac{y_m(\rho') y_m(\rho)}{||y_m||^2} e^{-\lambda_m^2(\xi-\xi')} \qquad (32)$$

Equation (31) is a Volterra integral equation which we solved by recognizing that $\Phi_1(\xi,\rho)$ should have a solution of the form

$$\Phi_1(\xi,\rho) \; = \; \sum_{m=1}^{\infty} h_m(\xi) y_m(\rho) \; , \tag{33}$$

The functions $h_m(\xi)$ have been obtained numerically, and the first three such functions are plotted in Figures 1-3. Approximate analytical solutions for $h_1(\xi)$ and $h_2(\xi)$ are given by

$$h_1(\xi) \; = \; -0.2632 \, k_o (1-e^{-\lambda_1^2 \xi}) \, e^{-\lambda_1^2 \xi} \tag{34}$$

and

$$h_2(\xi) \; = \; [-k_o(0.001022+0.0364k_o)(1-e^{-(2\lambda_1^2-\lambda_2^2)\xi})$$

$$+k_o(0.19316-0.0172k_o)(1-e^{-\lambda_1^2\xi})-0.00468k_o(1-e^{-\lambda_2^2\xi})$$

$$+0.00482k_o^2(1-e^{-(3\lambda_1^2-\lambda_2^2)\xi})$$

$$+0.00861k_o^2(1-e^{-2\lambda_1^2\xi})] \, e^{-\lambda_2^2\xi} \tag{35}$$

These approximations are also shown in Figures 1 and 2.

With the functions $h_m(\xi)$ determined, the nondimensional concentration is given by

$$\phi(\xi,\rho) \; = \; \sum_{m=1}^{\infty} [A_m e^{-\lambda_m^2\xi} + h_m(\xi)] y_m(\rho) \tag{36}$$

The appropriate average or "mixing cup" concentration is defined by

$$\overline{\phi}(\xi) \; = \; \frac{\int_o^1 \rho(1-\rho^2)\phi(\xi,\rho)d\rho}{\int_o^1 \rho(1-\rho^2)d\rho} \; = \; 4\!\int_o^1 \rho(1-\rho^2)\phi(\xi,\rho)d\rho \tag{37}$$

Substituting for $\phi(\xi,\rho)$ in Equation (37) from Equation (36) and integrating over ρ, and introducing numerical values of

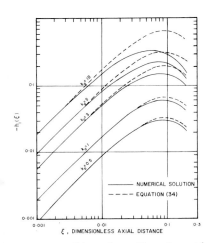

Fig. 1. The function
$h_1(\xi)$ of Equation (33).

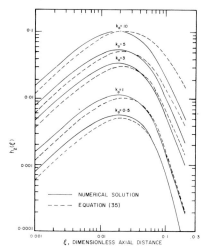

Fig. 2. The function
$h_2(\xi)$ of Equation (33).

the coefficients we obtain

$$\overline{\phi}(\xi) = 0.81095e^{-7.31359\xi} + 0.097528\,e^{-44.6095\xi}$$
$$+ 0.032504e^{-113.921\xi} + 0.01544\,e^{-215.241\xi}$$
$$+ \ldots + 0.55475h_1(\xi) - 0.12098h_2(\xi)$$
$$+ 0.055208h_3(\xi) - 0.032448h_4(\xi) + \ldots \qquad (38)$$

For no coagulation (for which $h_1(\xi) = h_2(\xi) = \ldots = 0$) Equation (38) reduces to the result of Gormley and Kennedy, although the present analysis and numerical values are more exact than the previous results.

Because of the rapid convergence of Equation (38) for $\xi > 0.01$, a simple, useful approximation for $\overline{\phi}(\xi)$ is obtained by applying Equations (34) and (35) to give

$$\overline{\phi}(\xi) \simeq [0.81905 - 0.1461k_o(1-e^{-7.31359\xi})]e^{-7.31359\xi}$$
$$+ 0.097528\,e^{-44.6095\xi} - 0.12098h_2(\xi) , \qquad (39)$$

where $h_2(\xi)$ is given by Equation (35).

127

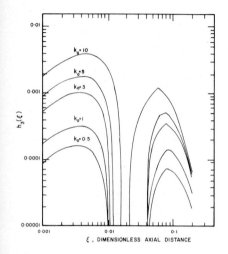

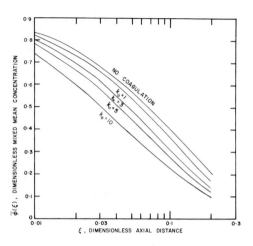

Fig. 3. The function
$h_3(\xi)$ of Equation (33).

Fig. 4. The mixed mean
concentration of particles as a
function of axial position
predicted by Equation (38).

The solution for simultaneous coagulation and convective
diffusion, Equation (38), is plotted in Figure 4 for various
values of the coagulation constant, k_0. The results shown are
based on the first three eigenvalues and eigenfunctions of the
system and are accurate for $\xi > 0.001$. For $\xi < 0.001$, more than
three terms should be used for good accuracy, but for $\xi > 0.1$
only the first eigenfunction is needed because of the rapid
convergence of the eigenfunction expansion. Figure 4 shows
that for $k_0 < 1$ the reduction of particle concentration due to
coagulation is less than 10% of the loss of particles due to
deposition on the tube wall.

V.   THE METHOD OF SUCCESSIVE APPROXIMATIONS

Because of the limitations inherent in perturbation
theory it is desirable to have an alternate technique to
solve a set of coupled equations of the form of (10). Although this technique suffers from some of the limitations

128

of perturbation theory, it is capable of extension to non-linear systems in a systematic manner.

To illustrate this technique let us apply it to Equations (15)-(18), dropping the subscripts on ϕ_1 and ξ_1 for convenience. Let us indicate the order of the approximation by means of a superscript, that is, $\phi^{(i)}$ refers to the ith approximation to ϕ. The zeroth approximation $\phi^{(0)}$ is the solution of no coagulation, given by Equation (5). Introducing $\phi^{(0)}$ into the sink term, $k_o\phi^2$, of Equation (15), we obtain an inhomogeneous equation for $\phi^{(1)}$, where the inhomogeneity involves the known function $\phi^{(0)}$. The solution for $\phi^{(1)}$ is then obtained by applying the Green's function theory of the previous section. The essential difference in the two approaches is that the method of successive approximations takes into account the nonlinear term neglected in the particular perturbation solution used above. Furthermore, the Volterra integral equations of the perturbation theory do not arise since the inhomogeneity associated with the nonlinear sink term involves the lower order solution. Repeated application of this method of successive approximations requires solution of equations of the form

$$(1-\rho^2)\frac{\partial\phi^{(i)}}{\partial\xi} - \frac{1}{\rho}\frac{\partial}{\partial\rho}\left(\rho\frac{\partial\phi^{(i)}}{\partial\rho}\right) = -k_o[\phi^{(i-1)}]^2 \qquad (40)$$

Applying the Green's function approach used in the previous section, we obtain the following approximations

$$\overline{\phi^{(1)}}(\xi) = (0.8191049-0.145408k_o)e^{-\lambda_1^2\xi} + 0.147309k_o\,e^{-2\lambda_1^2\xi} +$$

$$(0.097528-0.024504k_o)e^{-\lambda_2^2\xi} +\ldots+(0.0325044+0.002625k_o)e^{-\lambda_3^2\xi} +\ldots \qquad (41)$$

$$\overline{\phi^{(2)}}(\xi) = (0.819049-0.146026k_o+0.025876k_o^2-0.001532k_o^3)\,e^{-\lambda_1^2\xi} +$$

$$(0.147404k_o-0.052299k_o^2+0.004643k_o^3)\,e^{-2\lambda_1^2\xi} +$$

$$(0.026470k_o^2-0.004699k_o^3)e^{-3\lambda_1^2\xi} + 0.001518k_o^3\,e^{-4\lambda_1^2\xi} +$$

$$(0.097528-0.001236k_o-0.000047k_o^2+0.000070k_o^3)\,e^{-\lambda_2^2\xi} + \ldots \qquad (42)$$

$$\overline{\phi^{(3)}}(\xi) = (0.819049 - 0.146026k_o + 0.026100k_o^2 - 0.004630k_o^3 +$$

$$0.001095k_o^4 + \ldots)e^{-\lambda_1^2\xi} + (0.147513k_o - 0.052600k_o^2 + 0.014015k_o^3 -$$

$$0.002215k_o^4 + \ldots)e^{-2\lambda_1^2\xi} + (0.02662k_o^2 - 0.014222k_o^3 + 0.003372k_o^4 +$$

$$\ldots)e^{-3\lambda_1^2\xi} + (0.004762\,k_o^3 - 0.002259\,k_o^4 + \ldots)\,e^{-4\lambda_1^2\xi} + \ldots +$$

$$(0.097528 - 0.001236k_o - 0.000458k_o^2 - 0.0000066k_o^3 +$$

$$0.000006k_o^4 + \ldots)\,e^{-\lambda_2^2\xi} + \ldots + \qquad (43)$$

Several features of these solutions should be noted. The coagulation constant, k_o, appears as power series expansions multiplying exponential terms in the axial co-ordinate, ξ. The terms of order k_o^2 and higher are associated with the nonlinearity of the coagulation sink term. A comparison of the successive approximations indicates that the accuracy of the coefficients of the powers of k_o improves as higher approximations are calculated. For the third approximation $\overline{\phi^{(3)}}$, the coefficients up to k_o^3 are reasonably accurate, but the coefficients of k_o^4 are not. The range of values of k_o for which each approximation is valid depends on the convergence of the various power series in $k_o < 5$, neglecting terms of order k_o^4.

To order k_o Equations (41)-(43) are nearly identical to the approximate solution, Equation (39), obtained from perturbation theory, which is to be expected. It is now possible to estimate the range of validity of the perturbation analysis since the method of successive approximations incorporates the nonlinear terms neglected in the perturbation theory. Equations (38) and (43) give identical results for $\xi > 0$ and $k_o \sim 0(1)$, and results for $k_o = 3$ are compared in Figure 5. For $k_o = 10$ Equation (43) gives spurious results because the neglected terms of order k_o^4 become important. This comparison suggests that the perturbation analysis is accurate for $k_o \leq 3$ and can be used with caution up to $k_o \sim 0(10)$.

VII. THE ASYMPTOTIC SOLUTION FOR LARGE $K_{i,j}$

The analyses presented above are appropriate for $k_{i,j}^{(\ell)} < 1$, which includes the limiting case of no coagulation. For

130

$k_{i,j}^{(\ell)} \gg 1$ coagulation proceeds at such a fast rate that radial gradients in the concentration of particles can be neglected, and the population balances corresponding to Equation (10) simplify to ordinary differential equations integrated numerically by Runge-Kutta methods. Numerical solution yields $\phi_i = \phi_i(\xi)$ for a particular value of the radial coordinate, ρ, that is, the solution is carried out along a streamline in the annular region lying between $\rho + d\rho$. Typical results of this type of calculation are shown in Figure 6 for an initial monodispersion for which $k_o = 50$. The figure shows the nondimensional concentrations ϕ_1 and ϕ_2 as a function of ξ_1 for various radial positions. Not surprisingly, ϕ_1 decreases monotonically due to coagulation, and ϕ_2 increases near the tube inlet due to the generation of particles of size a_2 from the coagulation of particles of size a_1. As ξ_1 increases, ϕ_2 goes through a maximum and then decreases due to the coagulation of a_2 particles with a_1, a_2 and other particles. The solution for ϕ_1 corresponding to deposition with no coagulation is plotted on the figure for comparison. It is clear that the deposition rate is very much smaller than the coagulation rate in this case, so deposition can be neglected.

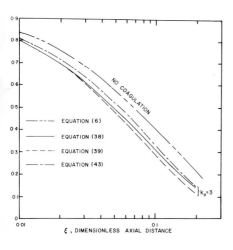

Fig. 5. A comparison between the results from perturbation analysis and the method of successive approximations.

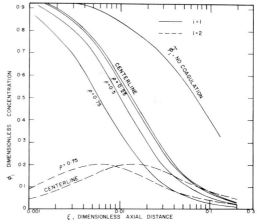

Fig. 6. Particle concentrations predicted for $k_0 = 50$ assuming no deposition compared with the theory for coagulation.

VIII. CONCLUDING COMMENTS

The present work should be useful to the experimentalist for assessing the relative importance of deposition and co-agulation in flow experiments. The experimentalist who wishes to extract information about coagulation constants or deposition rates from experiments involving simultaneous deposition and coagulation is advised to work under conditions such that the limiting cases apply. Otherwise the interpretation of experimental results becomes difficult owing to the intracta-bility of the governing equations. For deposition studies the coagulation constant, k_O, should be in the range $k_O < 0.1$ to permit the neglect of coagulation. For coagulation experiments it is estimated that for $\kappa_O > 50$ deposition can be neglected in the analysis and interpretation of results. Between these two extremes Equation (43) can be used to estimate the relative importance of coagulation compared to deposition.

IX. REFERENCES

1. Fuchs, N.A., "The Mechanics of Aerosols", Pergamon Press, New York, 1964.
2. Davies, C.N., "Aerosol Science," Academic Press, London and New York, 1966.
3. Hidy, G.M., and Brock, J.R., "The Dynamics of Aerocolloidal Systems," Pergamon Press, New York, 1970.
4. Fuchs, N.A., and Sutugin, A.G., "Topics in Current Aerosol Research," edited by G.M. Hidy and J.R. Brock, Pergamon Press, Oxford, 1971.
5. Nicolaon, G., and Kerker, M., Faraday Symposia of the Chemical Society, No. 7, 133 (1973).
6. Davis, E.J., Can. J. Chem. Eng. 51, 562 (1973).
7. Brown, G.M., AIChE J. 6, 179 (1960).
8. Sideman, S., Luss, D., and Peck, R.E., Appl. Sci. Res. A14 157 (1965).
9. Sellars, J.R., Tribus, M., and Klein, J.S., Trans. ASME 78, 441 (1956).
10. Liao, C.S., Ph.D. Dissertation, Clarkson College of Technology, (1974).
11. Lauwerier, H.A., Appl. Sci. Res. A2, 184 (1951).
12. Hsu, C.J., AIChE J. 17, 732 (1971).
13. Deavours, C.A., J. Heat Transfer 96, 489 (1974).
14. Gormley, P., and Kennedy, M., Proc. Roy. Irish Acad. 52A, 163 (1949).
15. Müller, H., Kolloidbeihefte 27, 223 (1928).
16. Smoluchowski, M. von, Phys. Z. 17, 557 (1916).
17. Smoluchowski, M. von, Z. Phys. Chem. 92, 129 (1918).
18. Davis, E.J., and Liao, S.C., J. Colloid Interface Sci. 50, 488 (1975).

ROTATIONAL RELAXATION OF THE BROWNIAN MOTION

T. S. Chow
Xerox Corporation

Encouraged by the excellent agreement between the hydrodynamic model and the molecular dynamic calculation of the linear velocity autocorrelation function of an atom translating in a simple fluid, a similar treatment is extended to the angular velocity autocorrelation function (AVCF) of rotating Brownian particles. Both the inertia and viscoelastic effects are simultaneously included in our hydrodynamic fluctuation model. An inertia dependent drag torque is used and a Maxwell viscoelastic relaxation time is introduced. The autocorrelation functions for the random torque and angular velocity are solved analytically and are given in closed form. By choosing physically reasonable values for the viscoelastic parameters, good agreement is obtained with the AVCF determined by molecular dynamic calculation.

I. INTRODUCTION

The inertia effect on the angular velocity autocorrelation function (AVCF) gives this function an asymptotic decay of $t^{-5/2}$ at long times.[1-2] These studies were based on the hydrodynamic theory of Brownian motions. Since the orientational relaxation time $(1/D_r = \gamma_0/kT$, D_r is the rotational diffusion coefficient, γ_0 is the frictional coefficient, k is Boltzmann's constant and T is the temperature) is much smaller than the angular velocity correlation time $(I/\gamma_0$, I is the moment of inertia), the slow decay at long times may not easily be observed.[2] Therefore, the short time behavior of the AVCF becomes more important. For the case of a linear velocity autocorrelation function (LVCF), a successful agreement between the molecular dynamic studies[3-5] and hydrodynamic models[6-8] was reached. We have established that the inertia effect causes the slow decay at long times and also concluded that the short time Enskog behavior of the LVCF can be explained adequately by including the viscoelastic relaxation.[7]

In this article, a similar treatment is extended to the AVCF. A Maxwell viscoelastic relaxation time τ is introduced in an isotropic incompressible viscous fluid. The constitutive relation of the local stress (σ_{ij}) and strain rate (e_{ij}) is

$$\sigma_{ij}(t) = - P \, \delta_{ij} + 2\int_{-\infty}^{t} \eta(t-s)e_{ij}(s)ds \tag{1}$$

where P is the pressure, δ_{ij} is the Kronecker delta, $\eta = (\eta_0/\tau) \exp(-t/\tau)$ and η_0 is the zero shear viscosity.

According to molecular kinetic theory, the distribution function f of a Brownian particle of density ρ and moment of inertia $I=8\pi\rho a^5/15$ in phase space obeys the equation[1,9]

$$\frac{\partial f}{\partial t} + \frac{\overline{J}}{I} \cdot \overline{\nabla}_\theta f = \int_0^t G(t-s)\overline{\nabla}_J \cdot (\overline{\nabla}_J + \frac{\overline{J}}{IkT})f(s)ds \tag{2}$$

Here $\overline{J}$ is the angular momentum. $\overline{\nabla}_\theta$ is the gradient with respect to the angular coordinates and $\overline{\nabla}_J$ that with respect to angular momentum. The correlation function for the random torque G is determined from hydrodynamic fluctuation theory[1,10] in the next section. The following equation for the AVCF can be obtained from Eq. (2).

$$IkT\dot{\phi}(t) + \int_0^t G(t-s)\phi(s)ds = 0 \tag{3}$$

The literature describes two alternative approaches, the Langevin equation and linear response theory,[11-12] of deriving Eq. (3). The random torque correlation function G has also been called a memory function.[6,11-13] An easy way of solving Eq. (3) is by the use of a Laplace transform.

II. MEMORY FUNCTION

The Laplace transform of the memory function is defined by

$$\overline{G}(p) = \int_0^\infty e^{-pt}G(t)dt \tag{4}$$

Based on the hydrodynamic fluctuation theory, the memory function has been calculated by considering the torque on a sphere. For $\tau=0$, we have obtained[1,14]

$$\frac{\overline{G}(p)}{\gamma_0 kT} = 1 + \frac{\alpha_0^2 p}{3(1+\alpha_0\sqrt{p})} \tag{5}$$

where $\gamma_0 = 8\pi\eta_0 a^3$, $\alpha_0 = a(\rho_0/\eta_0)^{1/2}$

and ρ_0 is the fluid density. When discussing a viscoelastic liquid, Eq. (1) gives

$$\overline{\eta}(p) = \eta_0/(1 + p\tau) \tag{6}$$

Replacing n_0 by $\bar{n}(p)$, Eq. (5) becomes

$$\frac{\bar{G}(p)}{\gamma_0 kT} = \frac{1}{1+\tau p} + \frac{1}{3} \frac{\alpha_0^2 p}{1+\alpha_0 \sqrt{p(1+\tau p)}} \tag{7}$$

and consequently[7]

$$\frac{G(t)}{\gamma_0 kT} = \frac{1}{\tau} \exp(-t/\tau) + \frac{\alpha_0}{3} [\frac{2}{\sqrt{\tau}} \delta(t) + \frac{d}{dt} g(t)] \tag{8}$$

where

$$g(t) = \frac{1}{\pi} \int_0^{1/\tau} \frac{\sqrt{r(1-\tau r)} \ \exp \ (-tr)}{\sqrt{r(1-\tau r)} + 1/\alpha_0^2} \ dr +$$

$$\frac{2}{\alpha_0 \sqrt{1+4 \ \tau/\alpha_0^2}} \exp \ [-t(1+\sqrt{1+4 \ \tau/\alpha_0^2})/2\tau]$$

III. THE AVCF

Substitution of the memory function, Eq. (8), into Eq. (3) leads to a complicated equation for the AVCF which can be solved analytically. For simplicity and convenience, we define the following parameters:

$$\beta = \gamma_0/I, \ x = \beta t, \ q = p/\beta$$

$$\varepsilon = \alpha_0 \sqrt{\beta} = \sqrt{15 \ \rho_0/\rho} \quad \text{(the inertia parameter)}$$

and $\delta = \tau\beta$ (the viscoelastic parameter).

Thus, the Laplace representation of the AVCF is

$$\frac{\bar{\phi}(q)}{\phi(0)} = \frac{(1+\delta q) \ (1+\varepsilon\sqrt{Q})}{\varepsilon Q^{3/2} + (1+\varepsilon^2/3)Q + \varepsilon Q^{1/2} + 1} \tag{9}$$

with $\phi(0) = kT/I$ and $Q = q(1+\delta q)$.

Before proceeding, we should note that the initial value of the AVCF for an incompressible fluid is not $\phi(o) = kT/I*$ as mentioned by Berne.[2] Here $I*$ is the effective moment of inertia. In discussions of translational Brownian motion, several

authors[2,15-16] have introduced an effective mass for the
initial value of the LVCF in an incompressible fluid, because
the drag on a sphere contains a bulk compression term.[14] When
a sphere rotates in an incompressible fluid, it is a shear
motion and the drag torque does not have a shear independent
term.[14] Therefore, no confusion should exist for an initial
value of the AVCF in which $\phi(o) = kT/I$. Of course, the moment
of inertia rather than the effective moment of inertia is required in statistical mechanics, Eq. (2-3).

   When we evaluate the complex integration required to
Laplace invert Eq. (9), the denomenator of Eq. (9) can be
put in the form

$$\varepsilon(\sqrt{Q} + d_1)(\sqrt{Q} + d_2)(\sqrt{Q} + d_3) \tag{10}$$

The above expression indicates there are two branch points at
q=o and $1/\delta$ and two sheets of the Riemann surfaces. We obtain

$$\frac{\phi(x)}{\phi(o)} = \frac{1}{2\pi i} \int_\Gamma \frac{\overline{\phi}(q)}{\overline{\phi}(o)} e^{qx}dq + \Sigma Res \left[\frac{\overline{\phi}(q)}{\overline{\phi}(o)} e^{qx}\right] \tag{11}$$

where Γ is the dumbell-shape contour (clockwise) surrounding
the branch cut between the two branch points along the real
axis.

   The integral term on the right hand side of Eq. (11)
reduces to

$$\frac{1}{\pi} \int_0^{1/\delta} e^{-xr} Im \left[\frac{\overline{\phi}(q)}{\overline{\phi}(o)} \exp(qx)\right]_{q=re^{-i\pi}} dr \tag{12}$$

Since only one sheet of the Riemann surfaces corresponding to
the function $\sqrt{Q} + d_j$ contributes to nontrivial residues, the
pertinent values for Q and q are

$$\sqrt{Q} = -d_j$$

and

$$q = (1+\sqrt{1+4\delta d_j^2})/2\delta \tag{13}$$

for j=1,2,3. The sum of residues term on the right hand side
of Eq. (11) becomes

$$2 \sum_{j=1}^{3} \frac{d_j}{\sqrt{1+4\delta d_j^2}} \psi_j (\sqrt{Q} \text{ and } q \text{ are given by Eq. (13))} \qquad (14)$$

where $\psi_j = (\sqrt{Q} + d_j) \frac{\overline{\phi(q)}}{\phi(o)} \exp(qx)$

We have one real root (d_1) and two complex conjugate (d_2 and d_3) for all values of ε and δ. Therefore, the final solution of Eq. (11) is

$$\frac{\phi(x)}{\phi(o)} = \frac{\varepsilon^3}{3\pi} \int_0^{1/\delta} \frac{(1-\delta r) \, S^{3/2} \exp(-xr) \, dr}{[1-(1+\varepsilon^2/3)S]^2 + \varepsilon^2 S(1-S)^2} +$$

$$\frac{2d_1 \psi_1}{\sqrt{1+4\delta d_1^2}} + 2\text{Re} \left[\frac{d_2 \psi_2}{\sqrt{1+4\delta d_2^2}} \right] \qquad (15)$$

where $S=r(1-\delta r)$. This equation contains both the inertia and viscoelastic effects on the AVCF. Three special cases of Eq. (15) are:

1) When $\delta=o$, Eq. (15) leads to the solution for the inertia effect

$$\frac{\phi(x)}{\phi(o)} = \frac{\varepsilon^3}{3\pi} \int_0^\infty \frac{r^{3/2} \exp(-xr) \, dr}{[1-(1+\varepsilon^2/3)r]^2 + \varepsilon^2 r(1-r)^2} \qquad (16)$$

2) When $\varepsilon=o$, the viscoelastic dependence for the AVCF has the relation

$$\frac{\phi(x)}{\phi(o)} = \frac{1}{\sqrt{1-4\delta}} [h(r_-)-h(r_+)] \qquad (17)$$

where

$$h(r) = (1-\delta r)\exp(-rx)$$

and

$$r_\mp = (1 \mp \sqrt{1-4\delta})/2\delta$$

Eq. (17) has the similar form obtained for the corresponding translational Brownian motion.[7]

3) Finally, the classical Stokes-Einstein result for δ=o and ε=o has the form

$$\frac{\phi(x)}{\phi(o)} = \exp(-x) \tag{18}$$

IV. NUMERICAL RESULTS

   To conclude this discussion, we present a numerical example of results obtained from Eqs. (15-18). The normalized AVCF is plotted as a function of the nondimensional time x=βt. The dotted curves in Figs. 1 and 2 are the classical exponential decay, Eq. (18). The solid curves in Fig. 1 include only the inertia effect calculated on the basis of Eq. (16). The AVCF is a monotonic function of time.

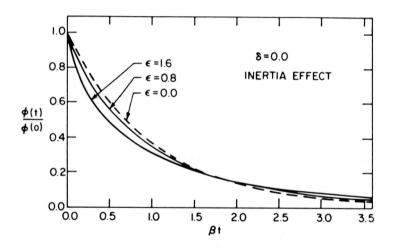

Fig. 1. The inertia effect based on Eq. (16).

In Fig. 2, the viscoelastic effect computed from Eq. (17) is indicated as solid lines. The AVCF behaves as a damping oscillatory function and has a negative backscattered region which has also been observed in molecular dynamic calculations.

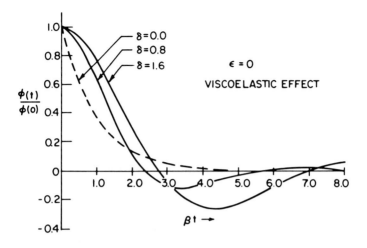

Fig. 2.   The viscoelastic effect
based on Eq. (17).

The simultaneous consideration of the viscoelastic and inertia
effects based on Eq. (15) should provide a better solution for
the AVCF especially at short times.[7]   They are shown as solid
lines in Fig. 3.   Because the viscoelastic property and molecular radius are not presented together with the corresponding

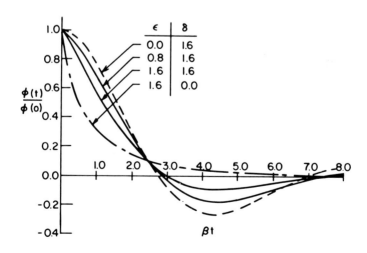

Fig. 3.   Viscoelastic and inertia
effects on the basis of Eq. (15).

139

molecular dynamic calculation of the AVCF in the literature, we will only make a qualitative comparison between these results obtained from Eq. (15) and those from a molecular dynamic calculation.

Consider a liquid at 70°K which has the following properties: the density is 1g/cc, viscosity is 10^{-3} poise, diffusion coefficient is 10^{-5} cm^2/sec and the viscoelastic relaxation time is 10^{-13} sec. A simple estimation leads to

$$a \sim 10^{-8} \text{ cm} \quad , \quad \beta \sim 10^{13} \text{ sec}^{-1}$$

$$\varepsilon \sim O(1) \text{ and } \delta \sim O(1) \tag{19}$$

We therefore conclude that Eq. (15), Fig. 3, is in qualitative agreement with Berne and Harp's molecular dynamic AVCF for liquid CO (Fig. 12 of Ref. 2). An exact comparison is straightforward when all the pertinent parameters are chosen.

V.   REFERENCES

1.   T. S. Chow, Phys. Fluids 16, 31 (1973).
2.   B. J. Berne, J. Chem. Phys. 56, 2164 (1972).
3.   B. J. Alder and T. E. Wainwright, Phys. Rev. A1, 18 (1970).
4.   A. Rahman, Phys. Rev. 136, A405 (1964).
5.   G. Subramanian, D. Levitt and H. T. David, J. Chem. Phys. 60, 591 (1974).
6.   R. Zwanzig and M. Bixon, Phys. Rev. A2, 2005 (1970).
7.   T. S. Chow, J. Chem. Phys. 61, 2868 (1974).
8.   T. S. Chow and J. J. Hermans, J. Chem. Phys. 56, 3150 (1972).
9.   S. Kim and I. Oppenheim, Physica 54, 593 (1971).
10.   T. S. Chow and J. J. Hermans, J. Chem. Phys. 59, 1283 (1973).
11.   R. Kubo, Rept. Progr. Phys. 29, 255 (1966).
12.   T. S. Chow and J. J. Hermans, Proc. Konink, Ned. Acad. Wetenschap. B77, 18 (1974).
13.   B. J. Berne and G. D. Harp, in "Advances in Chemical Physics" (I. Prigogine, and S. A. Rice, Eds.) Vol. 17, p. 63. Interscience, N. Y., 1970.
14.   L. D. Landau and E. M. Lifshitz, "Fluid Mechanics," Addison-Wesley, Reading, Mass., 1958.
15.   E. H. Hauge and A. Martin-Lof, J. Stat. Phys. 7, 259 (1973).
16.   R. Zwanzig and M. Bixon, J. Fluid Mech. 69, 21 (1975).

AEROSOL PARTICLE SIZE AND SHAPE MEASUREMENTS

THROUGH ASYMMETRIC LIGHT SCATTERING

J. Allen and R. B. Husar
Air Pollution Research Laboratory
Department of Mechanical Engineering
Washington University

ABSTRACT

An experimental particle counter is built around a low power helium-neon laser operated with a remote mirror in order to utilize the high intensity light within the standing wave cavity. The hydrodynamically focused aerosol stream passes down through the center of the laser beam and as each particle crosses the intersection, the scattered light pulse is detected by two photomultipliers, one on each side of the interaction zone. The signals are added and subtracted electronically. The sum gives the scattering cross section or optical size, but the difference pulse reveals to what extent, if any, the particle scatters light asymmetrically.

We find asymmetric scattering to be a specific property of non-spherical aerosols. Thus, for the configuration of our instrument, 0.365 micron polystyrene-latex spheres give an 8% probability of asymmetric scattering (1.25% if multiplets are excluded), dry sodium chloride shows 10%, while the same aerosol shows almost no asymmetry at relative humidities above the deliquescence point. Atmospheric samples generally have asymmetric scattering probabilities of 2 to 5%. The technique is to be used as a complement to existing methods for characterizing the physical properties of ambient aerosols.

I. INTRODUCTION

Aerosol particles in the size range from about 0.1 micron to 3 microns are of great importance in air pollution studies. While small enough to be respirable, they are still large enough to be the principal contributors to atmospheric turbidity through their ability to scatter light. Yet this very light scattering provides us with a practical means of counting and size spectrometry within the above size range. The Mie theory (1) provides the set of formulae relating a particle's light scattering cross section to its geometric size and refractive index (n). Thus if n is known, one can in principle measure aerosol sizes by passing the particles through a light beam, one at a time, and recording the intensity of scattered radiation from each. Such a method has the advantage of making the measurements in situ and in real time, while most impaction and filtration methods require subsequent laboratory analysis.

While the principle of an optical counter may be fairly simple, the design and construction of a practical instrument involves a number of complications, along with opportunities for innovation.

II. OBJECTIVE

Optical counters in various forms have existed for a number of years and several types are now sold commercially. Our project, therefore, is aimed at developing an instrument at reasonable cost with certain specific improvements.

A. Improved Sensitivity

Scattering theory shows that as particle sizes become small compared to the wavelength of light, the scattering cross section falls very much faster than does the geometric cross section. In the small-size limit, scattering cross section is proportional to the fourth power of the geometric area, or the sixth power of the diameter. This necessitates intense illumination and sensitive detectors, yet the very steepness of the scattering versus size function facilitates excellent size resolution.

B. Shape Discrimination

Not all aerosol particles are spherical; some are crystalline or irregularly shaped. It is possible to extract some shape information from the scattering pattern which permits one at least to discriminate between spherical and non-spherical aerosols.

III. APPARATUS

The present apparatus (our second prototype) consists of a low-power helium-neon laser operating with a remote mirror so that it is possible to exploit the high intensity light within the actual standing wave cavity. The aerosol to be analyzed is hydrodynamically focused into a narrow stream which intersects the light beam. The scattered light is collected on both sides of the beam by a lens and photo-multiplier on each side (Fig. 1). Thus whenever a particle crosses the laser beam, a pulse appears at the output of each detector. These pulses are processed electronically to give one signal equal to the sum and another equal to the differ-ence. The sum gives the effective optical size of the parti-cle, while the difference, called the asymmetric scattering gives information related to shape. A detailed description follows:

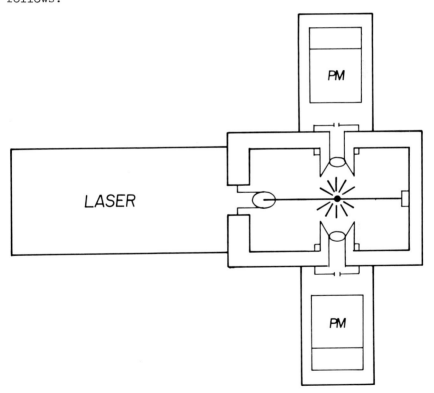

Fig. 1. The aerosol particles (moving into the page) cross the standing wave laser beam and scatter light which is collected by the lenses and detected by the two photo-multipliers.

A. The Laser

The decision to use a laser was made after studying the
properties of some more conventional light sources, and while
there are other devices which can provide comparable intensities, the laser offers the advantage of a well-defined
parallel beam. If constructed as an emitting laser, our unit
would have a nominal output of 2 milliwatts, but with a
totally reflecting mirror 7 cm from the Brewster window at the
tube end, the intensity within the cavity is at least 20 times
that great. In the TEM 00 mode, the beam has a Gaussian profile with a 2σ diameter of about 0.5 mm, and thus a central
intensity of nearly 50 watts per cm^2.

Operating a laser with a remote mirror is not without its
disadvantages. Maintaining the mirror in critical angular
alignment requires a very rigid structure, but a more serious
problem is the cavity's lengthwise dimensional stability, for
every time the mirror spacing (cavity length) changes by 1/2
wavelength, the laser must change to another longitudinal
mode. In the course of that transition, a phenomenon known
as mode-hopping occurs during which the laser oscillates
randomly between the two adjacent modes, with the result that
its light output is noisy and erratic.

The original model had an aluminum frame, in retrospect
a bad choice. The large thermal expansion of this metal
resulted in a total dimensional change (over the 38 cm cavity
length) of 30 half-wavelengths per °C temperature change,
and a mode hop at just the wrong time often spoiled an otherwise excellent count with a burst of random noise. We considered certain special materials such as invar, fused quartz,
and certain ceramics which have almost no expansion, but
rejected them as too costly and difficult to fabricate.

The frame of the present instrument was therefore made
of cherry wood. This material combines excellent mechanical
and working properties with a low thermal expansion in the
direction parallel to the grain (less than 1/10 that of
aluminum) and this small change is easily compensated by a
short length of brass. The structure has so far proved to
be entirely satisfactory, offering excellent stability,
stiffness and inherent vibration damping.

B. The Injector

As stated above, the laser has a 2σ beam diameter of
about 0.5 mm. In order to achieve a pulse height resolution
of 10%, the aerosol stream must cross the beam within a

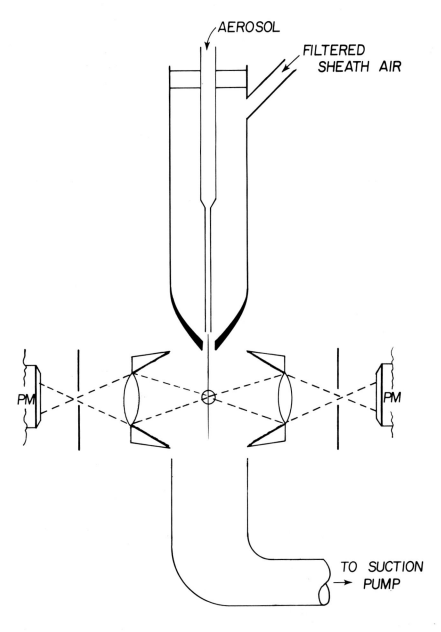

Fig. 2. Configuration of the injector, lenses and detectors. This is an end view looking along the laser beam.

region where the intensity is at least 90% of its central
value. This limits the stream diameter to 0.1 mm. This
order of focusing has been achieved by means of an aerodynam-
ically focused nozzle (see Fig. 2). The aerosol passes
through a series of telescopically fitted pieces of hypodermic tubing, the last being 0.25 mm I.D. The aerosol then
passes through a tapered nozzle where the accelerating flow
of filtered sheath air reduces its diameter to the required
fineness. The particles cross the laser beam at about
10 meters/sec, yielding pulse durations of 50 microseconds.
With an aerosol flow rate of .042 cm^3/sec, one can handle
aerosols of 48,000 particles per cm^3 before there is a 1%
probability of two particles being in the beam at one time.
Normal ambient aerosols have on the order of 1000 particles/
cm^3 in the countable size range of the instrument
(0.18 to 1.1 micron).

C. The Photodetectors

     The axes of the laser, the injector, and the optical
system are all mutually perpendicular. On each side of the
intereaction zone is a lens of 13 mm focal length and f 2.5
aperture. (These are standard D-mount lenses from 8 mm
moving picture cameras.) At the conjugate focal point of
each lens is a diaphragm stop and behind this is a 1.5 inch
end window photomultiplier with a type S-20 cathode for good
response to the red (6328 A) He-Ne laser light. In the housing of each photomultiplier is an op-amp follower to provide
low-impedance output for transmission to the signal processing amplifiers.

D. Signal Processing

     At each input to the signal processing amplifier, a
simple attenuator serves to adjust for incidental differences
in tube sensitivities (Fig. 3). The two signals then go to
operational amplifiers. The first of these is connected as a
summing amplifier so that the pulses from the right and left
tubes are added. Another operational amplifier operates in
its differential mode in order to subtract the two pulses.
If a particle scatters light symmetrically (equally to left
and right) there will be an output pulse, of course, in the
sum channel, but complete cancellation will result in no
output in the difference channel. On the other hand, if a
particle scatters asymmetrically, the cancellation will be
imperfect and output pulses will appear in both the sum and
the difference channels. The two signals are fed to a dual-
trace oscilloscope and to two electronic counters. Either
signal can be plugged into a logarithmic amplifier

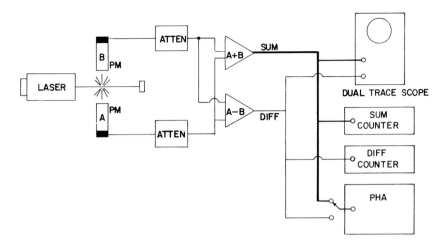

Fig. 3. Signal processing: The attenuators correct for incidental differences in photomultiplier sensitivities. The signals are then fed to two operational amplifiers. The upper adds the two signals to produce the sum pulse, while the lower, operating as a differential amplifier, yields a difference pulse whenever the A and B signals are unequal.

(to accommodate the more than 300:1 range in pulse heights between 0.18 and 1.1 micron particles) and a multi-channel analyzer. An audio amplifier and loudspeaker permit listening to the pulses. This is often helpful when making adjustments or estimating counting rates. Additional electronics includes a single-channel analyzer and a linear gate which permit the sum pulses to be gated by the difference pulses or vice versa. In this way, we can display the spectrum of only those sum pulses which are accompanied by difference pulses or else the spectrum of difference pulses associated with a restricted range of sum pulse heights.

IV. PERFORMANCE AND RESULTS

A. Small Particle Sensitivity

As the instrument is now constructed and operating, it is capable of resolving the signal produced by 0.188 micron polystyrene latex (PSL) clearly above the background noise. However, the present prototype was built less than a month ago and has thus not yet been tuned, adjusted, and modified to maximize its performance. Signal-to-noise ratio can

147

undoubtedly be improved by appropriately limiting the band-
width of the signal amplifiers, as well as by improvements in
the optics, such as faster lenses and better control of stray
light. The upper size limit of 1.1 micron results solely
from the restricted dynamic range of the logarithmic amplifier ahead of the multi-channel analyzer.

B. Shape Discrimination

    Research in this area has unquestionably yielded the most
interesting results. Our earliest experiments with the first
prototype did not employ the multi-channel analyzer, but
merely counted the total numbers of sum and difference pulses
obtained during a time interval for a variety of aerosols.
In taking the counts, the discriminators were set just above
the background noise levels. The ratio of difference pulses
to sum pulses counted represents the probability that a particle shows asymmetric scattering. Fig. 4 represents an average
derived from many such runs done on the new prototype.

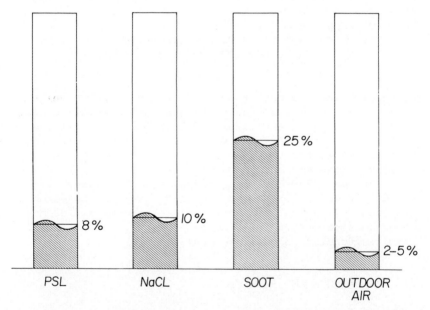

*Fig. 4. Asymmetric scattering probabilities for three
laboratory aerosols and atmospheric air. The wavy lines
suggest the spread in the data.*

The wavy line at the top of each bar suggests the approximate
scatter in the data. It is clear that the soot, which is
known to consist of highly irregular chainlike aggregates
(2) shows the most asymmetric scattering (25%). Sodium

chloride, made up of cubic crystals, shows less (10%), while polystyrene-latex spheres show less still (8%).

Why, however, do 8% of the PSL particles show asymmetric scattering? We hypothesized that this was due to doublets (or multiplets) consisting of two (or more) still stuck together while being counted. In support of this, it was noticed that the asymmetric counting ratio did increase when a thicker suspension was used in the nebulizer which generated the aerosol, thereby increasing the probability that a droplet would contain more than one sphere and thereby produce a multiplet upon evaporating. However, better evidence was needed, for it was crucial to determine whether the instrument could reliably discriminate between spherical and non-spherical aerosols.

As the instrument was improved and the multi-channel analyzer installed, it was possible to resolve clearly the singlet, doublet, and triplet peaks of 0.365 micron PSL. We then added a single-channel analyzer in the sum pulse line whose output triggered a gate in the difference line. The sum counter was tapped into the SCA output while the difference counter read the pulses at the gate output (Fig. 5).

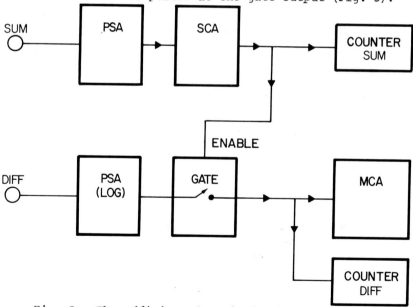

Fig. 5. The addition of a single-channel analyzer, a linear gate, and another pulse shaping amplifier makes it possible to select and count only those sum and difference pulses associated with a restricted range or 'window' of effective particle sizes.

Thus, the sum counter would see only those pulses whose heights fell within the SCA window, and the difference counter would see only those difference pulses coincident with (belonging to) the selected sum pulses. Three runs made with the SCA window set respectively at the singlet, doublet, and triplet peaks yielded the following asymmetric counting ratios: (Fig. 6)

```
Singlet   1.25%
Doublet   30 %
Triplet   34 %
```

We considered this sufficient confirmation that any asymmetric scattering from PSL spheres was due to multiplets whose collective shape was non-spherical.

Dry sodium chloride shows a sizeable (10%) asymmetric scattering probability. However, when the aerosol was aged in a humidity chamber, the asymmetric count ratio varied considerably as a function of humidity, but invariably fell to near zero at relative humidities greater than 75%. This is undoubtedly due to deliquescence of the salt crystals to form spherical solution droplets (3).

All inland atmospheric aerosols which have been sampled so far have had asymmetric counting ratios between 2 and 5%. This low value has two important consequences: first, the assumption that all the particles are spherical will not introduce statistically serious errors into the size spectrograms of common atmospheric aerosols obtained by optical counters. On the other hand, the appearance of a high asymmetric counting ratio in an environmental sample would indicate the presence of what is ordinarily a comparatively rare particulate species. And it would also serve as a warning that in analyzing particle sizes, it might not be safe to assume sphericity.

One further question was whether there exists any correlation between particle size and asymmetric scattering for heterodisperse aerosols. To determine this, the sum signal was fed to the multi-channel analyzer but its input was gated by the difference channel. It would thus tell us, "What is the size spectrum of those particles which exhibit asymmetric scattering?"

The resulting spectrum for NaCl showed a sharp drop below 0.3 micron and almost no particles below 0.25 micron appeared in the gated spectrum, even though these smaller sizes were

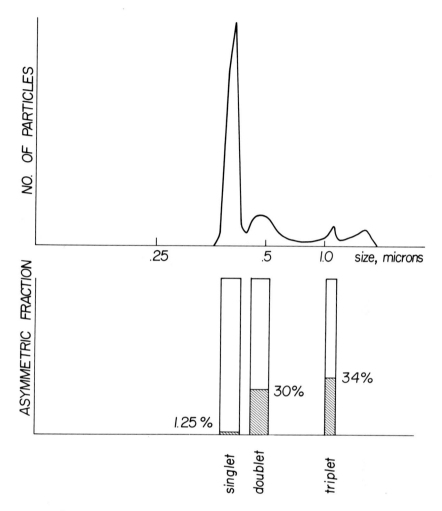

Fig. 6. Proof that multiplets cause whatever asymmetric scattering is observed from polystyrene-latex. Above is the multichannel analyzer spectrum of 0.365 micron PSL. The bar graphs below indicate the asymmetric scattering fraction obtained when the apparatus in Fig.5 restricted counting to each of 3 narrow windows in the size spectrum. The widths of the bars correspond to the widths of the windows. The vertical scale of the spectrum has been exaggerated to the right of the singlet peak in order to show the multiplet peaks more clearly.

very abundant in the ungated spectrum (Fig. 7). Outdoor air samples (collected near our St. Louis, MO laboratory) showed similar behavior as did an aerosol generated by atomizing

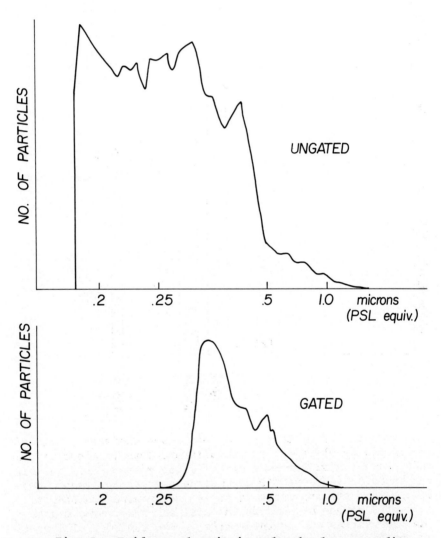

Fig. 7. Evidence that it is only the larger sodium chloride particles which produce asymmetric scattering. Above is the total multi-channel analyzer spectrum for dry NaCl aerosol. Below is the spectrum for the same aerosol, but gated by the difference channel so that only those particles with asymmetric scattering were counted.

sea water (obviously not collected near our St. Louis laboratory). Soot also showed a higher asymmetric scattering
probability associated with larger particles, but unlike the
other aerosols did not show a clear cutoff. Even particles
as small as 0.19 micron (PSL equivalent) showed a significant
asymmetric scattering probability. It should be noted, of
course, that soot, unlike the other aerosols, absorbs a sizeable amount of the incident light rather than scattering it
conservatively. Thus, a soot particle measured by scattered
light is probably much larger than its scattering intensity
could indicate. It may be significant that 0.3 micron, the
minimum size for efficient asymmetric scattering, is about
one half the wavelength of the incident light.

V. FUTURE PLANS

    Naturally, we hope to continue improving our instrument
and testing it on a variety of environmental and laboratory-
generated aerosols, including a number of smokes and dusts,
fibrous aerosols, and some crystalline pollutants such as
sulfates. In doing so, we hope to learn both about the apparatus and the optical properties of the aerosols in question.
We feel that our optical size spectrometer has the potential
to become a useful tool for aerosol analysis. Used in conjunction with other devices, it may prove valuable in determining the composition and origin of air pollutants without
having to make conventional chemical analyses.

VI. REFERENCES

1. Mie, G., Ann. Physik 25, 377 (1908).

2. Williamson, S.J., "Fundamentals of Air Pollution,"
        p. 245, Addison-Wesley Publishing Co., Reading, Mass.
        1973.

3. Friedlander, S.K., "Smoke, Dust and Haze," p. 424
        McGraw-Hill, New York, 1976 (In press).

VII. ACKNOWLEDGEMENT

    This research was supported by the U.S. Environmental
Protection Agency, Environmental Sciences Laboratory,
Grant No. R803115 to Washington University. The first
prototype laser was made available to us by Dr. R.Knollenberg.

APPROXIMATION FOR THE ABSORPTION COEFFICIENT OF ATMOSPHERIC AEROSOL PARTICLES IN TERMS OF MEASURABLE BULK PROPERTIES

by Gottfried Hänel and Ralph Dlugi, Institut für Meteorologie, Johannes Gutenberg-Universität, Mainz, West Germany

A simple approximation for the absorption coefficient of airborne aerosol particles is derived. Discussion yields, that the approximation is a good, economic equation for practic use.

The treatment of radiative transfer in the atmosphere requires among other informations the knowledge of the absorption coefficient of airborne aerosol particles. To get a simple analytical form of the absorption coefficient it is suitable to use the MIE (1908) theory, which presupposes homogeneous spheres, as a first approximation.

The absorption coefficient 6_A of N_K homogeneous spheres in a volume of air V_K is given by

$$6_A = \frac{1}{V_K} \sum_{j=1}^{j=N_K} \pi r_j^2 \, \mathcal{X}_{Aj}(\lambda, r_j, n_j, k_j) \qquad (1)$$

where λ is the wavelength of radiation, r_j the j-th sphere's radius, $n_j - i \cdot k_j$ the complex refractive index of the j-th sphere, and $\mathcal{X}_{Aj}$ its efficiency factor of absorption. For size parameters $\alpha = 2\pi r/\lambda \leq 0.8$, $1.25 \leq n \leq 1.75$, and $k \leq 1$ it is possible to derive an approximation

$$\mathcal{X}_A \cong \frac{24\alpha n k}{\left(n^2+2\right)^2 + k^2\left(2n^2+k^2+4\right)} + \dots \cong \frac{8\alpha k}{n^2+2} + \dots \qquad (2)$$

(PENNDORF 1962, HÄNEL 1976). To compare this approximation with the efficiency factor $\varkappa_{A,MIE}$, which is computed with the exact MIE theory, the generalized size parameter of absorption $\alpha_A = (8\alpha k)/(n^2+2)$ is defined. Computations of $\varkappa_{A,MIE}$ versus α_A for complex refractive indices close to those measured on samples of atmospheric aerosol particles yield the rough approximation $\varkappa_{A,MIE} \cong \alpha_A$ for $\alpha_A \lessgtr 1$ and $\varkappa_{A,MIE} \cong 1$ for $\alpha_A > 1$ (HÄNEL 1976). Assuming that the most important absorbing particles have $\alpha_A \lessgtr 1$, equation (1) reads

$$\delta_A \cong \frac{12\pi}{\lambda V_K} \sum_{j=1}^{j=N_K} \frac{k_j}{n_j^2+2} V_j = \frac{12\pi}{\lambda V_K} \sum_{j=1}^{j=N_K} \frac{k_j}{n_j^2+2} \frac{m_j}{\vartheta_j} \quad (3)$$

where $V_j = 4/3\pi r_j^3 = m_j/\vartheta_j$ is the volume, m_j the mass and ϑ_j the bulk density of the j-th sphere. This formula tells, that the absorption coefficient of a single particle is proportional to the product $k_j V_j$ or $k_j m_j$. For spheres of the same complex refractive index $n-i\cdot k$ and of the same bulk density ϑ the formula (3) reads

$$\delta_A \cong \frac{12\pi}{\lambda V_K} \frac{k}{\vartheta} \frac{1}{n^2+2} \sum_{j=1}^{j=N_K} m_j = \frac{12\pi}{\lambda} \frac{k}{\vartheta} \frac{1}{n^2+2} \frac{M}{V_K} \quad (4)$$

where M/V_K denotes the mass of the spheres per unit volume of air.

This formula can be applied to atmospheric aerosol particles, since the assumption, that all particles have the same bulk complex refractive index and the same bulk density is supported by coagulation processes in the atmosphere. Especially the most important absorbing particles - these are usually the smallest ones - can assumed to be of almost the same chemical composition (compare WHITBY 1974). This is the reason why the mean bulk values of the complex refractive index and the density coming from measurements on samples of atmospheric aerosol particles (VOLZ 1972, CHIN-I LIN et al. 1973, LINDBERG and SNYDER 1973, FISCHER 1973 and 1975, THUDIUM 1976, HÄNEL 1968 and 1976) can be used for computation of the absorption coefficient with the help of equation (4). The mass of particles per unit volume of air can be measured with the help of the well known and widely used filter techniques.

We have compared the approximation formula (4) with the results coming from the exact MIE theory (HÄNEL 1976). For both computations the same experimental aerosol data have been used. The comparison has been performed for three types of aerosol (clean air, urban, and maritime) in the wavelength region o.3 to 12.o μm of radiation as well as for relative humidities between o and o.95. The results yield that the approximation (4) gives errors smaller than about 4o % for the wavelengths o.3 to 2.5 μm and 9.25 to 12.o μm. The errors are smaller than about 8o % for the wavelength 3.o μm. These errors are of the same order like the sum of experimental errors inhibited using equation (4) on the base of experimental data. For these reasons we regard the approximation (4) as a good, economic first approximation to exact MIE theory for practical use.

REFERENCES

1. Chin-I Lin, Marcia Baker, R.J. Charlson, Applied Optics 12, 1356-1363 (1973).
2. Fischer, K., Beitr. Phys. Atm. 46, 89-1oo (1973).
3. Fischer, K., Applied Optics 14, 2851-2856 (1975).

4. Hänel, G.,Tellus XX, pp. 371-379 (1968).
5. Hänel, G.,Advances in Geophysics 19, 73-188 (1976).
6. Lindberg, J.D., D.G. Snyder, Applied Optics 12 573-578 (1973).
7. Mie, G., Ann. Phys. 25, 377-445 (1908).
8. Penndorf,R., J. Atm. Sci. 19, p. 193 (1962).
9. Thudium, J. Journal Aerosol Sci. 7, 167-173 (1976).
1o. Volz, F.E., Applied Optics 11, 755-759 (1972).
11. Whitby, K.T., Proc. of the "Jahreskongress 1974 der Gesellschaft für Aerosolforschung e.V.", Bad Soden, Germany (1974).

THE EFFECT OF AEROSOLES AND CLOUDS ON RADIATIVE TRANSFER PROBLEMS IN THE ATMOSPHERE

O. Theimer*

New Mexico State University, Arts & Sciences Research Center

The main subject of this paper are small scattering corrections to radiative transfer problems such as thermal sensing of the atmosphere. The quantities of main interest are the source function and the spectral intensity seen by a detector in elevation Z, facing the surface of the earth. These quantities are obtained as iterative solutions of the equation of transfer. Explicit formulas are given for layered atmospheres in which the characteristic parameters depend only on altitude. Numerical values for the scattering corrections in such an atmosphere are presented.

I.   INTRODUCTION

This paper is concerned with small scattering corrections to radiative transfer problems such as thermal sensing, which depend sensitively on the optical parameters of the atmosphere.

As shown in Fig. 1 we consider a detector $\bar{a}$ in altitude Z, accepting radiation within a solid angle Ω which determines the area A of earth surface seen by the detector. The beam of radiation from A to $\bar{a}$ has direction (θ,ϕ) where θ is the polar angle with the direction $\hat{z}$ normal to the earth surface, and ϕ is the azimuthal angle about $\hat{z}$. The spectral intensity in distance s from A is $I_\nu(s,\theta,\phi)$, and the spectral intensity $I_\nu(S,\theta,\phi)$ received by the detector is used for determining the temperature distribution in the atmosphere. To this purpose approximate solutions of the equation of transfer are needed which are the main subject of this paper.

---

* This work was supported by the U. S. Army Electronic Command, Atmospheric Sciences Laboratory, White Sands Missile Range, New Mexico

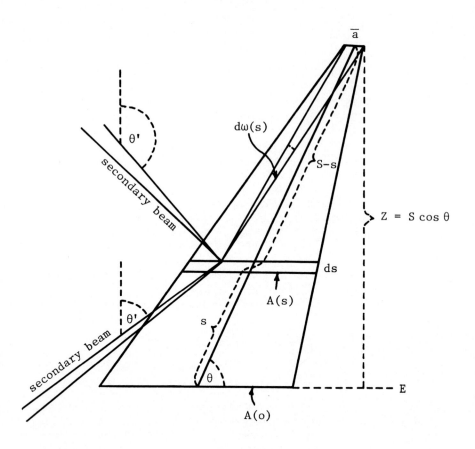

Fig. 1. *Primary beam from earth surface E to detector $\bar{a}$ in distance S. A(s) is area seen by the detector in distance s from earth surface.*

II. ATTENUATION AND AMPLIFICATION OF BEAMS

We consider attenuation of a beam due to absorbing or scattering particles, by a small amount (1)

$$dI_\nu^{ex}(s) = -I_\nu(s)\rho(s)\kappa_\nu^{ex}(s)ds, \qquad (1)$$

where κ_ν^{ex} is the mass extinction coefficient for the frequency ν; it is related to the cross-section c_ν^a for absorption, and c_ν^s for scattering by the equation

$$\rho\kappa_\nu^{ex} = \rho(\kappa_\nu^a + \kappa_\nu^s) = N(C_\nu^a + C_\nu^s), \qquad (2)$$

where $\rho(N)$ is the mass (number) density of the particles respectively.

Next, we consider amplification of a beam due to emission by particles in the beam, by the small amount (1)

$$dI_\nu^{em}(s,\theta,\phi) = j_\nu(s,\theta,\phi)\rho(s)ds, \qquad (3)$$

where $j_\nu(s,\theta,\phi)$ is the spectral emissivity due to spontaneous emission (j_ν^e) and scattering (j_ν^s). Thus,

$$j_\nu(\theta,\phi) = j_\nu^e + j_\nu^s(\theta,\phi). \qquad (4)$$

$j_\nu^s(s,\theta,\phi)$ represents radiation emitted into the primary beam in direction (θ,ϕ) due to scattering from secondary beams with all directions (θ',ϕ') and intensities $I_\nu(s,\theta',\phi')$. (See Fig. 1). Hence,

$$j_\nu^s(s,\theta,\phi) = \frac{\kappa_\nu^s(s)}{4\pi}\int_0^\pi \int_0^{2\pi} p(\theta,\phi,\theta',\phi')I_\nu(s,\theta',\phi')\sin\theta'd\theta'd\phi'. \qquad (5)$$

The "phase function" $p(\theta,\phi,\theta',\phi')$ is related to the differential scattering cross-section

$$c_\nu^s(\vartheta) = (C_\nu^s/4\pi)p(x^s), \quad x^s = \cos\vartheta, \qquad (6)$$

where ϑ is the scattering angle. It is convenient to expand $p(x^s)$ into a series (2) of Legendre's polynomials $P_n(x^s)$,

$$p(x^s) = \sum_{n=0}^{n_{max}} P_n P_n(x^s). \qquad (7)$$

Given the uncertainty in atmospheric composition and in the shape of the scattering particles, an operationally meaningful value for n_{max} will not exceed 3 in most practical cases. The $P_n(x^s)$ are related to Legendre's associated functions of the first kind, $P_n^m(x)$, $x = \cos\theta$, by the relation (2)

$$P_n(x^S) = P_n(x)P_n(x') + 2 \sum_{m=1}^{n} \frac{(n-m)!}{(n+m)!} P_n^m(x)P_n^m(x') \cos m(\phi-\phi').$$

$$(8)$$

III. THE GENERALIZED KIRCHOFF LAW

In a closed system in complete thermal equilibrium the spectral intensity I_ν becomes the blackbody intensity $B_\nu(T)$ at the temperature T. Furthermore dI_ν^{ex} and dI_ν^{em} of the Eqs. (1) and (3) have to be opposite equal. This gives the generalized Kirchoff Law

$$\frac{j_\nu^e + j_\nu^s}{\kappa_\nu^a + \kappa_\gamma^s} = \quad (j_\nu^e/\kappa_\nu^a) = B_\nu(T). \qquad (9a)$$

$$(j_\nu^s/\kappa_\nu^s) = B_\nu(T). \qquad (9b)$$

In the earth atmosphere Eq. (9a) is usually well satisfied, but Eq. (9b) breaks down more or less significantly. This is taken into account by the source function (1)

$$J_\nu \equiv \frac{j_\nu}{\kappa_\nu^{ex}} = \frac{j_\nu^e}{\kappa_\nu^a} \frac{(1 + j_\nu^s/j_\nu^e)}{(1 + \kappa_\nu^s/\kappa_\nu^a)} = B_\nu(T)(1 + \delta_\nu^s). \qquad (10)$$

An approximate calculation of the scattering correction δ_ν^s is our main concern.

IV. THE EQUATION OF TRANSFER

We consider a homogeneous atmosphere in which the parameters κ_ν^{ex}, J_ν and ρ depend on the altitude z and θ, but not on ϕ. Taking the difference $|dI_\nu^{em}| - |dI_\nu^{ex}|$ of the Eqs. (1) and (3) and integrating over z gives, for a beam pointing from the earth to the sky ($\theta < \pi/2$)

$$I_\nu^\uparrow(z,\theta) = I_\nu(0,\theta)e^{-\tau_\nu(\theta,z,0)}$$

$$+ \int_0^z [J_\nu(z',\theta)\kappa_\nu^{ex}(z')\rho(z')/\cos\theta]e^{-\tau_\nu(\theta,z,z')} dz',$$

$$(11)$$

where

$$\tau_\nu(\theta,z,z') = (1/\cos\theta) \int_{z'}^{z} \kappa_\nu^{ex}(z'')\rho(z'')dz'', \qquad 0 < \cos\theta < 1,$$

(12)

is the optical thickness in the interval between z' and z. For a beam pointing from the sky down to the earth $(\pi/2 < \theta < \pi)$

$$I_\nu^\downarrow(z,\theta) = -\int_{z}^{\infty} [J_\nu(z',\theta)\kappa_\nu^{ex}(z')\rho(z')/\cos\theta]e^{-\tau_\nu(\theta,z',z)} dz'$$

(13)

with

$$\tau_\nu(\theta,z',z) = (1/\cos\theta) \int_{z'}^{\infty} \kappa_\nu^{ex}(z'')\rho(z'')dz'', \qquad -1 < \cos\theta < 0,$$

(14)

The first term in Eq. (11) represents radiation emitted by the earth surface and attenuated during transfer to the point (z,θ). This term does not occur in Eq. (13) since sources outside the sky are neglected. The integral in the Eqs. (11) and (13) represents spectral intensity due to emission and extinction by the particles in the beam.

V. THE LAYERED ATMOSPHERE

We study an atmosphere in which κ_ν^{ex}, J_ν and ρ do not change in layers of constant altitude z. Such a situation is realized in atmospheres with or without layers of aerosoles or clouds. For convenience we treat the parameters κ_ν^{ex} etc. as piecewise constant step functions by subdividing the range from 0 to z_{max} into $\mathcal{N}$ intervals labeled μ and bounded by z_μ and $z_{\mu=1}$. $\mathcal{N}$ is of order 10-100 and z_{max} is an appropriate upper limit for the atmosphere. Eq. (11) can now be written in a form in which complicated integrals are replaced by sums. Thus, dropping the subscript ν

$$I^{\uparrow}(z,\theta) = I(0,\theta)e^{-\tau(\theta,z,0)}$$

$$+ \sum_{\mu=1}^{\mu_z} J_\mu(\theta)\left[e^{-\tau(\theta,z,z_\mu)} - e^{-\tau(\theta,z,z_{\mu-1})}\right]. \quad (15)$$

The optical thickness $\tau(\theta,z,z')$ of Eq. (11) with z' located in the μ-th interval, is also a sum of the form

$$\tau(\theta,z,z') = \tau(\theta,z_\mu,z') + \tau(\theta,z,z_\mu)$$

$$= (1/\cos\theta)\kappa_\mu^{ex}\rho_\mu(z_\mu - z')$$

$$+ \sum_{\mu'=\mu+1}^{\mu_z} (1/\cos\theta)\kappa_{\mu'}^{ex}\rho_{\mu'}(z_{\mu'} - z_{\mu'-1}). \quad (16)$$

Similarly

$$I^{\downarrow}(z,\theta) = \sum_{\mu=\mu_z+1}^{\mu_{\mathcal{N}}} J_\mu(\theta)\left[e^{-\tau(\theta,z_{\mu-1},z)} - e^{-\tau(\theta,z_\mu,z)}\right], \quad (17)$$

with the optical thicknesses given by an equation of type (16).

VI. THE SOURCE FUNCTION $J_\mu(\theta,\phi)$.

The source function $J(z,\theta,\phi)$ contains the scattering emissivity $j^s(z,\theta,\phi)$ of Eq. (5), which depends on the radiation coming from all directions (θ',ϕ') to the point at altitude z in a primary beam with direction (θ,ϕ). The approximate calculation of $j^s(z,\theta,\phi)$ is necessary for determining δ_ν^s of Eq. (10) and, hence, it is the main objective of this paper. Taking Eq. (5) as a starting point we note that in a layered atmosphere, the only term in Eq. (5) depending on ϕ' is the phase function $p(\theta,\phi,\theta',\phi')$. Now, if the integration over ϕ' in Eq. (5) is carried out the ϕ-dependent part of the phase function, as given by the Eqs. (7) and (8), cancels and we get

$$j^S(z,\theta) = \frac{\kappa^S}{2}(z) \sum_{n=0}^{n_{max}} p_n P_n(x) \int_{-1}^{+1} P_n(x')I(z,x')dx'. \tag{18}$$

For later convenience let

$$\sum_{n=0}^{n_{max}} p_n P_n(x) P_n(x') = \sum_{n=0}^{n_{max}} \bar{q}_n(x) x'^n, \tag{19}$$

then

$$j^S(z,\theta) = \frac{\kappa^S(z)}{2} \sum_n \bar{q}_n(x) \int_{-1}^{+1} x'^n I(z,x')dx'. \tag{20}$$

Next, we substitute $I(z,x')$ from the Eqs. (15) and (17) into Eq. (20) and note that these intensities contain the *unknown* source functions $J_\mu(\theta')$ and surface intensities $I(0,\theta')$ associated with the secondary beams. This situation suggests an iterative solution of Eq. (20) where, in the first stage of iteration, $J_\mu(\theta')$ and $I(0,\theta')$ are guessed on the basis of physical considerations. Referring to our earlier discussion of the generalized Kirchoff Law we propose the trial solution

$$J_\mu(\theta') = B(T_\mu), \quad I(0,\theta') = B(T_o), \tag{21}$$

where T_μ is the temperature in the μ-th interval, and T_o is the temperature of the earth surface. Eq. (21) is a good approximation if the scattering correction of Eq. (10) satisfies the inequality

$$\delta^S = [(j^S/j^e)-(\kappa^S/\kappa^a)]/[1+(\kappa^S/\kappa^a)] \ll 1. \tag{22}$$

Combining the relations (21), (20), (17) and (15) we get

$$j^S(z,\theta) = \left(\tfrac{1}{2}\kappa^S(z)\sum_n \bar{q}_n(x)\right) \times \left(\int_0^1 x'^n \left\{ B(T_o)e^{-\tau(z,0)/x'} \right.\right.$$

$$\left.\left. + \sum_{\mu=1}^{\mu_z} B(T_\mu)\left[e^{-\tau(z,z_\mu)/x'} - e^{-\tau(z,z_{\mu-1})/x'}\right]\right\} dx' \right.$$

$$+ \int_{-1}^{o} x'^n \left\{ \sum_{\mu=\mu_z+1}^{\mu_{\mathcal{N}}} B(T_\mu) \right.$$

$$\times \left. \left[e^{-\tau(z_{\mu-1},z)/x'} - e^{-\tau(z_\mu,z)/x'} \right] \right\} dx' \Bigg) . \tag{23}$$

It is seen that we are dealing with integrals of the form (3)

$$\int_o^1 x'^n e^{-a/x'} dx' = E_{n+2}(a), \quad n = 0,1,2,\cdots \tag{24}$$

where

$$\frac{e^{-a}}{a+n} < E_n(a) \leqslant \frac{e^{-a}}{a+n-1}, \quad a > 0, \tag{25}$$

is the exponential integral of n-th order. Substituting Eq. (24) into (23) gives

$$j^S(z,\theta) = \kappa^S(z)\overline{B(z,\theta)}, \tag{26}$$

where $\overline{B(z,\theta)}$ is an average blackbody intensity of the form

$$\overline{B(z,\theta)} = \left(\tfrac{1}{2}\sum_n \overline{q}_n(x) \right) \times \left(B(T_o)E_m[\tau(z,0)] \right.$$

$$\left. + \sum_{\mu=1}^{\mathcal{N}} B(T_\mu)\big|E_m[\tau(z,z_\mu)]-E_m[\tau(z,z_{\mu-1})]\big] \right) , \tag{27}$$

where $\tau(z,z_\mu)$ and $\tau(z,z_{\mu-1})$ represents the positive magnitude of τ and $m = n + 2$. Each term in the sum over μ represents the contribution to $j^S(z,\theta)$ of secondary radiation emitted by the whole extend of the μ-th atmospheric layer with temperature T_μ. Each term contains the difference of two quantities referring to the upper and lower boundary of the μ-th layer, and representing the average attenuation of radiation coming from this layer and being observed in altitude z.

The net result of our preceding analysis is as follows. In the equations of transfer (15) and (17) the source function $J_\mu(\theta)$ is approximated by the expression

$$J_\mu(\theta) = B(T_\mu)(\kappa_\mu^a/\kappa_\mu^{ex}) + \overline{B_\mu(\theta)}(\kappa_\mu^s/\kappa_\mu^{ex}) \quad , \tag{28}$$

where $\overline{B_\mu(\theta)} = \overline{B(z_{\mu,}\theta)}$.

VII. RADIATIVE TRANSFER IN A CLOUD STRATUM

We consider a cloud stratum of 10^5 cm thickness containing water droplets of constant number density N, droplet radius r, and temperature T. Neglecting scattering the source function $J = B(T)$ is also constant; however, if scattering is taken into account J is given by Eq. (28), and the second term in this expression depends on the altitude, i.e. on the subscript $\mu = 1,2,\cdots 10$, which labels the layers into which the cloud is formally divided. Assuming that the detector is at the upper edge of the cloud we use Eq. (15) with $J_\mu(\theta)$ given by Eq. (28). This yields, after some manipulation

$$I^\uparrow(z,\theta) = I(0,\theta)e^{-\tau(\theta,z,0)} + B(\tau)(\kappa^a/\kappa^{ex})[1-e^{-\tau(\theta,z,0)}]$$

$$+ \frac{\kappa^s}{\kappa^{ex}} \left\{ [\overline{B_{10}(\theta)} - \overline{B_1(\theta)}\, e^{-\tau(\theta,z,0)}] \right.$$

$$\left. + \sum_{\mu=1}^{q} e^{-\tau(\theta,z,z_\mu)}[\overline{B_\mu(\theta)} - \overline{B_{\mu+1}(\theta)}] \right\} , \quad z = z_{10} \ . \tag{29}$$

The curly bracket represents a scattering correction, proportional to (κ^s/κ^{ex}) and containing the quantities $\overline{B_\mu(\theta)}$ which depend on μ; they are given by Eq. (27) if z is taken to be z_μ, and if Σ_μ is replaced by $\Sigma_{\mu'}$. Thus

$$B_\mu(\theta) = \left(\tfrac{1}{2} \sum_n \overline{q_n}(x) \right) \times \left(B(T_0)E_m[\tau(z_\mu,0)] \right.$$

$$+ B(T) \left\{ 2E_m[\tau(z_\mu,z_\mu)] - E_m[\tau(z_\mu,0)] \right.$$

$$- E_m[\tau(z_\mu,z_{10})]\Big\}\bigg) \quad . \tag{30}$$

To find the coefficients $\bar{q}_n(x)$ we assume that the droplet radius r is small enough for Rayleigh's theory to be valid. In this case, and considering a perpendicular beam ($\theta=0$)

$$p(x^s) = (3/4)(1+x^{s^2}), \quad \bar{q}_o(x) = \bar{q}_2(x) = 3/4, \quad x = 1. \tag{31}$$

With this result we get finally, for $\theta = 0$,

$$\bar{B}_\mu = (3/8) \sum_{m=2,4} \bigg(B(T_o)E_m[\tau(z_\mu,0)] + B(T)\Big\{2E_m[\tau(z_\mu,z_\mu)]$$

$$- E_m[\tau(z_\mu,0)] - E_m[\tau(z_\mu,z_{10})]\Big\}\bigg) \quad . \tag{32}$$

VIII. NUMERICAL RESULTS

We consider water droplets with radius $r = 0.016$ cm, at $300°K$. The radiation wave length λ is 0.3 cm, and the parameter $y = 2\pi r/\lambda \simeq 0.32 < 1$ indicates that Rayleigh's theory is fairly appropriate. The optical cross-sections are (5) $c^s \simeq 0.9 \times 10^{-4}$ cm^2, $c^a \simeq 7.3 \times 10^{-4}$ cm^2, $c^{ex} \simeq 8.2 \times 10^{-4}$ cm^2. The number density of droplets is $N = 10^{-1}$ and N^{-2} corresponding to a mean free photon path (attenuation length) $\ell = (Nc^{ex})^{-1} \simeq 10^4$cm and 10^5cm respectively. This represents strong and weak attenuation by the cloud. The earth's surface temperature is assumed equal to the cloud temperature, and its surface intensity $I(0)$ is assumed equal to $B(T)$. With this simplifying assumption Eq. (32) reduces to

$$\bar{B}_\mu = B(T)\left\{1 - \frac{3}{8} \sum_{m=2,4} E_m[\tau(z_\mu,z_{10})]\right\} \begin{array}{l} = \frac{1}{2}B(T) \text{ if } \mu = 10 \\[12pt] \simeq B(T) \text{ if } \mu = 1 \end{array} \tag{33}$$

where

$$\tau(z_\mu,z_{10}) = Nc^{ex}(z_{10}-z_\mu) = \begin{array}{l} 0.82(10-\mu) \text{ if } N = 10^{-1} \\[12pt] 0.082(10-\mu) \text{ if } N = 10^{-2} \end{array} \tag{34}$$

It is seen that $\overline{B}_\mu$ varies with μ between the values $\frac{1}{2}B(T)$ and $B(T)$ in a simple manner determined by the function $E_m[\tau(z_\mu, z_{10})]$ which may be well approximated by the upper limit given in Eq. (25).

IX. REFERENCES

1. Chandrasekhar, S., "Radiative Transfer" Chapter I. Dover Publications, Inc., New York, 1960.
2. Jahnke, E., and Emde, F., "Tables of Functions," Fourth Ed. Chapter VII, Dover Publications, Inc., New York, 1945.
3. "Handbook of Mathematical Functions," (M. Abramowitz and I. A. Stegun, Eds.), Chapter V, National Bureau of Standards, Applied Mathematics Series 55, Sixth Printing, 1967.
4. Van de Hulst, M. C., "Light Scattering by Small Particles" Chapter 19 and 20, John Wiley & Sons, Inc. New York, 1962.
5. Ref. 4, p. 270ff.

CHEMICAL COMPOSITION OF ATMOSPHERIC AEROSOL POLLUTANTS
BY HIGH RESOLUTION X-RAY FLUORESCENCE SPECTROMETRY

Jack Wagman
U. S. Environmental Protection Agency
Environmental Sciences Research Laboratory
Research Triangle Park , North Carolina   27711

X-ray fluorescence spectrometry has a number of at-
tributes that make it especially attractive for the analysis
of airborne particulate matter.   These include direct analysis
of filter deposits without further sample preparation, non-
destructiveness, and fairly uniform detectability for all
elements from atomic number 9 upward.   Atmospheric aerosol
samples and samples from power plants, incinerators and other
emission sources, typically contain several dozen elements at
widely different concentrations, thus presenting many possible
interelement line interference problems for some of the x-ray
fluorescence analyzers in current use.   X-ray crystal (wave-
length dispersion) spectrometers with their high resolution
capabilities have been rediscovered as ideal instruments for
this application.   In addition to yielding elemental analyses
with minimal data manipulation, they have the potential for
identification and concentration measurement of elemental
valence states in aerosol samples.   A simultaneous multiwave-
length spectrometer, containing an array of 16 fixed monochro-
mators and a sequential channel, has been specially equipped
and adapted for rapid and routine analysis of large numbers of
air pollution samples.   The performance of this instrument and
results obtained with source and ambient aerosols are discussed.

INTRODUCTION

The effects of aerosols as air pollutants are broad in
range and include many types of physiological actions, soiling,
deterioration of materials, reduction of visibility and
insolation, and other atmospheric phenomena.   The ability to
assess and understand these effects, however, has been limited
to a large extent by the relatively slow though steady progress

171

that has been made through the years in the development of precise and dependable methods for determining the physical and chemical properties of airborne particulate matter.

Chemical characterization generally begins with elemental composition. X-ray fluorescence (XRF) spectrometry is becoming a broadly used procedure for the elemental analysis of filter-deposited airborne particles because of a number of attractive features. These include (a) the direct analysis of filter deposits with no need for sample preparation; (b) the non-destructiveness of the method permitting samples to be retained for further analysis or future reference; (c) the fairly uniform detectability across the periodic table with the ability to analyze all elements from atomic number 9 (F) upward; and (d) the availability of commercial instruments that permit the analysis of samples for a large number of elements in relatively short time intervals and at low cost.

Atmospheric particulate samples and samples from power plants, incinerators and other source emissions, typically contain many elements at widely different concentrations, thus presenting many possible interelement line interference problems. A practical solution to the routine analysis of large numbers of such samples has become possible with the recent availability of multichannel wavelength spectrometers that combine two important features, i.e., high spectral resolving power and simultaneous measurement of a large number of elemental concentrations.

An instrument of this kind, adapted for analysis of filter-deposited samples of particulate matter, has been set up and is in routine use at the EPA Environmental Research Center in North Carolina. A description of its essential features and some results obtained with it are given in this paper.

GENERAL DESCRIPTION OF INSTRUMENT

The EPA multichannel spectrometer (Figure 1) is based upon a newly-designed Siemens (Model MRS-3) simultaneous spectrometer[1] adapted for use with filter-deposited samples. The instrument includes automatic sample handling and computer-controlled operation and data processing.

The multichannel spectrometer accomplishes rapid high resolution multielemental analysis through the simultaneous use of 16 fixed monochromators and one scanner. Each fixed channel is optimally designed for the analysis of a particular

---

[1] Simultaneous x-ray multispectrometers are currently manufactured also by Applied Research Laboratories, Philips Electronic Instruments, and Rigaku Denki. Mention of trade names or commercial products does not constitute endorsement or recommendation for use.

Fig. 1. *Simultaneous multiwavelength x-ray spectrometer.*

element by the proper selection of the analyte line, crystal and detector. Table 1 lists the elements analyzed by the fixed channels along with the analyte lines, crystals and detectors employed for each channel. This list includes the nine elements with lowest atomic number that can be analyzed (i.e., F, Na, Mg, Al, Si, P, S, Cl, K) plus seven additional elements (i.e., Cr, Mn, As, Br, Cd, Hg, Pb).

The monochromators are located above the specimen and opposite the x-ray tube within the vacuum-tight spectrometer housing. Arranged in a semi-circle about the specimen, the monochromators are mounted at ten positions. At seven of the positions, a double monochromator is used while two single monochromators and the scanner occupy the remaining positions. A double monochromator is essentially two monochromators molded into a single frame. As shown schematically in

TABLE 1

Fixed Monochromators in EPA Spectrometer

Electronic Channel	Element	Line	Detector Type	Crystal	Window Thickness, μm	Spectrometer Position
1	Chromium	K_α	Scintillation	LiF(200)	- -	1
2	Lead	L_β	Scintillation	LiF(200)	- -	1
3	Manganese	K_α	Scintillation	LiF(200)	- - -	2
4	Arsenic	K_β	Scintillation	LiF(200)	- -	3
5	Mercury	L_α	Scintillation	LiF(200)	- -	5
6	Bromine	K_α	Scintillation	LiF(200)	- -	9
7	Phosphorus	K_α	Flow	PET	2	10
8	Silicon	K_α	Flow	PET	2	2
9	Cadmium	L_α	Flow	PET	6	3
10	Aluminum	K_α	Flow	PET	2	4
11	Sulfur	K_α	Flow	PET	2	5
12	Sodium	K_α	Flow	KAP	0.4	7
13	Fluorine	K_α	Flow	KAP	0.4	4
14	Magnesium	K_α	Flow	ADP	0.4	8
15	Potassium	K_α	Flow	PET	6	8
16	Chlorine	K_α	Flow	PET	6	9

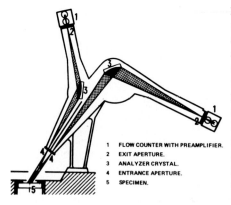

1 FLOW COUNTER WITH PREAMPLIFIER.
2 EXIT APERTURE.
3 ANALYZER CRYSTAL.
4 ENTRANCE APERTURE.
5 SPECIMEN.

*Fig. 2. Beam paths
in a double monochromator.*

Figure 2, each monochromator contains an inlet slit, crystal and outlet slit in a focusing arrangement. The crystals are curved in a logarithmic spiral in order to achieve a constant Bragg angle over the entire surface without grinding(1,2). Scintillation counters are used for high energy characteristic x-ray detection and flow proportional counters for low energy detection.

The scanning channel (Figure 3) is a curved-crystal spectrometer adjustable over a

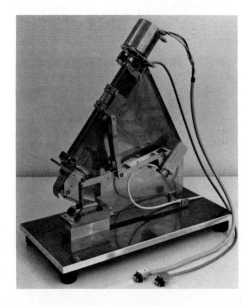

Fig. 3. Compact scanning spectrometer channel.

2θ range of 30 to 120°; it is compactly designed so that it occupies only one spectrometer position. The scanner, as shown schematically in Figure 4, is a fully-focusing(3) spectrometer, with its inlet slit, crystal and detector slit always on the Rowland circle. Since elements from atomic number 9 through 19 (fluorine through potassium) were included for measurement by fixed channels, the scanning spectrometer was equipped with a LiF (200) crystal thus making it effective for analysis of all higher atomic number elements.

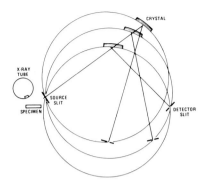

Fig. 4. Scanning spectrometer shown schematically at three wavelength settings.

The resolution of a spectrometer is an important factor in its capability for analyzing typical air pollution samples. A crystal spectrometer's resolution may be expressed as the full width at half maximum (FWHM) of the diffracted characteristic lines. Table 2 lists resolutions measured at four spectral lines with the scanning crystal spectrometer channel and with a state-of-the-art energy dispersion spectrometer, respectively. It is important to note here that, for the elements

TABLE 2

Resolutions Measured with Scanning Wavelength and Energy Dispersive Spectrometers

Element Line	2Θ, deg	Energy, keV	LiF (200) Crystal Spectrometer		Energy Dispersive[a] Spectrometer
			$\Delta\Theta$, deg	ΔE, eV	ΔE, eV
Pb L$_\alpha$	33.93	10.55	0.26	154	185
Zn K$_\alpha$	41.78	8.63	0.22	87	170
Mn K$_\alpha$	62.97	5.90	0.20	33	155
Ca K$_\alpha$	113.09	3.69	0.19	6	125

a. *Princeton Gamma - Tech Model PGT-1000.*

between atomic numbers 20 through 30 (Ca to Zn) where the K$_\alpha$ and K$_\beta$ lines of adjacent elements are difficult to separate, the crystal spectrometer has a resolution capability from 2 to 20 fold greater than the solid state detector. Figure 5 shows a comparison of spectra obtained on the same sample deposit of CaCl mixed with KCl using the scanning channel and the energy dispersion spectrometer. The K$_\beta$ potassium line is easily separated by the scanning channel, but is evident in the energy dispersive spectrum only as a slight distortion in the right side of the calcium K$_\alpha$ peak. Figure 6 shows the clear separation between the K lines for Mn, Fe, and Co that is achieved with the scanning channel. Energy spectrometers produce serious overlapping of the lines in this region, thus requiring data manipulations subject to significant errors for many types of samples.

JACK WAGMAN

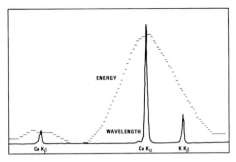

Fig. 5. *Comparison of x-ray spectra obtained with energy
and wavelength spectrometers for a calcium-potassium sample.*

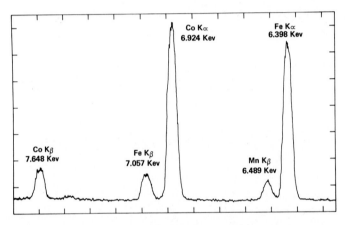

Fig. 6. *Separation of K lines of
Mn, Fe, and Co by the scanning channel.*

SAMPLE CONFIGURATION AND HANDLING

Filters used to collect air pollution particulate samples
for x-ray analysis should have the following characteristics:
(a) low mass to reduce scattering of incident radiation, hence
minimizing the background count, (b) low content of the elements
to be measured, and (c) minimal penetration of particles below
the surface. Also, filters composed of low atomic number
elements are desirable since these, in general, have lower
attenuation coefficients and therefore lower absorption effects
when penetration of the aerosol into the filter occurs. Filter
materials which have been found most suitable for x-ray fluorescence analysis are the thin membrane types, e.g., Millipore,
Fluoropore, and Nuclepore.

In collecting particulate samples for x-ray analysis, it is
also an advantage to work with thin aerosol deposits, inasmuch
as these greatly reduce matrix effects that would otherwise
require significant corrections for attenuation and enhancement.

When multi-element analyses are to be carried out, however, com-
promises must be made in selecting optimal deposit densities,
and matrix correction factors will often be required for the
lighter elements. The ideal aerosol deposit is one that has
reached maximum mass per unit area with no significant inter-
particle shadowing in either the incident beam or the measured
fluorescent beam. For such samples, fluorescent intensities
need only to be corrected for particle size effects where these
are significant.

To facilitate the processing of sample deposits in the
multichannel analyzer, special plastic frame mounts were
designed for 47mm filters as illustrated in Figure 7. A custom-
designed automatic sample loader has been fabricated to feed
filter samples to the inlet of the multichannel spectrometer
and remove the samples after they have been analyzed. It holds
up to 100 frame-mounted filters and may be operated manually,
but is normally controlled by the mini-computer program.

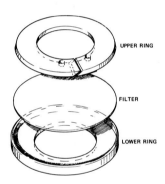

UPPER RING

FILTER

LOWER RING

Fig. 7. Exploded view of a frame-mounted filter.

CALIBRATIONS

Before elemental concentrations in particulate samples
can be determined, a set of calibration standards for the
elements to be analyzed must be inserted and processed in
the instrument. Through a computer-controlled calibration
program, sensitivity factors in cps per $\mu g/cm^2$ are determined
and stored in the computer memory.

Several types of thin deposited standards have been
employed to calibrate the spectrometer. These include:
(a) Thin films formed from vacuum-evaporated elements
and compounds. These were prepared at concentrations of
about 50 $\mu g/cm^2$ on Mylar substrate (3.8 μm thickness) with
an estimated accuracy of five percent.
(b) Aerosols generated, e. g., in a Collison atomizer,
and collected on membrane filters.
(c) Filter deposits of materials from solutions after

solvent evaporation. This method is subject to non-uniform deposits due to migration of ions during evaporation unless precautions are taken.

(d) Filter deposits of fine powdered materials from liquid suspension.

INSTRUMENT PERFORMANCE

The minimum detection limit for x-ray analysis is often defined as the concentration of analyte corresponding to a net count equal to three times the standard deviation, i.e., three times the square root of the background count. Table 3 lists sensitivities and detection limits for 30 elements determined in the EPA simultaneous spectrometer. The minimum detection limits for 100 seconds of count time are in the range of 2 to 10 ng/cm^2 for more than half of the elements measured including most of the low atomic number elements. The detection limit exceeds 30 ng/cm^2 for only four of these elements.

TABLE 3
EPA Multispectrometer XRF Analyzer Element Sensitivities and Detection Limits

Element	Sensitivity, counts/100 sec $\mu g/cm^2$	Detection Limit (100 sec, 3σ), ng/cm^2	Element	Sensitivity, counts/100 sec $\mu g/cm^2$	Detection Limit (100 sec, 3σ), ng/cm^2
F	220	149	Co	16540	3
Na	534	29	Ni	14504	10
Mg	10280	2	Cu	18880	43
Al	8074	3	Zn	21066	7
Si	11614	3	As	17125	10
P	13392	15	Se	22922	12
S	28013	9	Br	50340	28
Cl	25394	9	Cd	17303	2
K	121286	2	Sn	14800	2
Ca	87817	2	Sb	31100	4
Ti	85635	2	Ba	25000	7
V	18010	7	Pt	6812	20
Cr	7484	19	Au	8498	91
Mn	17522	14	Hg	5776	90
Fe	13300	18	Pb	16583	30

An estimate of the accuracy of the x-ray analytical procedure, as carried out with the simultaneous spectrometer, was determined in a comparison with gravimetric values for the mass of potassium sulfate aerosol generated in a Collison atomizer and deposited on a series of Nuclepore filters. The results (Figure 8) indicated very good agreement between the x-ray and gravimetric values.

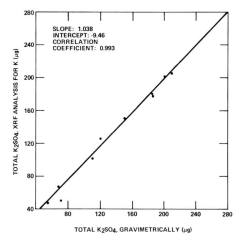

SLOPE: 1.038
INTERCEPT: -9.46
CORRELATION
COEFFICIENT: 0.993

TOTAL K$_2$SO$_4$, XRF ANALYSIS FOR K (μg)

TOTAL K$_2$SO$_4$, GRAVIMETRICALLY (μg)

Fig. 8. Comparison of x-ray fluorescence and gravimetric determinations of K_2SO_4 deposited on Nuclepore filters.

The results of another comparison, this time with atomic absorption analyses, is shown in Table 4. The samples consisted of aerosols collected on high purity quartz filters from a controlled combustion source fed with fuel oil spiked with known amounts of organometallic compounds. The calculated values for the five metals are based upon the amount of spiked fuel consumed and the known fraction of total emissions collected on the filter. The x-ray fluorescence concentrations are in remarkably good agreement with the calculated values with the exception of the value for cadmium which is about 10 percent low, probably because of some attenuation of the relatively low-energy Cd-L$_\alpha$ fluorescence resulting from partial penetration of collected aerosol into the interstices of the quartz fiber filter.

TABLE 4

Comparison of XRF and AAS Analyses of Particulate Samples from a Simulated Combustion Source

| | Total Collected, μg | | |
Element	Calculated Value[a]	Atomic Absorption	X-ray Fluorescence
Pb	202	228	212
Mn	182	195	183
Co	182	193	183
Cd	183	187	166
V	216	159	230

a. Based on analysis of fuel oil spiked with organometallic compounds.

ANALYSES OF SOURCE AND AMBIENT AEROSOLS

The EPA simultaneous multiwavelength spectrometer is being used to determine the elemental composition of large numbers of samples of particulate emissions collected from stationary sources that include power plants, incinerators, and chemical process plants; from mobile sources that include automobiles, diesel trucks, and aircraft; and from ambient air samples at various locations.

Table 5 shows striking differences in the elemental composition of particles emitted from coal-fired and oil-fired
power plants. The major components in coal combustion particles
are Si, Al, Fe, K, S, Ca, and Ti, which account for about half
of total mass, and probably over 90% if combined oxygen is
included. In contrast, oil combustion particles contain considerably more sulfur and substantial amounts of V and Ni, but
only trace quantities of the other elements measured. All of
the elements analyzed account for less than 15% of the total
mass of the oil combustion particles. However, sulfur alone,
expressed as sulfate, accounts for about 23% of the particle
mass.

TABLE 5

XRF Analysis of Particle Emissions from Coal- and Oil-fired Power Plants

	Concentration, %			Concentration, %	
Element	Coal-fired Samples[a]	Oil-fired Samples[b]	Element	Coal-fired Samples[a]	Oil-fired Samples[b]
Na	0.10	0.45	Fe	8.65	0.27
Mg	0.66	0.32	Ni	0.04	0.85
Al	12.46	0.05	Cu	0.13	0.12
Si	17.75	0.11	Zn	0.03	0.02
P	0.25	0.01	Br	0.14	
S	2.77	7.59	Cd	0.01	
K	4.57	0.04	Sn	0.03	
Ca	2.20	0.17	Sb	0.03	
Ti	1.32	0.01	Ba	0.25	0.01
V	0.14	4.08	Hg	0.13	0.03
Cr	0.41	0.13	Pb	0.70	0.07
Mn	0.17	0.05			

a. Mean values for 5 samples collected at Riverbend plant, Duke Power Co., Charlotte, N.C.

b. Mean values for 10 samples collected at Anclote plant, Florida Power Co., Tarpon Springs, Florida.

We have analyzed tail-pipe aerosol emissions from a variety
of light-duty passenger vehicles, including automobiles operated
on leaded and unleaded gasoline, and cars with diesel, lean-
burn, stratified-charge and rotary engines. Table 6 lists
elemental concentrations found in the emitted particles from
four different types of vehicles. The differences in composition are due not only to differences in the power or exhaust
systems, but also reflect differences in fuel composition. The
car operated on leaded gasoline emits mostly Pb, Cl, and Br in
proportions that are nearly stoichiometric for the compound
PbClBr. Cars equipped with oxidation catalysts convert some of
the sulfur in gasoline to SO_3 which combines with moisture
resulting in the emission of aerosol consisting almost entirely
of droplets containing approximately 50 percent (w/w) aqueous
sulfuric acid. Diesels emit a high rate of mostly carbonaceous

TABLE 6
XRF Analysis of Particle Emissions from a Variety of Automobiles

Element	Emission Rate[a], mg/km			
	1971 Camaro, leaded gasoline	1975 Granada (with catalyst), unleaded gasoline	1974 Mazda (rotary), unleaded gasoline	1975 Peugot, diesel
Na			0.004	
Mg	0.007	0.002	0.012	0.054
Al		0.024	0.024	
Si	0.039	0.008	0.043	0.25
P	0.005		0.14	0.090
S		2.17	0.38	1.49
Cl	1.96			
Ca	0.041	0.004	0.35	0.74
Ti		0.001		
V		0.006		
Mn	0.097			0.68
Fe	0.053	0.53	0.51	
Ni		0.004	0.045	
Cu	0.31		0.024	
Zn		0.014	0.053	
Br	5.06		0.033	
Cd	0.023		0.0006	
Ba			0.004	0.054
Hg	0.12			0.41
Pb	12.1	0.13	0.26	1.15
Total Mass	32.4	12.8	6.63	188

a. For vehicles operated on a dynamometer according to the 1975 U.S. Federal Test Procedure (Federal Register. 37(221), November 15 (1972).

aerosol, but carbon is too low in atomic number to be detected by x-ray fluorescence; sulfur as sulfate is a major noncarbonaceous constituent.

Aerosol samples collected from the ambient air contain the broadest range of elements and elemental concentrations and reflect contributions from both natural and anthropogenic sources. The x-ray fluorescence analyses for three samples listed in Table 7 are indicative of the high variability in composition that can be observed for urban air particulate samples collected at different locations and times. For example, the high sulfur content and relatively small amounts of aluminum and silicon in the sample represented by the data in the first column indicate a large contribution from fossil fuel combustion as compared to soil-erosion or reentrained dust particles. The analysis shown in the last column represents a sample with quite opposite characteristics, while the elemental concentrations listed in the middle column are typical of a

TABLE 7
XRF Analysis of Ambient Air Particulate Samples

	Concentration, $\mu g/cm^2$		
Element	Sample No. 7-19	Sample No. 7-24	Sample No. 10-04
F	-	-	0.09
Na	-	-	0.09
Mg	0.02	0.06	0.14
Al	0.28	1.20	5.30
Si	0.44	1.60	7.60
P	0.11	-	0.18
S	3.74	3.20	1.77
Cl	0.26	0.15	0.47
K	0.20	0.47	2.06
Ca	1.10	0.90	2.67
Ti	0.04	0.08	0.16
V	0.005	-	0.03
Fe	0.76	0.70	2.73
Zn	1.10	0.05	0.16
Br	0.40	-	0.75
Cd	0.04	0.007	0.04
Ba	-	0.04	0.10
Pb	1.80	0.23	3.0

sample with a more balanced contribution from these two types of sources.

ANALYSIS OF ELEMENTAL VALENCE STATES

The high resolution of crystal spectrometers has been used with some success to distinguish between different valence states of the same element. Using a primary beam collimator and a flat NaCl crystal, it is possible(4) to obtain the sulfur K_β x-ray emission fine structure shown in Figure 9 for a series of sulfur compounds. The peak positions are nearly the same for elemental sulfur, the sulfides, and thiourea, while for sulfite and sulfate the K_β peaks are at a lower Bragg angle (higher energy). Also, sulfate and sulfite yield a satellite K_β peak at a higher angle, while sulfite alone has a K_β satellite at a lower angle.

We are in the process of including this capability in the simultaneous multichannel instrument by adapting one of the fixed spectrometers. The possibility of distinguishing valence states of other elements is also being investigated in an effort to make fuller use of the potential of x-ray crystal spectrometers for characterizing particulate pollutants.

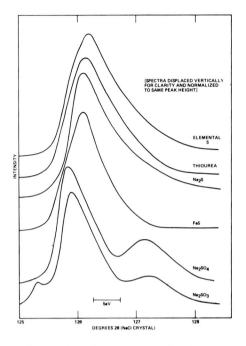

*Fig. 9. Kβ x-ray emission spectra
for various sulfur oxidation states.*

ACKNOWLEDGMENT

The author is pleased to acknowledge the participation of
L. S. Birks, J. V. Gilfrich, and J. W. Criss of the Naval
Research Laboratory, Washington, D. C., in the development of
specifications for the spectrometer and in optimizing its performance; K. T. Knapp and R. L. Bennett in instrument application; R. B. Kellogg and J. Lang in instrument operation and
data acquisition; and other Environmental Sciences Research
Laboratory personnel in the acquisition of aerosol samples.

REFERENCES

1. Barraud, J., C. R. Acad. Science, Paris 214, 795 (1942).
2. DeWolff, P. M., Appl. Sci. Res. B1, 119 (1950).
3. Johansson, T., Zeitschrift für Physik 82, 507 (1933).
4. Birks, L. S., Quarterly Progress Report, Interagency
    Agreement EPA-IAG-D5-0344 between U. S. Environmental
    Protection Agency and Naval Research Laboratory, May,
    1975.

SULFATE AEROSOL FORMATION UNDER CONDITIONS
OF VARIABLE LIGHT INTENSITY

D.L. Fox, M.R. Kuhlman, and P.C. Reist
Department of Environmental Sciences and Engineering
University of North Carolina at Chapel Hill

I. INTRODUCTION

In the presence of sunlight, sulfur dioxide in the ambient
air will be converted to form sulfate aerosol. This conversion
is thought to proceed through the formation of hydroxyl radicals which react with SO_2 to form SO_3 which then rapidly combines with water to yield H_2SO_4.

If SO_2 is introduced into a chamber containing "background" air with water vapor and few condensation nuclei and
the mixture is irradiated with a constant light source which
approximates natural mid-day sunlight, a burst of new condensation nuclei will be formed. The nuclei count will increase
dramatically, reach a peak and then rapidly decline, even
though the SO_2 and water vapor concentrations are essentially
constant. Figure 1 shows a typical number concentration vs.
time plot for this type of experiment.

The decrease in number concentration can be explained by
observing that nuclei grow by agglomeration and by condensation of freshly formed acid to form larger sized particles.
If agglomeration rates are faster than particle formation
rates the net particle number decreases. The changing aerosol
number size distribution is shown in Figure 2A. This is a
three dimensional plot of the number of particles per cm^3
within each of the 10 size ranges of an Electrical Aerosol
Analyzer (EAA). The lines represent scans of the analyzer
approximately every four minutes. As SO_2 is converted from
the gas phase to a liquid (H_2SO_4), total particle surface area
and volume increase as shown in Figures 2B and 2C. Surface
area appears to reach a steady-state or equilibrium value
whereas volume increases, and subsequently decreases as the
larger particles begin to be removed by gravity settling, or
grow out of the sensitive range of the mobility analyzer.

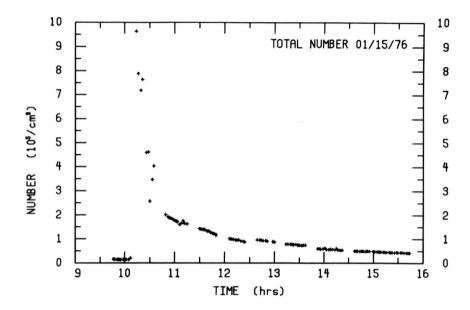

Fig.1. Total particle number measured by Electrical Aerosol Analyzer on January 15, 1976 under clear sky. $[SO_2]^o = 0.54$ *ppm in background air.*

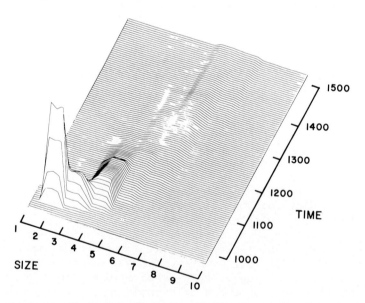

Fig.2. Aerosol number distribution on January 15, 1976 under clear sky. $[SO_2]^o = 0.54$ *ppm in background air. Size 1 = 0.004 μm; 2 = .007; 3 = .013; 4 = .024; 5 = .042; 6 = .075; 7 = .133; 8 = .237; 9 = .422; 10 = .750.*

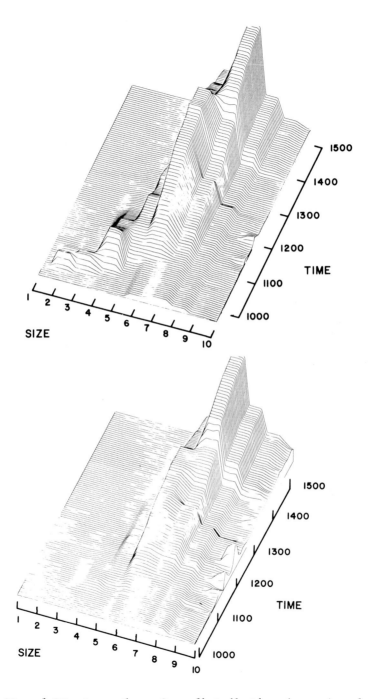

Figs.2B and 2C. Aerosol surface distribution (upper) and volume distribution (lower) on January 15, 1976.

187

With natural sunlight this simple picture can be greatly distorted. We have observed that when illumination of the reactive gas is intermittant the condensation nuclei count fluctuates dramatically, decreasing at a more rapid rate when illumination decreases and rising when the mixture is reilluminated. Certain patterns have become apparent even in the intermittant illumination case, however. In this paper we will discuss this phenomenon, offer an explanation for why it occurs, and discuss the implications of this explanation both as it pertains to current understanding of atmospheric aerosol formation and how this variable nuclei formation rate might affect the significance of measurements of ambient condensation nuclei.

II. EXPERIMENTAL

A series of experiments on aerosol growth mechanisms has been carried out in a 200 m^3 chamber located in rural North Carolina. In these experiments background air, either filtered or unfiltered, is admitted to the chamber in the early morning and the chamber is closed. This air may be either moist or dry but in either case contains a substantial quantity of water vapor. At some time a small amount of SO_2 is introduced into the chamber and rapidly mixed throughout the volume, at which point the mixing fans are turned off. SO_2 concentrations are typically in the range of 0.1 to 1 ppm. Details of the chamber design and its performance have been reported previously (1).

Almost immediately after injection of the SO_2, there is a rapid rise in the condensation nuclei count as measured by an Environment One Condensation Nuclei Counter and in the smaller size ranges of a TSI Electrical Aerosol Analyzer. These data indicate rapid formation of aerosol particles in the < 0.01 μM diameter size range. As mentioned previously, for a perfectly clear day the concentration of these small particles goes to a maximum in about 15 minutes, then rapidly declines in about two hours reaching an equilibrium value which seems to vary depending on such factors as initial SO_2 concentration, sunlight intensity, and relative humidity.

When the sunlight intensity is variable, such as on a partially cloudy day, small particle concentration peaks are noted at times when the sun reappears, these peaks sometimes rising much higher than would be expected in the case of constant irradiation. A typical evolving aerosol number distribution for this type of run is shown in Figure 3, and the same data viewed from the back are shown in Figure 4. Unlike the

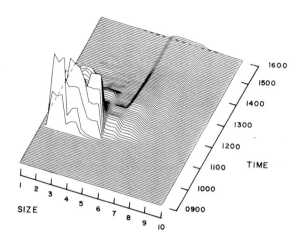

Fig.3. Aerosol number distribution on February 26, 1976 under partly cloudy sky. $[SO_2]^O = 0.20$ *ppm in background air. Size intervals are same as in Figure 2A.*

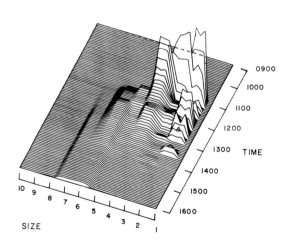

Fig.4. Aerosol number distribution on February 26, 1976. Projection viewing angle 180° from Figure 3.

clear day case in which the aerosol smoothly evolves to increasingly larger sizes, intermittant illumination produces a
transient bimodal size distribution. In this case nucleation
occurs even in the presence of an existing aerosol as can be
seen by the total surface area shown in Figure 5. The rapid
increases in the number of small particles in stages 2 and 3
of the Electrical Aerosol Analyzer correlate reasonably well
with the spikes in solar radiation for this day given the
limited time resolution of the EAA. It appears that the sudden increases in solar radiation bring about an increase in
the partial pressure of H_2SO_4 sufficient to cause the conversion process to change from diffusion controlled condensation
to binary homogeneous nucleation of new particles.

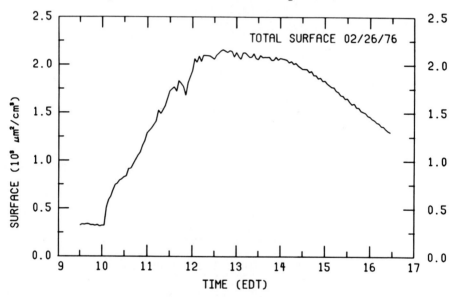

*Fig. 5. Total surface area measured by Electrical Aerosol
Analyzer on February 26, 1976.*

III. DISCUSSION

    For nucleation to occur two conditions must be met.
First, there must be sufficient number of H_2SO_4 molecules per
unit volume. According to Mirabel and Katz (2), this concentration must be somewhat greater than 3×10^{10} molecules per
cm^3. Figure 6 shows nucleation rates for various molecular
concentrations and relative humidities computed from Mirabel
and Katz's data. Above the critical minimum, relatively small
changes in molecular concentrations can result in enormous
differences in nucleation rates.

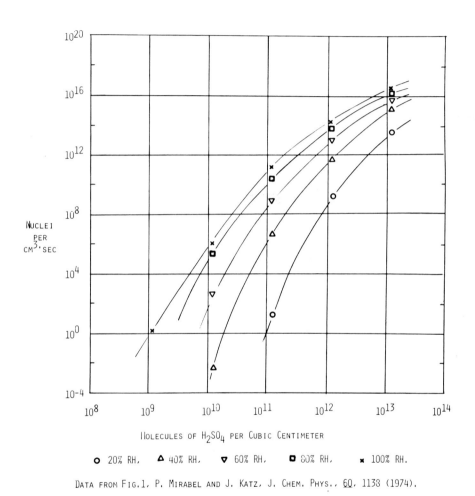

NUCLEI
PER
CM^3·SEC

MOLECULES OF H$_2$SO$_4$ PER CUBIC CENTIMETER

O 20% RH,   △ 40% RH,   ▽ 60% RH,   ◨ 80% RH,   ✕ 100% RH.

DATA FROM FIG.1, P. MIRABEL AND J. KATZ, J. CHEM. PHYS., 60, 1138 (1974).

Fig.6. Calculated rate of nucleation at various humidities as a function of H2SO4 molecular concentration (Air, NTP).

The second condition for nucleation is equally important.
There are a number of mechanisms for conversion of free molecules to aerosol particles of which nucleation is only one.
The most important other mechanism in the case of SO_2 conversion is diffusion of H_2SO_4 molecules to existing particles, or
in other words, condensation of the acid gas onto existing
particles. Condensation rates depend on the availability of
surface for condensation and the process tends to create more
surface. Thus condensation tends to a stable or equilibrium
condition and would damp out nucleation except that coagulation of the particles is also taking place. This agglomeration reduces the surface area available for condensation. For
similarly sized droplets, reduction of number concentration by
a factor of two due to coagulation reduces surface area by
about 20%.

In a "clean" system there is initially no surface available for condensation. Thus condensable molecules will build
up in concentration until nucleation starts. For H_2SO_4 this
nucleation will be rapid and will continue until sufficient
surface area is produced so that some condensation can take
place. Castleman et al. (3) estimated acid molecule concentration at this point by considering only nucleation and condensation to be occurring and equated the two processes. This
results in Equation (1),

$$[H_2SO_4]_{ss} = \frac{k[SO_2][OH][M]}{\pi\, r^2\, \bar{v}\, n} \tag{1}$$

where k is the rate constant for SO_2 + OH reaction, M is a
third body, r is particle radius, n is number of particles per
unit volume, and $\bar{v}$ is the velocity of the gas molecules relative to the particles. Equation (1) indicates that nucleation
will predominate over condensation when concentration of the
H_2SO_4 molecules exceeds the computed steady-state value, provided that condition one is also met. The terms in this
simple expression can be used to explain what occurs under
varying light intensity.

Field measurements and photochemical kinetics modeling
indicate that fluctuations occur in the concentration of OH
free radicals in a chemical system under variable solar radiation. Wang et al. (4) have measured the OH free radical concentration in the open atmosphere. They report OH levels as
high as 6 x 10^7 molecules cm^{-3} with a diurnal pattern peaking
at 1200-1400 hours on sunny days. They also report the OH
concentration would drop below the detection limit of 5 x 10^6
molecules cm^{-3} when it was cloudy or rainy. Their work gives

the approximate level present in the atmosphere and indicates a dependence on sunlight. This laboratory has used a photochemical model with initial conditions: $[C_3H_6]$ = .020 ppm, $[NO]$ = .002 ppm, $[NO_2]$ = .014 ppm and $[O_3]$ = .013 ppm and a variable light profile. The model exhibits fluctuations in OH concentration which correspond to solar radiation intensity changes. The fractional change in OH concentration is of the same order as that of solar radiation intensity. It seems reasonable, then, to expect $[H_2SO_4]_{ss}$ to vary with solar radiation intensity.

SO$_2$ concentrations are usually controlled by the experimenter in chamber studies, but may be quite variable in the atmosphere depending on rates of reaction with available species, proximity of sources, etc. Thus $[SO_2]$ variations could influence the value of $[H_2SO_4]_{ss}$ and hence the nuclei formation rate.

Factors which could lower gas phase acid concentrations even though the production rate remained constant would be increased particle size or increased particle number. Data from numerous investigators show that in closed systems at least, particle size increases with time (condensation and coagulation) whereas number concentration decreases (coagulation). Since a decrease in number concentration results in a decrease in surface area while continued condensation results in increased surface area, a point should be reached where these two factors are equal and an equilibrium surface would be observed. This has been noted by several investigators (5, 6, and 7). This equilibrium surface can provide sufficient area for condensation of H_2SO_4 molecules and thereby prohibit rapid increases in $[H_2SO_4]_{ss}$. Thus, measurable nucleation would not occur when $[SO_2]$ and $[OH]$ vary gradually.

Figure 7 shows the total number behavior as measured by the EAA with fluctuations in the number curve for a partly cloudy day. Similar spikes were observed with a Gardner small particle counter to provide verification. The number spikes show an increase to $\sim$ 250,000 nuclei cm^{-3} measured over 240 sec. between scans or a production rate of about 1000 nuclei cm^{-3} sec.$^{-1}$.

According to Mirabel and Katz (2) for a H_2O - H_2SO_4 system at 40% R.H. for the nucleation rate to increase from 1 nucleus cm^{-3} sec.$^{-1}$ to 1000 nuclei cm^{-3} sec.$^{-1}$, the activity of H_2SO_4 gas need only increase from 2 x 10^{-3} to 5 x 10^{-3}. This means the concentration of H_2SO_4 molecules must only change from 2.4 x 10^{10} molecules cm^{-3} to 5.9 x 10^{10} molecules cm^{-3}. If the chemical system develops a steady-state OH

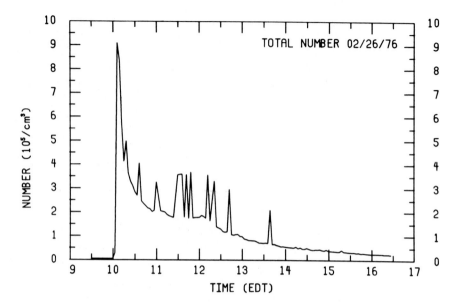

Fig.7. Total particle number measured by Electrical Aerosol Analyzer on February 26, 1976

concentration of 5×10^7 molecules cm^{-3} when the sun directly illuminates the chamber, equation (1) predicts $[H_2SO_4]_{ss} = 4.9 \times 10^{10}$ molecules cm^{-3} which corresponds to a production rate of 1000 nuclei $cm^{-3}sec.^{-1}$. If decreases in sunlight then permit the OH concentration to drop to 2×10^7-molecules cm^{-3}, the nucleation rate drops to < 1 nucleus $cm^{-3}sec.^{-1}$. Such changes in the number of H_2SO_4 molecules present would be most likely to result in nucleation.

IV.   IMPLICATIONS TO ATMOSPHERIC AEROSOL FORMATION

The conversion of sulfur dioxide in the gas phase to sulfuric acid in the aerosol phase in the ambient atmosphere is important for health, economic and aesthetic reasons. The mechanisms for conversion are complicated by a coupling of physical and chemical processes. A simplified model of SO_2 conversion may be discussed with two pathways - one pathway involving oxidation of SO_2 in the gas phase with intermediates proceeding to the formation of sulfuric acid vapor followed by nucleation and condensation on existing aerosols, and the other pathway involving the dissolution of SO_2 in aqueous aerosols with oxidation to sulfate occurring in solution. Many investigators feel that neither pathway predominates in the ambient atmosphere. Our data tends to indicate that conversion of SO_2 along the first pathway does occur at significant rates.

From our studies it is clear that H_2SO_4 nucleation can take place even with preexisting aerosol particles present. This nucleation does not contribute significant mass to the overall system and indeed, resembles a transient superimposed on an otherwise stable system. The possibility for nuclei formation suggests that caution should be used when interpreting condensation nuclei data collected near SO_2 sources or during periods of fluctuating solar radiation. On the other hand, even the transient spikes seemed to disappear as the concentration of larger sized particles (and hence, of surface area) built up. This tends to confirm Whitby and Husar's observations that in highly polluted areas such as the Los Angeles basin, nucleation is not a significant mechanism for the formation of H_2SO_4 aerosols (5).

A simple model estimates the steady-state concentration of H_2SO_4 to be on the order of 4×10^{10} molecules cm^{-3}. (3). This activity is in the range predicted by Mirabel and Katz (2) for nucleation rates of 1000 nuclei cm^{-3} $sec.^{-1}$. Nucleation has been observed under variable sunlight conditions and we suggest fluctuations in the production of H_2SO_4 molecules resulted in increases in the H_2SO_4 partial pressure sufficient to cause binary homogeneous nucleation with water.

V.   ACKNOWLEDGEMENTS

Financial support of the U.S. Environmental Protection Agency through Grant No. R802472 is gratefully acknowledged.

VI.   REFERENCES

1. Fox, Donald L., Joseph E. Sickles, Michael R. Kuhlman, Parker C. Reist and William E. Wilson "Design and Operating Parameters for a Large Ambient Aerosol Chamber", J. Air Pollution Control Assoc. 25, 1049-1053 (1975).
2. Mirabel, Philippe, and Joseph L. Katz, "Binary Homogeneous Nucleation as a Mechanism for the Formation of Aerosols", J. Chem. Phys. 60, 1138-1144 (1974).
3. Castleman, A.W. Jr., Richard E. Davis, H.R. Munkelwitz, I.N. Tang, and William P. Wood, "Kinetics of Association Reactions Pertaining to H_2SO_4 Aerosol Formation", Intern. J. Chem. Kinetics. Symposium No. 1 : 629-639, 1975.

4. Wang, Charles C., L.J. Davis, Jr., C.H. Wu, S. Japar, H. Niki and B. Weinstock, "Hydroxyl Radical Concentrations measured in Ambient Air", Science, 189, 797-800 (1975).
5. Husar, Rudolf B., and Kenneth T. Whitby, "Growth Mechanisms and Size Spectra of Photochemical Aerosols", Environ. Sci. Technol. 7, 241-247 (1973).
6. Clark, William Edgar, "Measurement of Aerosols Produced by the Photochemical Oxidation of SO_2 in Air", Ph.D. Thesis. University of Minnesota, August 1972.
7. McNelis, David N., Aerosol Formation from Gas Phase Reactions of Ozone and Olefins in the Presence of Sulfur Dioxide. EPA Report EPA-650/4-74-034. August 1974.

EFFECT OF GAS COMPOSITION ON THE COLLECTION EFFICIENCY OF MODEL GRID AND NUCLEPORE FILTERS

K.C. Fan, J. Lee, J.W. Gentry
Department of Chemical Engineering, University of Maryland

ABSTRACT

The particle collection efficiency and pressure drop across a nuclepore filter was measured as a function of time, particle size, and gas composition. Four particle sizes, two pore sizes, and two carrier gases were employed in this study. The scattering electron microscope was used to distinguish collection mechanisms.

Additional studies were carried out using model grid filters. These studies indicate that the pressure drop could be correlated with the Fuchs-Stechkina equation. Computer simulations indicate qualitative agreement with the experimental results but predict efficiencies lower than the experimental values. Electron micrographs provide some insight into the collection mechanisms.

INTRODUCTION

Collection mechanisms for nuclepore filters have been extensively investigated using monodisperse latex particles with air or nitrogen as a carrier gas (5,7,8,10,11). Spurny (9,7) has developed an expression for the pressure drop across nuclepore filter. Spurny and Lodge (9) have reported that the pores in the nuclepore filter have a tendency to fill uniformly. On the other hand, measurements by Fan (3) indicated the formation of caps rather than collection of particles within the pores. The present study differs in that the carrier gases with properties significantly different from air and nitrogen are employed.

Previous studies (4) with model filters have dealt primarily with single fibers rather than with the grids employed in this study. Choudhary (2) and Fan (4) have found that the Fuchs-Stechkina (6) equation was valid for air and argon. The present work demonstrates that the equation holds for gas mixtures as well.

EXPERIMENTAL PROCEDURE

A Dautrebande type generator was used to generate particles from aqueous dispersions of monodisperse latex particles using a procedure similar to that of Spurny (11) and Yeh (12). The particles were passed through two cold traps (the last stage was at − 77°C) in order to remove water

vapor. The particle collection efficiency was measured with
a Pheonix-Sinclair dust photometer (Model JM-4000). Since
the particles are monodisperse and have the same index of
refraction, the intensity of light is proportional to the
particle concentration.

Between the dust photometer and the cold traps was a
filter assembly in which the particles could either by-pass
the filter or pass through a filter. In these studies
nuclepore filters and model grid filters were used. A water
manometer with pressure taps across the filter measured the
pressure drop across the filter continuously (Figure 1).

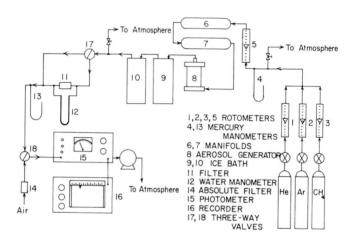

Fig. 1. Schematic diagram of aerosol detection system.

A detailed description of the fabrication of the model
filters has been reported previously (3). It should be
pointed out that the wires composing the filters are approximately 29μ and the interwire distance is approximately 34μ
in all three directions. The weave of the fiber provided a
complex geometry.

The carrier gas was mixed prior to entering the generator although the design provides the option of gas mixing
after the cold traps. The total pressure before the generator was continuously monitored.

PRESSURE DROP MEASUREMENTS

The pressure drop across the filters was measured as a function of gas velocity. Nuclepore filters with pore diameters of 3.0 and 5.0μ were used in the study. The experimental measurements were compared with the Spurny equation for pressure drop (5):

$$\Delta P = P_1 + \frac{C}{R} - [(P_1 + \frac{C}{R})^2 - \frac{1.14 \cdot 10^{-2} \, \eta \, L \, U \, P_1}{P \, R^2}]^{0.5}$$

where P_1 is the pressure before filter, R the pore radius, C a constant 0.0 305, P is the porosity, η the gas viscosity, L the filter thickness, U the gas velocity.

Measurements were made for two filter diameters, and for gas mixtures of argon and helium as well as methane. Typical results are shown in Table 1.

In this case, data for the viscosity of the gas mixture were taken from the Chapman and Cowling (1). Since all the variables in the Spurny equation are known, the ratio of the experimental to the theoretical pressure drop can be calculated. The results of our measurements indicate that this ratio of the theoretical to experimental pressure drop lies between 1.01 and 1.20. The basic conclusion of this analysis was that the Spruny equation is applicable to gas mixtures as well as air and nitrogen for nuclepore filters.

The study also revealed the Fuchs-Stechkina equation could be used to correlate the pressure drop across the grid filter with a value of C_0 between 0.514-0.53 compared with the value of the Happel constant of 0.5. The results confirmed for argon, and mixtures of argon and helium appear to be generally valid (Table 2).

ELECTRON MICROGRAPHS OF NUCLEPORE FILTERS

A sequence of photographs of the clogging of the filter was taken at 2, 4, and 10 minutes (Figures 2, 3). The particle diameter was 1.01μ, the pore diameter of the filter was 3.0μ, the carrier gas was argon, the flow rate was 0.35 CFM, the cross section was 5.07 cm^2 and the particle concentration was 8×10^2 particles/cm^3. Pictures were taken with both the smooth and the rough side of the filter facing the flow. The pressure drop was measured simultaneously.

Two points are indicated by the photographs. First, there were very few particles deposited within the pores or at distances more than one particle diameter from the pore surface. Specifically, the rate of deposition at the rims of the pores was 12 times as great as the deposition rate within the outlying area and 9 times as large as the deposition rate within the holes (Table 3).

TABLE 1
PRESSURE DROP AS A FUNCTION OF VELOCITY

Pore Diameter μ	Linear Velocity cm/sec	Pressure Drop in H_2O	Ratio Experimental to Theoretical Pressure Drop
Carrier Gas: Ar and He Mixture (2:1)			
5	32.6	6.38	1.06
5	44.9	9.25	1.10
5	57.9	12.24	1.14
5	69.9	15.03	1.15
5	84.2	17.87	1.13
3	24.4	6.39	1.01
3	34.2	9.12	1.03
3	46.0	12.64	1.05
3	55.4	15.93	1.09
3	70.0	19.68	1.06
Carrier Gas: Ar and CH_4 Mixture (6:4)			
5	31.8	5.76	1.16
5	42.2	7.93	1.20
5	52.6	10.03	1.20
5	63.6	11.93	1.19
3	21.7	4.85	1.08
3	42.6	9.87	1.10
3	54.6	13.07	1.14
3	64.7	15.51	1.14

TABLE 2
PRESSURE DROP (GRID FILTER)

$$P = (A_1 + A_2 \exp(A_3 U))U$$

Carrier Gas	A_1 $CmH_2O/cm/sec$	A_2 $CmH_2O/cm/sec$	A_3 Sec/cm	Fuchs-Stechkina Constant
Argon	9.06×10^{-3}	4.96×10^{-4}	9.05×10^{-3}	.52
Argon: Helium(1:1)	8.41×10^{-3}	7.59×10^{-4}	1.34×10^{-3}	.52
Argon: Helium (1:2)	1.01×10^{-2}	7.01×10^{-4}	2.71×10^{-3}	.53
Argon: Helium (2:1)	8.81×10^{-3}	2.22×10^{-4}	8.16×10^{-3}	.514

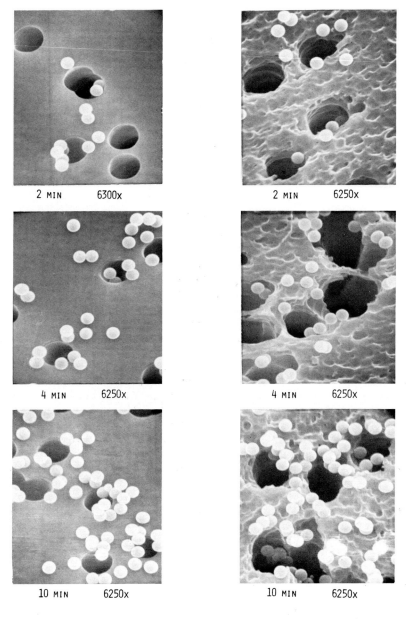

Fig. 2. Electron micrographs of 1.01μ particles captured by 3μ nuclepore filters as a function of time (smooth side facing the flow).

Fig. 3. Electron micrographs of 1.01μ particles captured by 3μ nuclepore filters as a function of time (rough side facing the flow).

TABLE 3
PARTICLE DEPOSITION RATES
(Particles/cm^2-min)

	Total Particles	Hole Area	Rim Area	Outlying Area
Smooth Side				
	12	.13	.83	.05
	28	.15	.76	.1
	49	0.0	.94	.06
Rough Side				
	12	.17	.7	.13
	27	.14	.72	.15
	73	.15	.75	.11

It seems reasonable to assume that the collection within the pores can be assumed to be due to diffusion while the collection at distances far from the pores is primarily due to inertial impaction. The particles in this study are 1.01μ and are unlikely to be deposited by diffusion. It is our belief that the principle collection mechanism is direct interception. The second point that should be noted is that the particles first collect on the rim of the pore, and then form a cap finally sealing the pore.

In a sequence of electron micrographs taken using methane as a carrier gas (Figures 4,5), filters with pore diameters of 3.0 and 5.0μ and particle diameters of 0.79μ and 0.36μ show the following results. The 0.79μ particles collect more uniformly in the pores of the 5.0μ filters and form caps for the 3.0μ particles. The 0.36μ particles appear to form caps for both 3 and 5μ pores. A possible explanation for this result is that for the 0.79μ particles, direct interception is the dominant mechanism for the cap formation since formation occurs at large values of r/R, where r is particle radiu. For the smaller particles, diffusion appears to be more important than previously suspected. This is indicated by the cap formation as well as the greater number of particles collected in the open spaces away from the pores.

ANALYSIS OF THE EFFICIENCY AND PRESSURE DROP MEASUREMENTS (NUCLEPORE FILTERS)

The measurements of the collection efficiency with pressure drop during clogging can be correlated as a function of ΔP^* where $\Delta P^* = \Delta P_t - \Delta P_o$, ΔP_o is the pressure drop at the beginning of the experiment (Figure 6).

PORE DIAMETER : 3.0 μ
MAGNIFICATION:12000

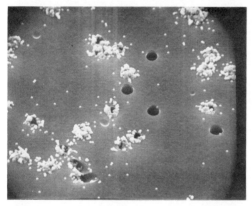

PORE DIAMETER : 3.0 μ
MAGNIFICATION: 2390

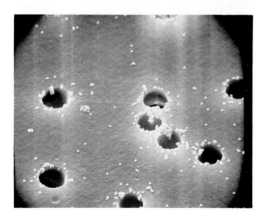

PORE DIAMETER: 5.0 μ
MAGNIFICATION: 2650

Fig. 4. Electron micrographs of 0.36 μ particles captured by 3 and 5μ nuclepore filters with methane at 0.4 cfm.

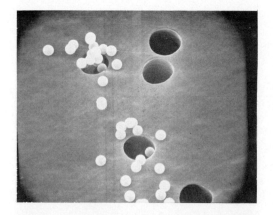

**PORE DIAMETER : 3.0 μ
MAGNIFICATION : 5800**

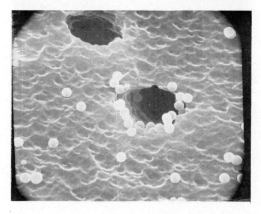

**PORE DIAMETER : 5.0 μ
MAGNIFICATION : 5700**

*Fig. 5. Electron micrographs of 0.79μ particles captured
by 3 and 5μ nuclepore filters with methane at 0.4 cfm.*

In order to compare the data with theory, the following
model was adopted. A uniform pore filling model was assumed
(this assumption could be modified to account for cap formation) and the pressure drop was used to determine the
effective pore radius. Spurny's equation which had been independently verified above was used to determine the radius.
Secondly, a semi-empirical equation of Spurny (7) was used to
calculate a theoretical efficiency which was compared with
experiments.

The result of this comparison was qualitatively in
agreement with theory, but quanitatively the efficiency increases more rapidly than the uniform pore model predicts as
shown in Figure 7, where $R' = R_o - R_t$, R_o is original pore
radius, R_t is pore radius at time t. This is consistent with
the cap formation observed in the electron micrographs. It
is interesting to note that the data for 0.79μ particles

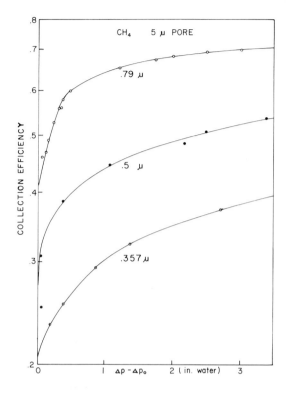

Fig. 6. Experimental collection efficiency of nuclepore filters as a function of pressure drop.

and a 5μ pore diameter give better agreement with the theory which is consistent with the electron micrographs for these conditions.

MODEL GRID FILTERS

The collection efficiency was measured for a filter assembly consisting of six successive screens. Three particle sizes of 1.01, 0.79, and 0.36μ were used with gas mixtures of argon and helium. The typical plot of collection efficiency as a function of velocity is shown in Figure 8. Since the density of the gas mixtures varies by a factor of 10 while the viscosity is nearly constant, the Stokes number can be varied independently of the Reynolds number.

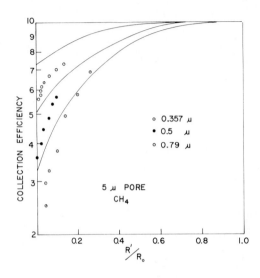

Fig. 7.   *The comparison of experimental efficiency with theoretical efficiency as a function of* **effect***ive pore radius.*

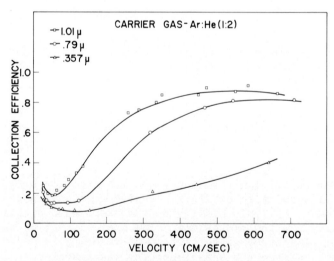

Fig. 8.   *Experimental efficiency of grid filters as a function of velocity with different sizes of particles.*

The effect of the gas mixtures can be shown dramatically when the efficiency is plotted as a function of the Reynolds number and then was a function of the Stokes number. If in-

206

ertial impaction is dominant, the efficiency for the gas mixtures will be represented by a universal function of the Stokes number for a given particle size and filter assembly. On the other hand, if direct interception is dominant, the efficiency should be determined by the Reynolds number. Electrostatic attraction would show similar behavior to direct interception.

Using the Kuwabara flow field, the theoretical efficiency was calculated for 1.01µ and 0.79µ particles for different gas mixtures. The numberical simulations indicate that for the larger particles, the efficiency can be represented by a function of the Stokes number alone. These experimental measurements indicate a qualitative agreement with theory, but the experimental efficiencies are less than those predicted by theory (Figures 9, 10).

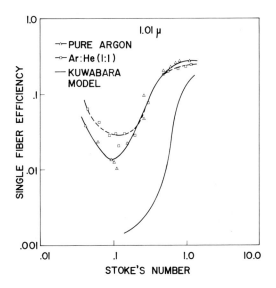

Fig. 9. Single fiber efficiency as a function of Stokes number.

Electron micrographs indicate that for the larger particles, most collection occurs within a narrow band of the fiber with a few particles collected at the edges (Figure 11). Such a deposition pattern is consistent with collection by impaction.

207

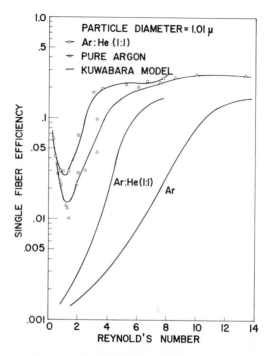

Fig. 10. *Single fiber efficiency as a function of Reynolds number.*

Fig. 11. *Electron micrograph of 1.01μ particles ca captured by grid filters at velocity of 224 cm/sec.*

CONCLUSIONS

The experimental results of this study indicate that the pressure drop for gas mixtures across nuclepore filters can be described by the Spurny equation and for grid filters by the Fuchs-Stechkina equation.

Experimental measurements of the collection efficiency can be correlated as a function of the pressure difference for filters undergoing clogging. Evidence indicates that diffusional processes play an important role for small particles, that cap formation is more prevalent than uniform pore filling, and that diffusion plays a significant role in cap formation. The Spurny equation for collection efficiency gives qualitative agreement with the experimental measurements during clogging.

Experimental measurements for the grid filter indicate that for 1.01μ particles collection is primarily a function of the Stokes number. The result has been confirmed for gas mixtures as well as pure gases. Qualitative agreement with the Kuwabara method has been found. The deposition pattern of 1.01μ particles is consistent with collection processes dominated by inertial impaction.

ACKNOWLEDGEMENT

K. C. Fan and J. Lee wish to acknowledge the support of the Minta Martin Foundation, the Center for Materials Research, and the Computer Center of the University of Maryland.

REFERENCES
1. Chapman, S. and T.G. Cowling, "The Mathematical Theory of Non-uniform Gases", p. 232, University Press, Cambridge, 1952.
2. Choudhary, K.R., Ph.D. Thesis, University of Maryland, College Park, Maryland (1974).
3. Fan, K.C., K.R. Choudhary, and J.W. Gentry, Submitted to Canadian J. Chemical Engineering (1976).
4. Gentry, J.W., and K.C. Fan, Powder Technolgy, 14, 253 (1976).
5. Machacova, I., J. Hrbek, V. Hampl, and K. Spurny, Collection Czechoslov. Chem. Commun., 35, 2087 (1970).
6. Pich, J., "Aerosol Science" (C.N. Davis, ED.), p. 223, Academic Press, New York, 1966.
7. Spurny, K.R., J. Havlova, J.P. Lodge Jr., and B. Wilder, Staub, 35, 77 (1975).
8. Spurny, K.R., and G. Madelaine, Collection Czechoslov. Chem. Commun., 36, 2857 (1971).
9. Spurny, K.R., J.P. Lodge Jr., E.R. Frank, and D.C. Sheesley, Evironmental Science and Technology, 5, 453 (1969).

10. Spurny, K.R., G. Pfefferkorn, and R. Blaschke, <u>Staub</u>, 31
     5 (1971).
11. Spurny, K.R., J. Havlova, J.P. Lodge Jr., E.R. Ackerman,
     D. C. Sheesley, and B. Wilder, <u>Evironmental Science
     and Technology</u>, 8, 759 (1974).
12. Yeh, Hsu-Chi, and B.Y.H. Liv, J. <u>Aerosol Science</u>, 5, 205
     (1974).

CALCULATIONS OF UNIPOLAR AEROSOL CHARGING[I]

W. H. Marlow
Brookhaven National Laboratory

Abstract

By using the best available data for charging rates for small spheres with arbitrary dielectric constants, the time-dependent charge distributions of two unipolar polydisperse aerosols is calculated. The behavior of this charging is studied as a function of polydispersity in both aerosol and ion distributions. Particles ranging from 0.005 μm to 1.0 μm radius are included and ions typical of the atmosphere and atmospheric applications are used.

I. INTRODUCTION

Unipolar aerosol charging is the acquisition of electric charge by aerosols in the presence of ions of a single sign. Under essentially all conditions, electric charges in gaseous media serve as nuclei for the rapid clustering or attachment of gas phase molecular species. The composition distribution of these cluster ions is characteristic of the composition of the gas and, in important cases, gaseous trace constituents may even dominate in determining the properties of the cluster ions. As we are all well aware, aerosols too exhibit a wide variety of properties in both their size and composition distributions. The question therefore arises as to how the variabilities in both cluster ions and aerosol will affect the nature of aerosol charging.

Previous calculations of bipolar charging[1] indicate that both cluster ion and aerosol polydispersity characteristics can interact to change the nature of the charge carried by very fine particles from what they are in the monodisperse case.

[I]This work was performed under the auspices of the U.S. Energy Research and Development Administration.

Since bipolar charging is a process that reaches a steady
state, it is to be expected that large differences from the
monodisperse situations could not occur because of effects of
compensation by ions of opposing sign. However, for unipolar
charging where a steady state does not occur, aerosol and ion
polydispersities could be expected to play important roles in
aerosol charging.

To examine this conjecture realistically, I postulate a
system consisting of an aerosol of six different radius particles (0.005, 0.01, 0.05, 0.1, 0.5, and 1.0 micrometers) and of
six different cluster ions with electrical mobilities of 2.2,
2.1, 2.0, 1.8, 1.5, and 1.0 cm^2/(volt-sec) which have masses
respectively of 69, 73, 88, 109, 148, and 279 a.m.u. The
aerosol covers the dominant range of atmospheric particulate
matter while the ions span the range of positive atmospheric
ion mobilities cited in the literature. Masses are derived
from the empirical curve of Huertas, Marty, and Fontan.[2]
Such a system is mathematically described by the following
equations:

$$\frac{dA_{Os}}{dt} = G_{Os} - A_{Os} \sum_{i=i}^{6} I_i f_{iOs} \qquad S = 1, 2, ---, 6$$

$$\frac{dA_{ns}}{dt} = G_{ns} + A_{n-1\,s} \sum_{i=i}^{6} I_i f_{i\,n-1\,s} - A_{ns} \sum_{i=i}^{6} I_i f_{i\,ns}$$

$$S = 1, 2, ---, 6, \quad n = 0, 1, ---, N(s) -1,$$

$$\frac{dA_{N(s)s}}{dt} = A_{N(s)-1\,s} \sum_{i=i}^{6} I_i f_{i\,N(s)-1\,s}$$

$$\frac{d I_i}{dt} = E_i - I_i \sum_{s=1}^{6} \sum_{n=0}^{N(s)-1} f_{i\,ns} A_{ns} \qquad i = 1, 2, ---, 6$$

G_{ns}: generation rate for particles of size-class s with n
elementary charges

E_i: generation rate for ions of variety i

A_{ns}:   number density of particles of size-class s carrying n
             charges

I_i: number density of ions of variety i

f_{ins}:   flux rate per ion of species i to a particle of size s
             carrying n charges

t: time

The ion fluxes f_{ins} have been chosen in accordance with
the best available data. Recent measurements by Pui[3] indicate that the free molecular and near free molecular fluxes of
Marlow and Brock[4] faithfully predict the charging rates for
neutral particles of 0.005 and 0.01 micrometers radius though
they underestimate the fluxes for larger particles. For the
0.1 micrometer and perhaps 0.5 micrometer particles, collision-
dominated flux rates are generally considered accurate. However, for the middle of the transition region, neither extra-
polations of the first-order corrections to the free molecular
flux nor the continuum data are very good, but the latter
appear somewhat better (i.e., half an order of magnitude too
high) than the former (almost one order of magnitude low).
Therefore, these continuum fluxes were used for the 0.05
micrometer and larger particles. In all cases, a particle
dielectric constant of 80 was assumed for the image force.

II. COMPUTATIONAL RESULTS

A. Constant Total Ion Number Density

For the purpose of comparison, the charging of the
particles by two different, constant $\frac{d\,I_i}{dt} = 0$ cluster ion
distributions was computed. It is the model used in most
unipolar computations. Due to the constancy of the small ions,
the aerosol particles all charge independently of particles of
other sizes and therefore the calculation actually represents
six monodisperse aerosols. Figure 1 presents the results of
the calculations. Both cases include the same total number of
cluster ions. Comparison of the two graphs reveals that care
must be exercised in using the commonly employed assumption
that the "N_o t" product (N_o = total number of ions and t = time)
uniquely characterizes unipolar charging processes. Quantitative examination of the data shows the average charge for the
continuum formula cases (0.05-1.0 μm) to be about 20% higher
in the high mobility case than in the low mobility case whereas

W. H. MARLOW

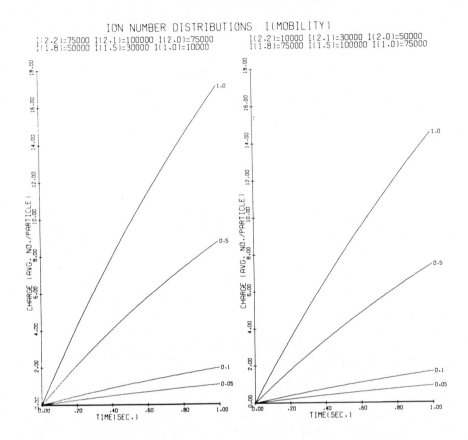

Fig. 1. Monodisperse aerosol charging: average particle
charge as a function of time. Particle radius (in micrometers) at right of line for its average charge; 0.005 and
0.01 micrometer particle charges too small to appear on graph.

it is on the order of 80% higher when the same comparison is
made for the smallest particles described by the 1st order
correction to free molecular flow. Such a result for the
larger particles is to be expected since dA_i/dt is proportional
to the _conductivity_ of the gas when only continuum ion fluxes
are used, as can be seen in the unipolar charging equations.
No such uniform relationship exists for the finest particles
and in fact it is reasonable to attribute the 50% greater
charge increase of the 0.005 μm and 0.01 μm particles to the
substantially greater number of high mobility cluster ions in
the high conductivity example. Particles in the transition
region which are described in these studies by the continuum
formulas (0.05 and 0.1 μm) would likely also show a charge

ratio higher than the conductivity ratio were better charging coefficients available.

B. Constant Total Charge

The opposite extreme is that represented by a fixed total number of charges. This is a situation wherein the effects of aerosol and ion polydispersities should be evident. The results of these calculations are presented in Figure 2.

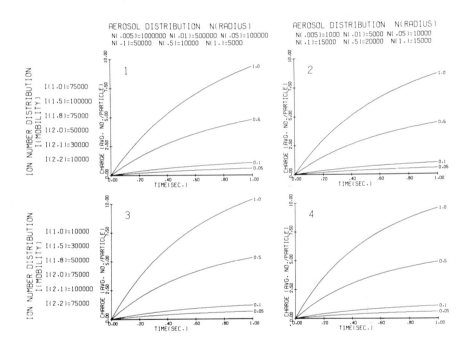

Fig. 2. Polydisperse aerosol charging: average particle charge as a function of time. Particle radius (in micrometers) at right of line for its average charge; 0.005 and 0.01 micrometer particle charges too small to appear on graph.

For a given ion mobility distribution, the average charge acquired by a particle is dependent upon the aerosol polydispersity. This can be seen by comparing graphs 1 and 2 or 3 and 4 in Figure 2. Note that these results imply the numbers of larger particles are principally responsible for controlling the extent of charging for all particles despite the much greater change in small particle numbers. In both cases, the charging difference is about 10%.

215

As in the constant ion density cases, the difference in charging between the high and low conductivity examples is considerable for the finest particles (comparing graphs 1 and 3 or 2 and 4 in Figure 2). This indicates that largely the same mechanisms pertain as was discussed in the constant ion density case.

III. CONCLUSIONS

At this stage of these investigations, several observations on unipolar polydisperse aerosol charging can be made:

(a) If conductivity replaces total ion number, then the product of time and conductivity can be used to accurately characterize the unipolar charging of aerosols of particles larger than 0.5 micrometers radius.

(b) Charging of particles smaller than 0.5 micrometers is much more dependent upon the availability of the most highly mobile cluster ions than upon total conductivity due to cluster ions.

(c) Overall aerosol charging is strongly affected by polydispersity with the number density of larger particles playing the dominant role.

IV. REFERENCES

1. Marlow, W. H. and Brock, J. R., J. Colloid Interface Sci. 51, 23 (1975).
2. Huertas, M. L., Marty, A. M., and Fontan, J., J. Geophys. Res. 79, 1737 (1974).
3. Pui, David Y. H., Experimental Study of Diffusion Charging of Aerosols, Ph.D. thesis, University of Minnesota, 1976.
4. Marlow, W. H. and Brock, J. R., J. Colloid Interface Sci. 50, 32 (1975).

COMPUTERIZED TREATMENT OF HANAI'S FORMULA, AND ITS APPLICATION TO THE STUDY OF THE DIELECTRIC BEHAVIOUR OF EMULSIONS.

M. Clausse* and *R. Royer***

*Laboratoire de Thermodynamique
**Département de Mathématiques

Institut Universitaire de Recherche Scientifique
Université de Pau et des Pays de l'Adour
B.P. 523 "Pau-Université"
64010 Pau (France)

ABSTRACT

The authors propose a computerized numerical analysis of Hanai's formula

$$\left(\frac{\varepsilon^{*}_{1} - \varepsilon^{*}_{2}}{\varepsilon^{*}_{1} - \varepsilon^{*}}\right)^{3} \frac{\varepsilon^{*}}{\varepsilon^{*}_{2}} = \frac{1}{(1 - \Phi)^{3}}$$

that gives the complex permittivity ε^{*} of a spherical dispersion system as a function of the complex permittivities ε^{*}_{1} (disperse phase) and ε^{*}_{2} (continuous phase) and of Φ, the volume fraction of the disperse phase.
    Using this method, it is possible to give an analytical representation of the diagrams of dielectric absorption occuring in spherical dispersion systems and especially in emulsions.

I. INTRODUCTION

It is well known that a composite system made of a dispersion of particles inside a continuous phase exhibits, when submitted to an alternating electric field, a specific bulk dielectric absorption.

When both components of the system are conductive but loss-free in the range of frequencies considered, in the most general case, a Cole-Cole type absorption occurs (Maxwell-Wagner effect), accompanied by a bulk conduction in the lower frequency range , (1) , (2) , (3) , (4) , (5) , (6) , (7).

When at least one of the components is polar and exhibits, besides a conduction, dipolar losses in the frequencies range concerned, it has been experimentally observed, (7) , (8) , (9) , that an other absorption, connected with the dipolar losses, arises in the system. Depending on the values of the permittivities and conductivities of both components and on the value of the critical frequency of the dipolar relaxation, this new absorption in the composite system can overlap or not the former one due to the Maxwell-Wagner effect , (5) , (7).

In the case of emulsions of the water-in-oil (w/o) type, experiments show that, in the kilocycle frequency range, the complex permittivity (1) $\varepsilon^* = \varepsilon' - j\,\varepsilon''$ of an emulsion is well represented by the following formula :

$$\varepsilon^* = \varepsilon_h + \frac{\varepsilon_l - \varepsilon_h}{1 + (j\omega\tau)^{(1-\alpha)}} + \frac{\chi_l}{j\omega} \qquad [1]$$

where ε_h and ε_l are respectively the limiting permittivities of the emulsion at high and low frequencies , τ its relaxation time and χ_l its steady bulk conductivity, ω being the angular frequency of the applied alternating electric field and $j = \sqrt{-1}$, (10) , (5) , (6).

---

(1) : To avoid the introduction of ε_o , the absolute permittivity of free space, in formulae, all permittivities referred to are absolute permittivities.

In order to link his experimental results to a theoretical model, Hanai , (11) , (12) , proposed the following formula for the complex permittivity ε^* of an emulsion :

$$\left(\frac{\varepsilon^*_1 - \varepsilon^*_2}{\varepsilon^*_1 - \varepsilon^*}\right)^3 \frac{\varepsilon^*}{\varepsilon^*_2} = \frac{1}{(1 - \phi)^3} \qquad [2]$$

ε^*_1 and ε^*_2 represent respectively the complex permittivity of the constituent phase number 1 (disperse phase) and of the constituent phase number 2 (continuous phase) , ϕ being the volume fraction of the disperse phase , $(0 \leqslant \phi \leqslant 1)$. Following an integration method first introduced by Bruggeman , (13) , and assuming an emulsion to be a dispersion of spherical particles inside a continuous phase, Hanai derived his formula from the well known Wagner's equation (14)

$$\frac{\varepsilon^* - \varepsilon^*_2}{\varepsilon^* + 2 \varepsilon^*_2} = \frac{\varepsilon^*_1 - \varepsilon^*_2}{\varepsilon^*_1 + 2 \varepsilon^*_2} \phi \qquad [3]$$

that holds for small values of ϕ only. Formula $[2]$ has been proposed as well by Clausse (5) who, after having established a general theorem for the substitution of complex permittivities for static permittivities in mixture formulae, derived it as an extension of Bruggeman's equation

$$\left(\frac{\varepsilon_1 - \varepsilon_2}{\varepsilon_1 - \varepsilon}\right)^3 \frac{\varepsilon}{\varepsilon_2} = \frac{1}{(1 - \phi)^3} \qquad [4]$$

that holds for static permittivities. This general procedure of substitution has been proposed also by Dukhin , (15) .

The validity of formula $[2]$ and its applicability to emulsions have been tested with more or less success by different authors. In particular , Hanai , Koizumi and Gotoh , (16) , showed , for o/w type emulsions , that their experimental results concerning permittivities and conductivities fit well formula $[2]$. In the case of w/o type emulsions which exhibit striking dielectric absorptions of the Cole-Cole type in the kilocycle frequency range, Hanai, (10), found that the experimental values of the high frequencies limiting permittivity ε_h of emulsions of water in a nujol-carbon tetrachloride mixture using the emulsifying agents Arlacel 83 , Span 20 and Span 60 can be well represented by the following formula

$$\left(\frac{\varepsilon_1 - \varepsilon_2}{\varepsilon_1 - \varepsilon_h}\right)^3 \frac{\varepsilon_h}{\varepsilon_2} = \frac{1}{(1 - \Phi)^3} \qquad [5]$$

where ε_1 is the static permittivity of water and ε_2 the permittivity (constant in the range of frequencies considered) of the oil phase. For the low frequencies limiting permittivity ε_l , Hanai found an influence of the shearing stress applied to his emulsions, the values of ε_l at rest lying far beyond the theoretical values given by :

$$\frac{\varepsilon_l}{\varepsilon_2} = \frac{1}{(1 - \Phi)^3} \qquad [6]$$

Using emulsions of water in a mixture of vaseline oil and wool-wax, Clausse , (5) , (17) , observed a good agreement of his experimental values of ε_l with formula $[6]$ and of those of ε_h with formula $[5]$ as well. As regards the steady bulk conductivity χ_l of w/o emulsions , which should be given by

$$\frac{\chi_l}{\chi_2} = \frac{1}{(1 - \Phi)^3} \qquad [7]$$

there is no definite conclusion, owing to the weak values of the conductivity χ_2 of oil phases.

However satisfactory may be the agreement of the theory with the experimental results reported above, there is still a need for a close and thorough analysis of formula $[2]$ and its comparison with experiments. For example, not only the limiting values ε_h and ε_l of the complex permittivity ε^* of an emulsion should be found to agree with equations $[5]$ and $[6]$ respectively, but the whole dielectric absorption phenomenon should be related to Hanai's general formula.

Therefore, it would be of great interest to manage to put formula $[2]$ in the form of equation $[1]$. This kind of transformation, that has been achieved for Wagner's equation, **cannot** be carried out in a simple way for Hanai's formula, because of the intricate form of that complex cubic equation.

To solve this problem, Clausse, (5), (18), proposed a method based on a numerical analysis of formula [2] by means of a computer, and applied it to the study of the dielectric properties of emulsions.

In the following sections, this method will be exposed and some general results reported.

II. THEORETICAL APPROACH

Let us consider Hanai's equation [2] as a formula generating in the complex plane a family of geometric transformations $\mathcal{H}$ with parameter Φ, volume fraction of the disperse phase.

At a given Φ, the point representative of the complex permittivity ε^* of the binary system can then be considered as the image by $\mathcal{H}$ of the points representative of the complex permittivities ε^*_1 and ε^*_2 of the disperse phase and the continuous one, as it is shown on figure 1.

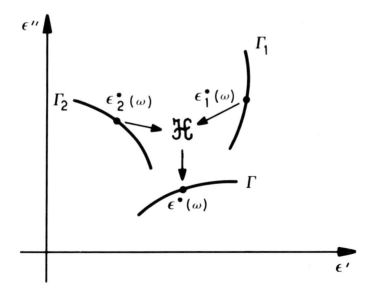

Fig. 1. Relation between ε^*_1 (ω), ε^*_2 (ω) and ε^* (ω) through transformation $\mathcal{H}$ in the complex plane.

If ε^*_1 and ε^*_2 are functions of the angular frequency ω of the applied electric field, the points representative of $\varepsilon^*_1 (\omega)$ and $\varepsilon^*_2 (\omega)$ circulate respectively on curves Γ_1 and Γ_2 as ω varies. Then their image by $\mathcal{K}$, which is the point representative of ε^* , circulates on a curve Γ which is the diagram of the dielectric properties of the binary system.

By a mathematical analysis of the curve Γ it is possible to obtain its analytical representation and therefore the function $\varepsilon^*(\omega)$ describing the dielectric behaviour of a binary system at a given Φ . Repeating this process for other values of Φ enables to obtain, at a given ω , the dependency of ε^* upon Φ .

Finally , formula [2] can be rewritten in the following form :

$$\varepsilon^* (\omega , \Phi) = \mathcal{K} \left\{ \varepsilon^*_1 (\omega) , \varepsilon^*_2 (\omega) , \Phi \right\} \qquad [8]$$

that shows clearly the dependency of ε^* upon the angular frequency ω and the volume fraction Φ of the disperse phase.

Calculating numerically from equation [8] by means of a computer, for different values of Φ ranging from 0 to 1 , the values of the complex permittivity ε^* corresponding to standard functions $\varepsilon^*_1 (\omega)$ and $\varepsilon^*_2 (\omega)$ (representing for example conduction absorptions or Debye's type dipolar absorptions), it is possible to determine models of dielectric behaviour for binary systems and especially for emulsions, and to build abaci in order to compare computer data with experimental results.

III. MATHEMATICAL PROCEDURE AND COMPUTER ROUTINE

Automatic solving of Hanai's equation gives rise to difficulties when dealing with values of ε^*_1 and ε^*_2 with real and imaginary parts of various magnitudes and when varying $(1-\Phi)^3$ in a wide range.

For instance, rewriting equation [2] as a polynomial of degree three

$$\frac{\varepsilon^* (1 - \Phi)^3}{\varepsilon^*_2} = \left(\frac{\varepsilon^*_1 - \varepsilon^*}{\varepsilon^*_1 - \varepsilon^*_2} \right)^3 \qquad [9]$$

makes it obvious that, for large values of Φ (namely, Φ near 1), all three solutions differ very little from ε^*_1, hence from each other. Therefore it is not possible to select the proper one while computing. Physical considerations (such as the signs and relative magnitudes of real and imaginary parts, and continuity with respect of ω) are required to obtain such a selection. For the same reason, simple iterative methods are expected to be badly convergent (perhaps not convergent at all). Besides, sophisticated iterative algorithms, (19), are time consuming though they provide a high reliability. We us instead a most classical direct process after standardization of the equation.

Let us choose $u = 1 - (\varepsilon^*/\varepsilon^*_1)$ for the new unknown; with $a = \varepsilon^*_1 - \varepsilon^*_2$ and $c = (1 - \Phi)^3 / 3 \varepsilon^*_2$, equation [9] becomes

$$3 c (1 - u) = \varepsilon^*_1{}^2 u^3 / a^3 \qquad [10]$$

or, in standard form :

$$u^3 + 3 b u - 3 b = 0 \qquad [11]$$

with $b = c a^3 / \varepsilon^*_1{}^2$.

If we try to get u in the form $u = z_1 z_2 (z_1 + z_2)$, we obtain

$$u^3 = z_1{}^3 z_2{}^3 (z_1{}^3 + z_2{}^3 + 3 u) \qquad [12]$$

which fits exactly [11] if $z_1{}^3 z_2{}^3 = - b$ and $z_1{}^3 + z_2{}^3 = - 3$

So, $b_1 = z_1{}^3$ and $b_2 = z_2{}^3$ must be the solutions of the quadratic equation $x^2 + 3 x - b = 0$, given by $-1.5 \pm (2.25 + b)^{1/2}$.

Depending on $(1 - \Phi)$ and $\varepsilon^*_1 - \varepsilon^*_2$, the real part of b may be quite small. To avoid a castastrophic loss of precision within subtraction, we calculate only

$$b_1 = -1.5 - (2.25 + b)^{1/2} \qquad [13]$$

with positive choice for the real part of the square root, while b_2 is gained by $b_2 = -b/b_1$.

After computation of any determination of z_1 and z_2, we know that $z_1 z_2 (z_1 + z_2)$ does not change when multiplying simultaneously z_1 and z_2 by $\lambda = \exp(2j\pi/3)$ or by $\bar{\lambda}$, $(j = \sqrt{-1})$. Then it is sufficient to consider z_1 in association with z_2, λz_2, $\bar{\lambda} z_2$.

The resulting algorithm is written as a routine in FORTRAN IV language, using double precision (long format of floating-point values) for the sake of precision. It is currently executed with good results on an IBM 360/65 machine, by means of the level G compiler under control of the 360 Operating System (release 21.7).

The FORTRAN names used are similar to those occuring in the present text, except for the greek letters; E1, E2, E for ε^*_1, ε^*_2, ε^*, PHI for Φ. The entry point is labelled HANAI; the calling statement may look

CALL HANAI (E1, E2, PHI, E)

with mandatory declarative instructions in the calling unit of program   CØMPLEX * 16 E1, E2, E (3)

REAL * 8 PHI

The data E1, E2, PHI are not changed by subroutine HANAI. The three results are returned in array E.

The routine itself is given as an annex, at the end of this paper.

IV. GENERAL RESULTS AND DISCUSSIONS

By using the HANAI computer routine, it is possible to solve Hanai's equation with in view a comparison of computer data with experimental results.

In the case of spherical dispersion systems whose both components are conductive only, formula [2] can be rewritten like formula [1], namely like a Cole-Cole formula with an additional term of conduction, for values of Φ , the volume fraction of disperse phase, as high as about 0. 70 , (5) . This result generalizes for higher values of Φ that obtained with Wagner's formula which is valid only for $\Phi < 0.10$. It explains qualitatively the dielectric relaxations exhibited in the kilocycle frequency range by many emulsions systems, (4) , (5) , (6) , giving so a theoretical representation of the so-called "interfacial polarization" , (4) , or "migration polarization" , (15) , occuring in emulsions with high concentrations of disperse phase.

As an illustration, the case of emulsions, either of the "water-in-oil" type (w/o) or of the "oil-in-water" type (o/w) using non-polar and non-conductive oil phases, is considered further below. Computer calculations were carried out for values of Φ ranging from 0 to 1 by steps of 0. 05 each. The complex permittivity ε^*_w of the aqueous phase was assumed to be given by

$$\varepsilon^*_w = \varepsilon_{ws} + \frac{X_w}{j\omega} \qquad [14]$$

ε_{ws} being the static permittivity and X_w the steady conductivity whose values were chosen after compilation of experimental data concerning the dielectric and conductive properties of water, (20) , (21) . To comply with the hypothesis that the oil phase was non-polar and non-conductive , the relative complex permittivity $\varepsilon^*_{oil}/\varepsilon_0$ of the oil phase was assumed to be equal to its real part $\varepsilon_{oil}/\varepsilon_0$ which was given a constant value of 2.50, in accordance with experimental data concerning the dielectric constant of many "oil" type liquids (liquid paraffin, vaseline oil, benzene, carbon tetrachloride, etc ...) .

Figure 2 gives an example of the Cole-Cole plots obtained in the case of w/o type emulsions , ($\varepsilon^*_1 = \varepsilon^*_w$ with $\varepsilon_1 = \varepsilon_{ws}$ and $X_1 = X_w$, $\varepsilon^*_2 = \varepsilon^*_{oil}$ with $\varepsilon_2 = \varepsilon_{oil}$).

An analysis of this diagram shows that the complex permittivity ε^* of the emulsion can be represented by a Cole-Cole formula

$$\varepsilon^* = \varepsilon_h + \frac{\varepsilon_l - \varepsilon_h}{1 + (j\omega\tau)^{(1-\alpha)}} \qquad [15]$$

225

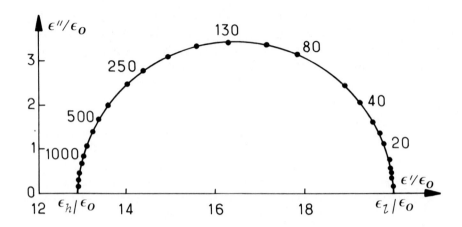

Fig. 2. Cole-Cole diagram obtained from computer data for a w/o emulsion with $\Phi = 0.50$. $\varepsilon_{ws}/\varepsilon_o = 80.40$, $\chi_w = 10^{-3}\ \Omega^{-1}.m^{-1}$, $\varepsilon_{oil}/\varepsilon_o = 2.50$. (The frequencies are given in KHz).

where ε_h and ε_l are given respectively by [5] and [6] , the additional ohmic term of formula [1] being null, as it could be expected from the assumption that the continuous phase (oil phase) was non conductive. This theoretical result, that remains valid for values of Φ as high as about 0.70 , is consistent with the experimental observations concerning the dielectric properties of w/o emulsions as it is illustrated by figure 3 which gives the experimental Cole-Cole diagram of an emulsion of water in a blend of 75% (wt/wt) vaseline oil and 25% (wt/wt) lanolin.

Experiments show that the limiting permittivities ε_h and ε_l of these emulsions are given within an uncertainty of 2% by [5] and [6] respectively, in accordance with the results of the computer treatment of Hanai's general formula [2] . For values of Φ greater than 0.70 , the Cole-Cole diagrams obtained from the computer data become more depressed and cannot be identified with circular plots , ε_h and ε_l however being still given respectively by [5] and [6] , (22) . This theoretical result is consistent with Hanai's experimental observations concerning emulsions of water in a mixture of nujol and carbon tetrachloride, (4) .

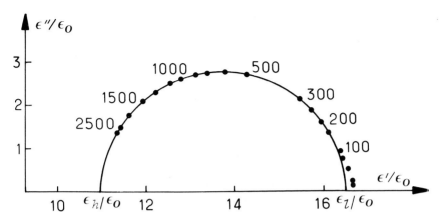

Fig. 3. *Cole-Cole diagram of an emulsion of water (Φ = 0.47) in a blend of 75% (wt/wt) vaseline oil and 25% (wt/wt) lanolin, at 23°C. (The frequencies are given in KHz).*

As concerns o/w type emulsions, ($\varepsilon^*_1 = \varepsilon^*_{oil}$ with $\varepsilon_1 = \varepsilon_{oil}$, $\varepsilon^*_2 = \varepsilon^*_w$ with $\varepsilon_2 = \varepsilon_{ws}$ and $\chi_2 = \chi_w$), the analysis of the computer data shows that, for values of Φ up to 0.70, the complex permittivity is given by formula [1], with ε_h, ε_l and χ_l fitting respectively the following formulas :

$$\left(\frac{\varepsilon_1 - \varepsilon_2}{\varepsilon_1 - \varepsilon_h}\right)^3 \frac{\varepsilon_h}{\varepsilon_2} = \frac{1}{(1-\Phi)^3} \qquad [16]$$

$$\varepsilon_l = \frac{3}{2}\varepsilon_1 + (\varepsilon_2 - \frac{3}{2}\varepsilon_1)(1-\Phi)^{3/2} \qquad [17]$$

$$\chi_l = \chi_2 (1-\Phi)^{3/2} \qquad [18]$$

As in the case of w/o emulsions, for values of Φ higher than 0.70, the Cole-Cole diagrams corresponding to the dielectric relaxation itself cannot be identified with circular plots, formulae [16], [17] and [18] being however still valid. Anyway, whatever the value of Φ, the magnitude of the dielectric relaxation, which can be evaluated by the dielectric increment ($\varepsilon_l - \varepsilon_h$), is very tiny in comparison with the values of ε_h, ε_l and χ_l/ω, as it is shown by the Table that gives, for w/o and o/w emulsions, the numerical values of $\varepsilon_l/\varepsilon_o$, $\varepsilon_h/\varepsilon_o$ and $(\varepsilon_l - \varepsilon_h)/\varepsilon_o$, ε_o being the permittivity of free space.

TABLE

Computed values of $\varepsilon_l/\varepsilon_o$, $\varepsilon_h/\varepsilon_o$ *and* $(\varepsilon_l - \varepsilon_h)/\varepsilon_o$ *for w/o and o/w emulsions.*

Volume fraction of water	w/o emulsions			o/w emulsions		
	$\varepsilon_l/\varepsilon_o$	$\varepsilon_h/\varepsilon_o$	$(\varepsilon_l-\varepsilon_h)/\varepsilon_o$	$\varepsilon_l/\varepsilon_o$	$\varepsilon_h/\varepsilon_o$	$(\varepsilon_l-\varepsilon_h)/\varepsilon_o$
0.00	2.500	2.500	0	3.750[a]	2.500	1.250[a]
0.10	3.429	3.322	0.107	6.174	5.730	0.444
0.20	4.883	4.514	0.369	10.606	10.372	0.234
0.30	7.289	6.279	1.010	16.345	16.201	0.144
0.40	11.574	8.936	2.638	23.141	23.045	0.096
0.50	20.000	12.971	7.029	30.850	30.783	0.067
0.60	39.062	19.066	19.996	39.374	39.327	0.047
0.70	92.592	28.069	64.523	48.641	48.610	0.031
0.80	312.490	40.861	271.629	58.596	58.577	0.019
0.90	2494.200	58.162	2436.038	69.195	69.186	0.009
1.00	∞	80.400	∞	80.400	80.400	0

a. This value has only the significance of a mathematical limit.

Consequently, the complex permittivity ε^* of o/w type emulsions can be written with a good approximation

$$\varepsilon^* = \varepsilon + \frac{\chi_l}{j\,\omega} \qquad [19]$$

ε being given by either formula [16] or formula [17] , both formulae differing very little form each other, owing to the small value of ε_1 in comparison with that of ε_2 . The experimental observations concerning the dielectric properties of emulsions of non-polar and non-conductive oil phases in water are consistent with those theoretical results. For instance, systematic data were reported by Hanai, Koizumi and Gotoh, (4), (16), who studied over the frequency range [20 Hz - 5 MHz] and up to $\Phi = 0.85$ the dielectric properties of emulsions in water of a nujol-carbon tetrachloride mixture. No dielectric relaxation could be put into evidence and the complex permittivity of the emulsions studied was given by formula [19], the experimental values of ε and χ_l fitting well formulas [17] and [18] respectively. Similar conclusions were reached by Clausse, Sherman and Sheppard, (23), (24), who investigated the dielectric behaviour of benzene-in-water microemulsions.

Because of the good agreement existing between the theoretical model derived from the computer analysis of Hanai's formula and the experimental data concerning the low frequency dielectric properties of w/o or o/w type emulsions, the possible extension of the method to more complicated systems has been investigated.

For instance, the method has been applied to the theoretical study of the dielectric behaviour of systems whose disperse phase is polar only, such as w/o emulsions in the gigacycle frequency range, or polar and conductive at the same time, such as dispersions of ice globules in the kilocycle frequency range, (5). In particular, it has been shown, (25), that the complex permittivity of emulsions or dispersions whose disperse phase exhibits a Debye type dielectric relaxation while the continuous one is non absorbing can be represented by a Cole-Cole formula for any value of Φ. Using this result in combination with that reported above concerning the dielectric behaviour of w/o emulsions, Clausse and Lachaise, (26), (27), built a theoretical model for the dielectric study of the monothermal freezing of supercooled water droplets dispersed within emulsions. The application of this model to the relevant experimental data led to the conclusion that the time distribution of the breakdowns of supercooling of the individual water droplets dispersed within an emulsion held at a subzero temperature is of the gaussian type, the mean lifetime of the droplets in the supercooled state being the most probable one as well, (28), (29).

In an attempt to explain the dielectric behaviour of dispersions of ice globules obtained from the breakdown of supercooling of emulsions of water, the method has been extended to dispersions of particles covered with shells, with satisfactory results,

(B. Lagourette, Thesis, 1976, to be published). The method has proved itself to be effective as well when applied to the study of the dielectric and conductive properties of pure or doped ice crystals which were assumed to be representable by a matrix containing inclusions of ionic impurities, (30), (31), (32), (33), (34).

V. REFERENCES

1. Sillars, R.W., J. Inst. Elect. Engrs. (London), 80, 378 (1937).
2. Dryden, J.S., and Meakins, R.J., Proc. Phys. Soc., 66 B, 427 (1953).
3. Hamon, B.V., Austral. J. Phys., 6, 304 (1953).
4. Hanai, T., in "Emulsion Science" (P. Sherman, Ed.) Chapt. 5, p. 353. Academic Press, London, 1958.
5. Clausse, M., Thesis, Université de Pau, France, 1971, B.S. CNRS, 32 160 17334.
6. Lafargue, C., Clausse, M., and Lachaise, J., C.R. Acad. Sci., 274 B, 540 (1972).
7. Beek (van), L.K.H., in "Progress in Dielectrics" (J.B. Birks, Ed.), vol. 7, p. 69 Heywood, London, 1967.
8. Lafargue, C., and Babin, L., Proc. of the 12th Colloq. AMPERE (Bordeaux, France), 12, p. 374, 1963.
9. Chapman, I.D., J. Phys. Chem. (USA), 721, 33 (1963).
10. Hanai, T., Kolloid-Z., 177, 57 (1961).
11. Hanai, T., Kolloid-Z., 171, 23 (1961).
12. Hanai, T., Kolloid-Z., 175, 61 (1961).
13. Bruggeman, D.A.G., Annln. Phys., 24, 636 (1935).
14. Wagner, K.W., Arch. Elektrotech., 2, 371 (1914).
15. Dukhin, S.S., in "Surface and Colloid Science" (E. Matijevic, Ed.) Vol. 3, p. 83, Wiley Interscience, New-York, 1971.
16. Hanai, T., Koizumi, N., and Gotoh, R., Kolloid-Z. 167, 41 (1959).
17. Clausse, M., C.R. Acad. Sci., 277 B, 261 (1973).
18. Clausse, M., C.R. Acad. Sci., 274 B, 649 (1972).
19. Dejon, P., and Nickel, K., in "Constructive Aspects of the Fundamental Theorem of Algebra" (P. Dejon and P. Henrici, Eds.), Wiley and Sons, New-York, 1969.
20. Hasted, J.B., in "Prògress in Dielectrics" (J.B. Birks and J. Hart, Eds.) Vol. 3, p. 101, Wiley and Sons, New-York, 1961.

21. Hippel (von) , A. , "Les Diélectriques et leurs applications",
Dunod , Paris , 1961 .
22. Clausse, M. , C. R. Acad. Sci. , 274 B , 887 (1972) .
23. Clausse, M. , and Sherman, P. , C. R. Acad. Sci. , 179 C ,
919 (1974) .
24. Clausse, M. , Sherman, P. , and Sheppard, R. J. , J. Colloid
Interface Sci. (1976) (in print) .
25. Clausse, M. , C. R. Acad. Sci. , 275 B , 427 (1972) .
26. Clausse, M. , and Lachaise, J. , C. R. Acad. Sci. , 275 B ,
797 (1972) .
27. Lachaise, J. , and Clausse, M. , C. R. Acad. Sci. , 276 B ,
287 (1973) .
28. Lachaise, J. , and Clausse, M. , J. Phys. D : Appli. Phys. ,
8 , 1227 (1975) .
29. Lafargue, C. , Lachaise, J. , and Clausse, M. , Proc. of the
8th I. C. P. S. (Giens , France) , Vol. I , p. 492 , 1975 .
30. Boned, C. , C. R. Acad. Sci. , 275 B , 801 (1972) .
31. Lafargue, C. , and Boned, C. , C. R. Acad. Sci. , 276 B ,
315 (1973) .
32. Boned, C. , C. R. Acad. Sci. , 276 B , 539 (1973) .
33. Boned, C. , C. R. Acad. Sci. , 282 B , 125 (1976) .
34. Boned, C. , J. Physique , 37 , 165 (1976) .

VI . ANNEX : *HANAI* ROUTINE

```
SUBRØUTINE  HANAI  (E1 , E2 , PHI , E)
REAL*8 PHI, DELR, RHØ1, RHØ2, ALP1, ALP2, C1(2), C2(2)
CØMPLEX*16  E1, E2, E(3), A, B, C, B1, B2, Z1, Z2, DEL
EQUIVALENCE (DEL, DELR), (B1, C1(1)), (B2, C2(1))
A  =  E1  -  E2
C  =  DCMPLX((1. DO-PHI)**3/3. DO, O. DO)/E2
B  =  C*A*A*A/E1/E1
DEL  =  CDSQRT(B + (2. 25DO, O. DO))
IF(DELR. LT. O. ) DEL  =  - DEL
B1  =  (-1. 5DO, O. DO)-DEL
B2  =  -B/B1
RHØ1  =  CDABS(B1)**0. 3333333333333333DO
IF(C1(1)) 1, 2, 3
1  ALP1  =  (DATAN(C1(2)/C1(1))+3. 141592653589793DO)/3. DO
GØ TØ  4
```

```
3   ALP1 = DATAN(C1(2)/C1(1))/3. DO
    GØ TØ 4
2   ALP1 = 0.5235987755982988DO
    IF(C1(2)) 5, 4, 4
5   ALP1 = - ALP1
4   Z1 = DCMPLX(RHØ1*DCØS(ALP1), RHØ1*DSIN(ALP1))
    RHØ2 = CDABS(B2)**0.3333333333333333DO
    IF(C2(1)) 6, 7, 8
6   ALP2 = (DATAN(C2(2)/C2(1)) + 3.141592653589793DO)/3. DO
    GØ TØ 9
8   ALP2 = DATAN(C2(2)/C2(1))/3. DO
    GØ TØ 9
7   ALP2 = 0.5235987755982988DO
    IF(C2(2)) 10, 9, 9
10  ALP2 = - ALP2
9   DØ 11 KSØL = 1, 3
    ALP2 = ALP2 + 2.094395102393195DO
    Z2 = DCMPLX(RHØ2*DCØS(ALP2), RHØ2*DSIN(ALP2))
11  E(KSØL) = ((1. DO, O. DO) - Z1*Z2*(Z1 + Z2))*E1
    RETURN
    END
```

Acknowledgements

The authors wish to express their thanks to Miss G.Trouilh and Mrs.M.Anglichau for the great care they have taken during the typing of this paper and to Mrs.M.Bernade and Mr.C.Ceresuela who prepared the illustrations.

DIELECTRIC   STUDY
OF
WATER–IN–HEXADECANE   MICROEMULSIONS.

A   PRELIMINARY   REPORT.

M. Clausse* , R.J. Sheppard* , C. Boned* , C.G. Essex*

*Laboratoire de Thermodynamique
Institut Universitaire de Recherche Scientifique
Université de Pau et des Pays de l'Adour
B.P. 523   "Pau-Université"
64010   Pau   (France)

*Dielectrics Group
Physics Department
Queen Elizabeth College
University of London
Campden Hill Road
Kensington, London W8 7AH,   (England)

ABSTRACT

        Microemulsions of the   "water–in–oil" type were ob-
tained from water and a mixture of 59% (wt/wt) hexadecane
(oil phase), 15% (wt/wt) potassium oleate (surfactant) and
25% (wt/wt) hexanol (cosurfactant). The microemulsions were
prepared at room temperature by shaking gently both phases
together in a glass bottle. It has been possible to obtain
stable transparent microemulsions for mass fractions p of
water as high as a critical value p_c (p_c # 0.32) above which
turbidity and instability was observed. The complex permitti-
vity $\varepsilon^* = \varepsilon' - j\varepsilon''$ of transparent microemulsions was studied
over the frequency range [100 kHz – 1000 MHz] . At a given
frequency, it was found that both ε' and ε'' increase
sharply when the mass fraction p of water approaches the
critical value p_c . A dielectric absorption was observed for
microemulsions with values of p close to p_c while none
could be put into evidence for lower values of p .

I . INTRODUCTION

Heterogeneous systems of high stability can be obtained by mixing, within a definite range of proportions, two normally non miscible liquids in presence of a suitable combination of surface active agents. Most of such systems involve water, an oil type liquid such as benzene or hexadecane, a surfactant of the ionic type, an alkali metal soap for instance, and a cosurfactant, often an alcohol that may be somewhat miscible with water or with the oil type liquid.

After having thoroughly investigated the structure of these fluid optically isotropic transparent systems by using different methods such as X‑rays and light scattering, ultracentrifugation, electron microscopy, rheology and NMR, Schulman and his collaborators, (1) , (2) , (3) , (4) , (5) , (6) , (7) , (8) , concluded that they consist of dispersions of uniform spherical droplets less than 1000 Å in diameter of either water or "oil" in the respective continuous phase, and consequently called them microemulsions.

The formation of the so-called microemulsions, either of the water-in-oil (w/o) or of the oil-in-water (o/w) type, occurs spontaneously or requires only a gentle shaking, in contrast with ordinary emulsions that need most often a vigorous homogenization. To explain this spontaneous emulsification process, Schulman and his co-workers, (8) , (9) , suggested that the surfactant and cosurfactant mixed in the right ratio give rise to a metastable negative interfacial tension so that a spontaneous breakdown of the water-oil interface occurs that leads to the formation of minute droplets ; then the resulting interfacial area increase develops until a zero net interfacial tension is reached. In addition, interactions between the "oil" phase and the cosurfactant are required so that the "oil" phase interpenetrate and associate with the complex interfacial film surrounding the dispersed droplets. These views have been endorsed by other authors, for some additions or variations , (10) , (11) , (12) , (13) , (14) , (15) , (16) , (17) .

A different opinion has been expressed by Ekwall , Mandell and Fontell, (18) , who stressed the similarity between Schulman's so-called microemulsions and isotropic micellar solutions obtained from ternary systems involving sodium di-2-ethylhexylsulfosuccinate, water and p-xylene or caprylic acid or n-decanol and concluded that "the term microemulsion is a misnomer" that should not be used in connection with transparent w/o or o/w systems. Their views have been supported by Shinoda, Kunieda and Friberg, (19) , (20). These authors pointed out that the transition to an ordinary emulsion occuring when an excess of water or "oil", depending on the case, is added to a microemulsion is similar to that occuring when a micellar solution is concerned and remarked that many micellar solutions are not different from microemulsions, regarding their composition and components proportions. Consequently, they proposed to call Schulman's microemulsions "swollen micellar solutions", what is very akin to the term "micellar emulsions" used by Adamson, (21) .

It could be possible to derive at least a partial answer to the problem from a study of the dielectric behaviour of microemulsion systems. The dielectric properties of emulsions have been submitted by many authors to thorough investigations, either experimental of theoretical, from which reliable general conclusions have been established, (22) , (23) , (24) . For instance, it has been shown experimentally that emulsions of water in non polar oil phases exhibit striking dielectric relaxations in the kilocycle frequency range, (22) , (24) , (25) , contrary to emulsions of non polar oil phases in water , (22) , (24) . These experimental results are consistent with the conclusions derived from a theoretical analysis of the dielectric behaviour of emulsion systems, (24) , (26) , (27) , (28). As concerns micellar systems or microemulsions, only a few attempts have been made to study their dielectric properties , (29) , (30) , (31) , (32) , although it was suggested by O'Konski, (33) , some fifteen years ago, that dielectric experiments could be of interest in investigating their physico-chemical properties.

II. MATERIALS AND METHODS

A. Preparation of samples

Water-hexadecane systems were prepared from distilled water and Baker[1] "very pure" n-hexadecane , using Fluka[2] dried potassium oleate as the surfactant and "Baker grade" 1-hexanol as the cosurfactant.

---

1 : J.T. Baker Chemicals N.V., Holland.
2 : A.G. Fluka, Switzerland

59 % (wt/wt) hexadecane, 25 % (wt/wt) hexanol and 15 % (wt/wt) potassium oleate were mixed together for 10 minutes at 50°C by means of a magnetic agitator, 1 % (wt/wt) distilled water being added to facilitate the obtention of a clear liquid phase of low viscosity. Amounts of this mixture and of water were weighed and mixed together in a test tube, only a gentle manual shaking being required to achieve the emulsification of both phases and the formation of a w/o type disperse system, for mass fractions of water less than 0.32 .

The values of p , the mass fraction of water ($0 \leqslant p \leqslant 1$), were determined within ± 0.002 from weight measurements made by means of a precision balance, the 1 % amount of water already present in the mixture of hexadecane and surface active agents being taken into account.

B. Dielectric measurements

Up to 5 MHz, the conductivities and relative real permittivities of the samples were measured by means of a parallel-plate condenser cell connected to a Wayne Kerr B 201 bridge used in conjunction with a Wayne Kerr SR 268 oscillator-detector unit, and following a method that has been described elsewhere , (34) , (35) , (32) . Over the frequency range [1 MHz - 100 MHz] , that overlaps the former one, a Boonton 33 A admittance bridge was used. The design and features of the cells, (one for frequencies up to 50 MHz , the other one for the 100 MHz frequency), have been given in full details in a recent paper from Essex et al. , (36) . The results at frequencies of 400 MHz and above were taken with an automated coaxial line system that has been designed by Sheppard and Grant , (37) , (38) .

ε' and ε'' , respectively real and imaginary part of the relative complex permittivity ε^* , were measured to ± 0.1 when using the Wayne Kerr apparatus and with about twice this uncertainty when using the Boonton bridge. The results taken with the automated coaxial line system were determined with an average uncertainty of 5 % for ε' and of 10 % for ε'' .

III. EXPERIMENTAL RESULTS

Following the procedure described in the preceding section, it was possible to obtain optically isotropic transparent fluid disperse systems, for values of the mass fraction p of water up to a critical value p_c which was found to be very close to 0.32. Usual dilution tests showed these systems to be of the w/o type. Held at room temperature during several weeks, they did not exhibit any noticeable alteration. For mass fractions of water slightly greater than 0.32, turbid unstable systems were obtained that break rapidly into two phases of which one looks like being of the previous type. Further increases of the water content lead to the formation of slightly turbid gels of high viscosity, then to optically isotropic transparent fluid disperse systems of the o/w type, and at least to milky fluid emulsions of the o/w type. For some details, these observations are consistent with results of previous works such as those of Shah et al., (16), and Rosano, (17).

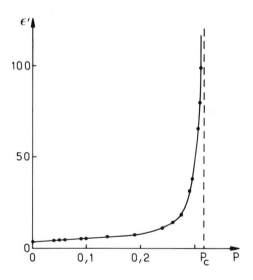

Fig.1. Variations of ϵ' versus p, at 1 MHz. (The dashed vertical straight line indicates the transition from transparency to turbidity).

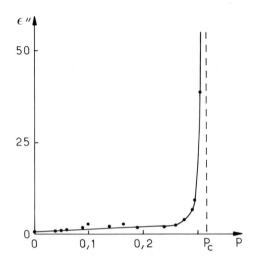

*Fig. 2.   Variations of ε" versus p , at 1 MHz.
(The dashed vertical straight line indicates the transition
from transparency to turbidity).*

     Figure 1   shows the variations of ε' at 1 MHz   when the mass fraction p of water increases from zero to the critical value p_c , and figure 2 those of ε'' under the same conditions. In both cases the diagram is a smooth curve with a sharply ascending branch as p approaches p_c .

     The values of ε' and ε'' were found to remain unchanged after a few days preservation of the samples at room temperature. On the contrary, tentative investigations of the dielectric behaviour of systems with mass fractions of water slightly greater than p_c revealed a change in the values of ε' and ε'' during the lapse of time necessary to carry out the experiments. This is consistent with the visual observation reported in the first paragraph of this section concerning the instability of these systems that break rapidly into two phases of which one looks like being of the subcritical type. A study of some of such phases readily proved that their dielectric properties are stable with time and similar to those of subcritical systems with mass fractions of water close to p_c . As for turbid gels, transparent o/w type systems and milky o/w type emulsions, because of their great conductivity, no reliable results could be established with the apparatus used, apart that the conductivity increases as the water content increases , what is qualitatively satisfactory.

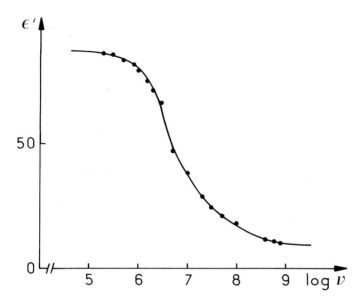

Fig. 3. Variations of ϵ' versus frequency, for $p = 0.308$. (The frequency ν is given in Hz).

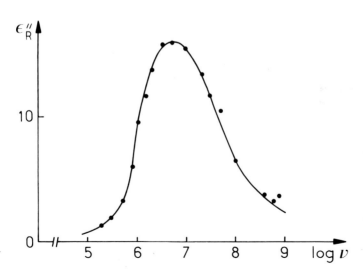

Fig. 4. Variations of ϵ''_R versus frequency, for $p = 0.308$. (The frequency ν is given in Hz).

The investigation of the dielectric behaviour of water-hexadecane systems with values of p lower than p_c proved that they exhibit striking dielectric relaxations, for values of p close to p_c . As an example, figures 3 and 4 show the features of the dispersion and absorption diagrams of the dielectric relaxation observed for $p = 0.308$. At each frequency the value of the relaxation loss factor ε''_R was obtained by subtracting from the value of the global loss factor ε'' that of the conduction loss factor ε''_C deduced from low frequency conductance measurements .

No relaxation phenomena could be put into evidence for values of p lower than approximately 0.20 .

IV. DISCUSSIONS

As reported by several authors, (22) , (23) , (24) , (25) , w/o type emulsions involving non polar oil phases exhibit along with a conduction absorption a dielectric relaxation of the Cole-Cole type related to interfacial polarization phenomena, whatever be the proportion of the disperse aqueous phase. This behaviour can be satisfactorily explained from the "spherical dispersion" model and the formula proposed by Hanai (22) to describe it,

$$\left(\frac{\varepsilon^*_1 - \varepsilon^*_2}{\varepsilon^*_1 - \varepsilon^*}\right)^3 \frac{\varepsilon^*}{\varepsilon^*_2} = \frac{1}{(1-\Phi)^3} \qquad [1]$$

where ε^*_1 , ε^*_2 and ε^* represent respectively the complex permittivities of the disperse phase, the continuous phase and the emulsion , Φ being the volume fraction of the disperse phase . From computer calculations, Clausse , (24) , (26) , (28) , (39) , proved that, in the case of w/o type emulsions, the complex permittivity ε^* given by [1] can be written like a Cole-Cole type formula with an additional ohmic term

$$\varepsilon^* = \varepsilon_h + \frac{\varepsilon_l - \varepsilon_h}{1 + (j\,\omega\tau)^{(1-\alpha)}} + \frac{\chi_l}{j\,\omega\,\varepsilon_o} \qquad [2]$$

ε_l is the low frequency limiting permittivity of the emulsion ,

ε_h the high frequency one and X_l the steady conductivity (equal to zero in the case of non conductive oil phases), ε_o being the absolute permittivity of free space, ω the angular frequency of the applied sinudoïdal electrical field and $j = \sqrt{-1}$.

From the results reported in the preceding section, it appears that the dielectric behaviour of transparent water-in-hexadecane systems differs from that of ordinary w/o emulsions. While no relaxation phenomena are noticeable for values of the mass fraction p of water smaller than about 0.20 , the features of the dielectric relaxation found for greater values of p do not fit Hanai's formula. In particular, for values of p close to the critical value p_c , the dielectric increment $\varepsilon_l - \varepsilon_h$ is much greater than that which can be predicted from formula [2] . Moreover , the transparent water-in-hexadecane systems are fairly conductive in spite of the low conductivity exhibited by the mixture of hexadecane, hexanol and potassium oleate.

A tentative explanation of the discrepancy existing between the actual behaviour of transparent water-in-hexadecane systems and the theoretical model that has proved suitable for ordinary emulsions could be the following one. For lower water proportions, i.e for values of p smaller than about 0.20, most of the water molecules are engaged into complexes aggregating surfactant molecules. So, the disperse water can be considered as being solubilized within the systems that consequently should be called solutions of hydrated micelles rather than microemulsions. For mass fractions of water greater than about 0.20, most of the water molecules gather into minute droplets surrounded by an interfacial film and dispersed within the systems that can then be defined as w/o type microemulsions, these microemulsions exhibiting a dielectric relaxation related to interfacial polarization phenomena as ordinary emulsions do. In both situations, the fact that the water-in-hexadecane systems are fairly conductive seems to indicate that part of the water molecules remain "free".

The conclusions of this study are consistent with those derived by Prince (40) who examined the molecular interactions taking place in the interphase of very small droplets and concluded that, in the case of water-in-hexadecane systems, a distinction based on aggregate size can be made between micellar solutions and microemulsions. To ascertain these results, a more detailed study is to be undertaken which concerns the investigation of the dielectric behaviour of water hexadecane systems in connection with the quantitative determination of their phase equilibria diagram.

V. REFERENCES

1. Hoar, T.P., and Schulman, J.H., <u>Nature</u> (London) , 152 , 102 (1943).
2. Schulman, J.H., and Mc Roberts, T.S., <u>Trans. Faraday Soc.</u> , 42 B , 165 (1946).
3. Schulman, J.H., and Riley, D.P., <u>J. Colloid Sci.</u> , 3 , 383 (1948).
4. Schulman, J.H., and Friend, J.A., <u>J. Colloid Sci.</u> , 4 , 497 (1949).
5. Bowcott, J.E., and Schulman, J.H., <u>Z. Elektrochem.</u> , 59 , Heft 4 , 283 (1955).
6. Schulman, J.H., Stoeckenius, W., and Prince, L.M. , <u>J. Phys. Chem.</u> , 63 , 1677 (1959).
7. Stoeckenius, W., Schulman, J.H., and Prince, L.M. , <u>Kolloid-Z.</u> , 169 , 170 (1960).
8. Cooke, C.E., and Schulman, J.H., "Proc. 2nd Scandinavian Symp. Surface Activity" , p. 231 , Stockholm , 1965 .
9. Schulman, J.H., and Montagne, J.B., <u>Ann. N.Y. Acad. Sci.</u> , 92 , 366 (1961).
10. Osipow, L.I., <u>J. Soc. Cosmetic Chemists</u> , 14 , 277 (1963).
11. Prince, L.M. <u>J. Colloid Interface Sci.</u> , 23 , 165 (1967).
12. Prince, L.M. <u>J. Colloid Interface Sci.</u> , 2 , 216 (1969).
13. Prince, L.M. <u>J. Soc. Cosmetic Chemists</u> , 21 , 193 (1970).
14. Gerbacia, W., and Rosano, H.L., <u>J. Colloid Interface Sci.</u> , 44 , 242 (1973).
15. Shah, D.O., and Hamlin, R.M., <u>Science</u> , 3970 , 483 (1971).
16. Shah, D.O., Tamjeedi, L.W., Falco, J.W., and Walker, R.D., <u>AIChE Journal</u> , 18 , 1116 (1972).
17. Rosano, H.L., <u>J. Soc. Cosmetic Chemists</u>, 25 , 609 , (1974).
18. Ekwall, P., Mandell, L. and Fontell, K., <u>J. Colloid Interface Sci.</u> , 33 , 215 (1970).
19. Shinoda, K., and Kunieda, H. <u>J. Colloid Interface Sci.</u> , 42 , 381 (1973).
20. Shinoda, K., and Friberg, S., <u>Advances in Colloid and Interface Sci.</u> , 4 , 281 (1975).
21. Adamson, A.W., <u>J. Colloid Interface Sci.</u> , 29 , 261, (1969).
22. Hanai, T. in "Emulsion Science" (P. Sherman, Ed.) , Chapt. 5 , p. 353 , Academic Press , London , 1968 .
23. Dukhin, S.S., in "Surface and Colloid Science" (E. Matijevic, Ed.) Vol. 3 , p. 83 , Wiley Interscience, New-York , 1971.

24. Clausse, M., Thesis, University of Pau, France, 1971
    B.S. CNRS, 32 160 17334.
25. Lafargue, C., Clausse, M., and Lachaise, J.,
    C.R. Acad. Sci., 274 B, 540 (1972).
26. Clausse, M., C.R. Acad. Sci., 274 B, 887 (1972).
27. Clausse, M., Colloid and Polymer Sci., 253, 1020 (1975).
28. Clausse, M., and Royer, R., Proc. of the I.C.C.S.
    (Puerto-Rico, U.S.A.), 1976.
29. Eicke, H.F., and Shepherd, J.C.W., Helvetica Chimica
    Acta, 57, 1951 (1974).
30. Beard, R.B., Mc Master, T.F., and Takashima, S.,
    J. Colloid Interface Sci., 48, 92 (1974).
31. Clausse, M., and Sherman, P., C.R. Acad. Sci., 279 C,
    919 (1974).
32. Clausse, M., Sherman, P., and Sheppard, R.J.,
    J. Colloid Interface Sci., 1976 (in press).
33. O'Konski, C.T., J. Phys. Chem., 64, 605 (1960).
34. South, G.P., Ph. D. Thesis, University of London, 1970.
35. Grant, E.H., South, G.P., Takashima, S. and Ichimura, H.,
    Biochem. J., 122, 691 (1971).
36. Essex, C.G., South, G.P., Sheppard, R.J., and Grant, E.H.,
    J. Phys. E : Sci. Inst., 8, 385 (1975).
37. Sheppard, R.J., J. Phys. D : Appl. Phys., 5, 1576 (1972).
38. Sheppard, R.J., and Grant, E.H., J. Phys. E : Sci. Inst.,
    5, 1208 (1972).
39. Clausse, M., C.R. Acad. Sci., 274 B, 649 (1972).
40. Prince, L.M., J. Colloid Interface Sci., 52, 182 (1975).

Acknowledgements

     The authors wish to express their thanks to Miss G.Trouilh and Mrs.M.Anglichau for the great care they have taken during the typing of this paper and to Mrs.M.Bernade and Mr.C.Ceresuela who prepared the illustrations.

THE STABILITY OF MICROEMULSION SYSTEMS

W. Gerbacia
Chevron Oil Field Research Company

H. L. Rosano and J. H. Whittam[1]
The City University of New York

Microemulsions are considered as thermodynamically unstable systems and distinct from thermodynamically stable swollen micellar solutions. The stability of microemulsion systems is assumed to arise from the presence of the surfactant-cosurfactant-solvent film surrounding the oil droplets. An osmotic effect is assumed to be operable and the energy barriers to coalescence are calculated by using regular solution concepts to describe the interactions in the interphase between two droplets in close approach. The energy barriers calculated by this model are seen to be high enough to explain the long term stability of these systems. The separation of one microemulsion system was observed for verification.

I. INTRODUCTION

Mixtures of an oil and water in the presence of a surface active agent usually form coarse emulsions which are optically opaque or nearly so, and usually separate on standing. In some cases, transparent mixtures of oil and water can be prepared with the proper combination of surface active materials.[1]

These systems which have been labeled microemulsions,[2] micellar emulsions[3] or swollen micellar solutions,[4] are considered to be thermodynamically stable.[3-7] The mixtures are transparent due to the small particle size of the dispersed components, which is usually less than 1400 A. Several investigators of the shape of the droplets have concluded that the droplets are spherical in shape.[8-10]

Recently, evidence has been presented that some transparent mixtures of this type are not thermodynamically stable.[11] This conclusion was based on the differences in the observed results when the order of mixing of the compo-

---

[1]Presently with The Gillette Company.

nents was changed. It was proposed that there may be two inherently different classes of transparent oil and water mixtures.[12] One class consists of thermodynamically stable micellar systems which form spontaneously, and the other consists of kinetically stable dispersions which can be stable and transparent for long periods of time. The latter systems will be labeled microemulsions here and the former, mixed micellar solutions[12] or swollen micellar solutions.

This investigation presents an initial attempt to account for the long term stability of oil external microemulsion systems. The formation of these systems has been discussed elsewhere.[11] It is assumed that the most stable state is represented by the separate phases and, that once dispersed, there is an energy barrier to coalescence of the dispersed droplets. The barrier originates because of the interfacial film surrounding the dispersed phase. A model is used to calculate the size of the energy barrier.

II. THEORY

In order to calculate the energy barrier to coalescence between two stabilized microemulsion droplets it is necessary to describe the interaction between the surface active films surrounding the two droplets. If there is an increase in free energy as the two films contact, there will be a net repulsion between the two droplets. Bagchi[13] described the interaction between colloid particles of AgI stabilized by polyvinyl alcohol by using a Flory-Huggins approach. The repulsion arose through an osmotic effect[14,15] or a volume restriction effect.[14] The repulsion energy was partially offset by the van der Waals attractive potential.

A somewhat different approach is taken here. The interfacial monolayer consists of the cosurfactant, surfactant and associated solvent. For an oil external microemulsion the solvent is the oil phase. The cosurfactant and surfactant are assumed to be anchored at the interface. As two droplets approach each other, the interfacial films surrounding the drops begin to mix as depicted in Figures la and b.

As the monolayers mix, solvent is displaced from the interfacial layers, causing an increase in free energy as the film becomes locally more concentrated in surfactant and cosurfactant. This is termed the osmotic effect.[13]

The energy change in the interfacial film is accounted for by using Scatchard's cohesive energy equations for three component systems,[16,17]

$$-E_M = (X_A \bar{V}_{AL} + X_S \bar{V}_{SL} + X_O \bar{V}_O) (C_{AA}\phi_{AL}^2 + C_{SS}\phi_{SL}^2$$

$$+ C_{OO}\phi_O^2 + 2C_{AS}\phi_{AL}\phi_{SL} + 2C_{AO}\phi_{AL}\phi_O + 2C_{SO}\phi_{SL}\phi_O) \quad (1)$$

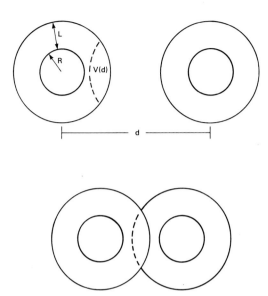

Fig. 1a.   Interaction between two microemulsion droplets,
$L \leq d \leq 2L$.

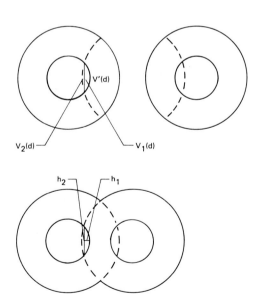

Fig. 1b.   Interaction between two microemulsion droplets,
$0 < d < L$.

(A list of symbols is given in Appendix I.) This energy term only considers interactions among the hydrophobic components. The aqueous phase is assumed not to interact with the oil phase. The change in energy on mixing is given by,

$$\Delta E_M = E_{Mi} - E_{Mf}$$ (2)

where E_{Mi} and E_{Mf} are calculated from equation 1 for the initial and final compositions, respectively.

In this model, volume changes and orientational effects on mixing are not considered. The change in free energy on mixing is therefore given by

$$\Delta G_M = \Delta E_M - T\Delta S_M$$ (3)

where ΔS_M is the regular solution entropy of mixing,

$$\Delta S_M = - RT \left[(x_O^f \ln \phi_O^f + x_A^f \ln \phi_{AL}^f + x_S^f \ln \phi_{SL}^f) \right.$$
$$\left. - (x_O^i \ln \phi_O^i + x_A^i \ln \phi_{AL}^i + x_S^i \ln \phi_{SL}^i) \right]$$ (4)

The only other energy term considered is the van der Waals interaction energy. In this model it is assumed that the water droplets are interacting through a medium composed of the surfactant and cosurfactant chains with the associated oil. The van der Waals energy for this interaction is calculated by,[18]

$$\Delta G_v = - \frac{N}{12} (A_3^{1/2} - A_1^{1/2})^2 \left[4R^2/(d^2 + 4Rd) \right.$$
$$\left. + 4R^2/(d^2 + 4Rd + 2R^2) + 2\ln (d^2 + 4Rd/(d^2 + 4Rd + 4R^2)) \right]$$ (5)

where the A's are the Hamaker constants. A_1 is the Hamaker constant for water, taken as 4.15×10^{-12} ergs.[18] The Hamaker constant for the interfacial film is composed of contributions from the three components. It is assumed to be a molar average of the contributions from individual components under the initial conditions. A_3 is calculated from,

$$A_3 = x_A^i A_{AL} + x_S^i A_{SL} + x_O^i A_O$$ (6)

It does not vary significantly as the oil is displaced. The individual terms are calculated from the approximation,[18]

$$A = 24 \pi a^2 \gamma$$ (7)

The surface tension of the analogous hydrocarbons are used for the γ's (eg. for a C_{12} amphiphile the surface tension of dodecane was used for γ). The separation distance a is taken as 1.2 A.[18]

The total interaction energy is given as,

$$\Delta G_T = \Delta G_M + \Delta G_\upsilon \qquad (8)$$

If expressions for the mole fractions and volume fractions of the components are known as a function of the separation distance, these terms can be calculated from the change in the interaction volume with distance.[13,19] The equations are given in Appendix II.

III. RESULTS

Figures 2 through 5 show the calculated energy curves for microemulsion systems in which the oil is hexadecane and the cosurfactant has a linear five carbon chain. The ratio of cosurfactant to surfactant (I) had been determined previously[11] by a titration procedure. The values of all parameters used were for T = 25°C.

Figure 2 shows the effect of varying the dispersed phase volume while keeping the droplet radius at 25 A, with a surfactant chain length of twelve carbons (L = 30 A), and a constant amount of surfactant. The small change in dispersed phase volume from 1.0 to 1.44 ml caused a change in the

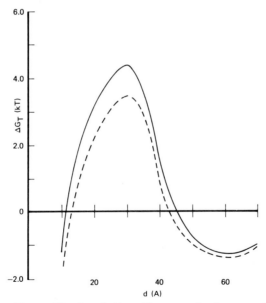

Fig. 2. The effect of the dispersed phase volume on the variation of ΔG_T (V_w), R = 25 A and surfacetant chain length is 12 carbon atoms; (——) V_w = 1.0 (---) V_w = 1.44 ml.

energy maximum of about +2 kT. There was approximately a 0.9 kT difference in the calculated maxima for the two cases. The system with the smaller dispersed phase volume had the higher energy barrier since the concentration of surface active components in the interphase was higher in that system.

In Figure 3 it can be seen that increasing the concentration of surfactant from 0.0017 moles to 0.0032 moles caused an increase in the energy maxima by 1.3 kT. The energy barrier did not change greatly as the droplet radius was varied from 25 to 35 A as shown in Figure 4. Calculations could not be carried out at radii greater than 40 A without displacing some of the surfactant or cosurfactant from the interface. Under these conditions the volume of the chains of the surface active components was greater than that available to the system.

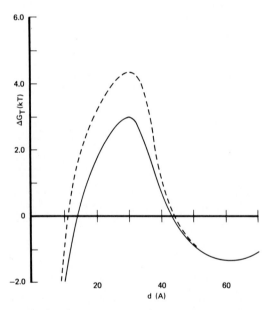

Fig. 3. Variation of ΔG_T with the number of moles of surfactant, R = 25 A and surfactant chain length is 12 carbon atoms; (- - -) n_S = 0.0032 moles, (——) n_S = 0.0017 moles.

Figure 5 shows the effect of changing the surfactant chain length and cosurfactant concentration in the interphase. The shorter chain length surfactant had a higher energy barrier. This was mainly due to the much larger amount of cosurfactant at the interface as determined by titration.

An oil external microemulsion system was prepared and its stability was observed for almost two years. The aqueous phase contained a red dye, rose bengal, in order to make any sedimentation and separation visible.

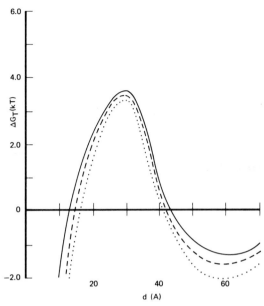

Fig. 4. The effect of R on the variation of ΔG_T, $V_w = 1.44$ ml, C_{12}; (——) R = 25 A, (---) R = 30 A, (····) R = 35 A.

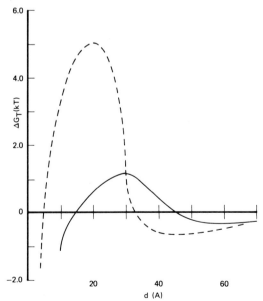

Fig. 5. The effect of surfactant chain length on variation of ΔG_T, R = 10 A, and $V_w = 1.44$ ml; (——) C_{12}, I = 7.5; (---) C_8. I = 24.

The microemulsion consisted of: 2.75 gm of sodium dodecyl sulfate (A & S Corporation, New York), 3.00 ml of a 1% rose bengal (Fischer Scientific Company, New Jersey) solution in distilled water, 120 ml of n-hexadecane (Baker Chemical Company, New Jersey), and 19.3 ml of 1-pentanol (Baker Chemical Company, New Jersey).

After 6-1/2 months sedimentation was observed. The red color had moved several centimeters down the tube. The original fluid height was 17.3 cm. After approximately 20 months the phases had separated completely. The system redispersed upon shaking, but was translucent. These experiments are continuing.

IV. DISCUSSION

In our model, discussed here, the energy barriers are due to the change in concentration of the interfacial film when solvent is displaced. When such a change in composition increases the free energy, the drops will be stabilized against coalescence. This type of model points out the significance of the solvent in stabilizing dispersed systems of this type.[13]

The demixing of the interfacial film can explain the stability of microemulsion systems. The energy barriers that have been calculated are significant and imply long term stability for these dispersed systems.

The minima in the free energy functions are not large under the conditions assumed in the calculations. Using the criteria elaborated by Hesselink, et al,[15] and Bagchi,[13] none of the systems would flocculate irreversibly. Some of the systems with minima greater than -1 kT should be indefinitely stable.[13,15]

The separation of the transparent microemulsion system demonstrated that some of these systems, at least, can be inherently unstable under certain conditions. Since the particle size was not known as a function of time, we were not able to calculate the actual coalescence energy barrier. If it is assumed that the droplets did not start settling out until they reach a diameter of 1500 A, the calculated energy barrier would be about 7 kT. This value was calculated using the equation for coalescence given by Davies and Rideal,[20]

$$V_t - V_O = \frac{4\pi \ kT}{3\eta_O} \ (1 - \phi_O) \ t \ e^{-\Delta G_T/kT} \tag{9}$$

where the initial radius was 25 A, and η_O = .033 poises, and the phase volume of the oil, is 0.98 for the microemulsion studied. The 7 kT value is considerably different from the calculated value of about 3.5 kT. However, both calculations can only be regarded as approximate in view of the assumptions made.

This model presents a basis for explaining the long term stability of thermodynamically unstable micellar systems. A more complete understanding will be achieved when orientational effects are considered and after more exhaustive stability measurements are made.

APPENDIX I

X_A, X_S, X_O = Mole fraction of alcohol, surfactant and oil respectively.

$\bar{V}_{AL}$, $\bar{V}_{SL}$, $\bar{V}_O$ = Molar volume of alcohol chains, surfactant chains and oil, respectively.

C_{AA}, C_{SS}, C_{OO} = Cohesive energy density of alcohol chains, surfactant chains and oil, respectively.

C_{AS}, C_{AO}, C_{SO} = Cohesive energy density of alcohol-surfactant chains, alcohol chains - oil, and surfactant chains - oil, respectively; e.g. $C_{AS} = (C_{AA} \cdot C_{SS})^{1/2}$.

ϕ_{AL}, ϕ_{SL}, ϕ_O = Volume fraction of alcohol chains, surfactant chains and oil, respectively.

f, i = Refer to the final state and initial state, respectively.

A_{AL}, A_{SL}, A_O = Hamaker constants for alcohol chains, surfactant chains, and oil respectively.

η_O = Viscosity of the oil phase

I = Ratio of cosurfactant to surfactant.[11]

V_t, V_O = Droplet volume at time t and t=0.

k = Boltzman's constant.

T = Temperature.

t = Time (sec).

APPENDIX II

The initial and final mole fraction for $L \leq d < 2L$ were calculated from;

$$X^i_A = (C \bar{V}_O I)/B$$

$$X^i_S = X^i_A/I$$

$$X^i_O = \phi^i_O/B$$

where

$$B = \phi^i_O + CV_O(I+1)$$

and C is the concentration of surfactant in the interphase;

$$C = \frac{3n_s}{4\pi m \ (3R^2 L + 3L^2 R + L^3)}$$

where n_s is the number of moles of surfactant and m is the number of droplets given by

$$m = \frac{3V_w}{4\pi R^3}$$

The final mole fractions are,

$$X^f_A = 2C \ \bar{V}_O \ I/D$$

$$X^f_S = X^f_A \ /I$$

$$X^f_O = \left[1 - 2(\phi^i_{AL} + \phi^i_{SL})\right] \ /D$$

where

$$D = \left[1 + 2C \ V_O \ (I + 1)\right] - 2 \ (\phi^i_{AL} + \phi^i_{SL})$$

The volume fractions are given by;

$$\phi^i_{AL} = CI \ \bar{V}_{AL}$$

$$\phi^i_{SL} = C \ \bar{V}_{SL}$$

$$\phi^i_O = 1 - (\phi^i_{AL} + \phi^i_{SL})$$

254

$$\phi_{AL}^{f} = 2\phi_{AL}^{i}$$

$$\phi_{SL}^{f} = 2\phi_{SL}^{i}$$

$$\phi_{0}^{f} = 1 - 2\ (\phi_{AL}^{i} + \phi_{SL}^{i})$$

For $0 \leq d < L$ the volume fraction and mole fractions are given by,

$$\phi_{AL}^{f} = 2CI\ \overline{V}_{AL} \cdot \left[v(d) - \left(v_{1}(d) + v_{2}(d)\right)\right] / E$$

$$\phi_{SL}^{f} = \phi_{AL}^{f} / I$$

$$\phi_{0}^{f} = 1 - (\phi_{AL}^{f} + \phi_{SL}^{f})$$

where the initial volume fractions are the same as above and

$$E = v(d) - \left(v_{1}(d) + v_{2}(d)\right)\left[2 - C\ (I\ \overline{V}_{AL} + \overline{V}_{SL})\right]\ ;$$

$$x_{A}^{f} = 2C\ \overline{V}_{0}\ I\ \left[v(d) - \left(v_{1}(d) + v_{2}(d)\right)\right] / F$$

$$x_{S}^{f} = x_{A}^{f} / I$$

$$x_{0}^{f} = v(d)\left[1 - 2C\ (\overline{V}_{SL} + I\ \overline{V}_{AL})\right] + \left[v_{1}(d) + v_{2}(d)\right]$$
$$\cdot \left[3C\ (\overline{V}_{SL} + I\ \overline{V}_{AL}) - 2\right] / F$$

where

$$F = v(d) \circ \left(1 + 2C\ (I + 1)\ V_{0}\right) - 2C\ (\overline{V}_{SL} + I\ \overline{V}_{AL})$$
$$+ \left(v_{1}(d) + v_{2}(d)\right) \cdot \left[3C\ (\overline{V}_{SL} + I\ V_{AL})\right.$$
$$\left. - \left(2 + 2C\ (I + 1)\ V_{0}\right)\right] .$$

The volume elements are given by

$$v(d) = \frac{2}{3} \pi \ (L - d/2)^2 \cdot (3R + 2L + d/2)$$

$$v_1(d) = \frac{\pi}{3} \ (H_1)^2 \cdot (3R - H_1)$$

$$v_2(d) = \frac{\pi}{3} \ (H_2)^2 \cdot (3R + 3L - H_2)$$

$$H_1 = R - \left[\frac{(d + 2R)^2 - (2RL + L^2)}{2(2R + d)} \right]$$

$$H_2 = L - (d + H_1) \quad .$$

V. REFERENCES

1. Hoar, T.P. and Schulman, J.H., Nature, 152, 102 (1943).
2. Stoeckenius, W., Schulman, J.H., and Prince, L.M., Kolloid-Z., 169, 170 (1960).
3. Adamson, A.W., J. Colloid Interface Sci., 29 (2), 261 (1969).
4. Shinoda, K., and Kunieda, H., ibid, 42 (2), 381 (1973).
5. Schulman, J.H. and McRoberts, T.S., Trans. Faraday Soc., 42B, 165 (1946).
6. Gillberg, G., Lehtinen, H., and Friberg, S., J. Colloid Interface Sci., 33 (1), 40 (1970).
7. Attwood, D., Currie, L.R.J., and Elworthy, P.H., ibid., 46 (2), 249 (1974).
8. Schulman, J.H., Stoeckenius, W., and Prince, L.M., J. Phys. Chem., 63, 1677 (1959).
9. Cooke, C.E., Jr. and Schulman, J.H., Second Scand. Symp. Surface Activity, p. 231, 1965.
10. Schulman, J.H. and Riley, D.P., J. Colloid Interface Sci., 3, 383 (1948).
11. Gerbacia, W. and Rosano, H.L., ibid., 44, 242 (1973).
12. Gerbacia, W.E., Rosano, H.L., and Zajac, M., J. Amer. Oil Chem. Soc., 53 (3), 101 (1976).
13. Bagchi, P., J. Colloid Interface Sci., 47 (1), 86 (1974).
14. Fischer, E.W., Kolloid-Z., 160, 120 (1958).
15. Hesselink, F.Th., Vrij, A., and Overbeek, J.Th.G., J. Phys. Chem., 75, 2094 (1971).
16. Scatchard, G., Chem. Rev., 8, 321 (1931).
17. Scatchard, G., Trans. Faraday Soc., 33, 160 (1937).
18. Vold, M.J., J. Colloid Sci., 16, 1 (1961).
19. Fischer, E.W., Kolloid-Z., 160, 120 (1958).
20. Davies, J.T., and Rideal, E.K., "Interfacial Phenomena", p. 366, Academic Press, N.Y., (1963).

RELATION OF STABILITY OF O/W EMULSION TO INTERFACIAL TENSION AND HLB OF EMULSIFIER

Yasukatsu Tamai
Chemical Research Institute of Non-Aqueous Solutions, Tohoku University
Shosuke Ebina and Kameo Hirai
Research Laboratory, Daido Chemical Industries

I.   INTRODUCTION

It is very important technology in the oil industry to prepare stable emulsion of a given oil in water, and also to control the stability of an emulsion, that is, stable in some range of temperature but unstable out of the required range. For this purpose, in general, the HLB value is considered to be a powerful measure which characterizes the ability of the emulsifier.   However, in some cases, especially with mixture of emulsifiers, HLB fails to predict the optimum blending composition.

In this work it is aimed first to examine the relation between HLB and the emulsion stability experimentally by observing the rate of oil separation with the automatic device developed, and second to compare the results with the measured interfacial tension of the corresponding system.   The effect of temperature on the emulsion stability is also studied and discussed.   With the mixed emulsifier, the predominancy is estimated of the components.

II.   EXPERIMENTALS

A.   Materials

1.   Liquid System
As the dispersed phase, non-polar liquid paraffin and partly polar butyl stearate are employed.   These organics are commercial and of reagent grade, and used as received. The continuous phase is water which is distilled in usual way of laboratory.

2. Emulsifier

Polyoxyethylene-nonylphenyl esters are applied for the emulsifier and listed in Table 1, in which the number of ethylene oxide in a molecule is indicated in parentheses.   There

TABLE 1
Nature of the Employed Emulsifier

		Solubility		in			
		Liquid Paraffin		Butyl Stearate		Water	
Emulsifier	HLB	0.1%	1.0%	0.1%	1.0%	0.1%	1.0%
POE(2)NP	5.7	o	o	o	o	x	x
POE(4)NP	8.9	o	o	o	o	o	x
POE(6)NP	10.9	o	x	o	o	o	o
POE(8.5)NP	12.6	x	x	o	o	o	o
POE(13)NP	14.5	x	x	o	o	o	o
POE(16)NP	15.2	x	x	o	x	o	o

HLB varies from 2 to 13, corresponding the number of ethylene oxide.   In some cases, the mixture is also applied and list-

TABLE 2
Nature of the Mixed Emulsifier

Mixing Rate		Solubility		in			
		Paraffin		Stearate		Water	
POE(x)NP:POE(y)NP	HLB	0.1%	1.0%	0.1%	1.0%	0.1%	1.0%
63(2):37(13)	8.9	o	o	o	o	x	x
41(2):59(13)	10.9	o	x	o	o	x	x
46(4):54(8.5)	10.9			x		o	x
22(2):78(13)	12.6	x	x	o	o	o	x

ed in Table 2.   In these Tables HLB values and solubility tendency are shown.   o and x mean soluble perfectly and imperfectly, that is, some portion remains insoluble.

3. Emulsion

Emulsion is made of 90vol% water and 10vol% oil.   Emulsifier is added by 0.1wt% as a whole.   In case emulsifier is not perfectly soluble in oil, it is desolved in water.   As is seen in Table 2, the mixtures of HLB 8.9 and 10.9 are soluble only imperfectly in both paraffin(1.0%) and water(0.1%). Therefore these systems are not examined.

To emulsify the system, 10 minutes agitation is applied

at the rate of 1580rpm with 7 sec pause in every 2 min.   It
is confirmed that after 6 minutes agitation the measured stability is always reproducible.

B.   Oil Separation Measurement

1. *Device*
     The oil separation is observed automatically with the two
electrodes immersed in the emulsion settled in a cylinder.
Fig. 1 shows the electrodes schematically.   The measuring principle is to monitor the cur-
rent between the electrodes.
When the circuit is open,
that is, the platinum electrode is in the oil phase,
the set of pair electrodes
is moved downward by a servo
motor until the circuit is
closed, that is, the both
electrodes come into the
emulsion phase.   The position of the electrodes are
recorded electronically.
The tinny platinum electrode
surrounded by teflon sheet
as shown in Fig. 1 works
very satisfactorily to avoid
the entrainment of oil which
is the cause of inaccuracy.

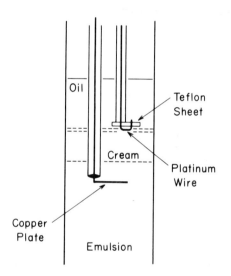

2. *Measurement*
     The prepared emulsion   *Fig.1.   The electrode assem-*
is poured into the measur-   *bly to monitor oil separation*
ing cylinder in which the electrodes are set.   Within 15 min.
about 5% of emulsion becomes creamy to the top layer, which
contains oil particles of more than 10 μm in diameter.   And
then the oil layer is separated from this creamy layer.
     The measurement is repeated at least three times under
the same conditions.   The scattering of the data at 60 min of
measurement is, in most cases, less than ± 20%.

C.   Interfacial Tension Measurement

     Interfacial tension is measured by Wilhelmy method at 25
± 0.5°C.   In this experiment emulsifier is also added by 0.1
wt% in the oil phase, except imperfectly soluble.   Wilhelmy

plate is made of platinum.   The position is kept constant at the interface by controlling electronically the counterbalance force.   It takes more than 30 min. to get equilibrium.

III.   RESULTS

A.   Stability and Interfacial Tension

In Fig. 2 and Fig. 3 both interfacial tension and oil sep-

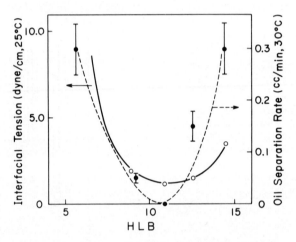

Fig. 2.   *Interfacial tension and oil separation rate of paraffin/water system*

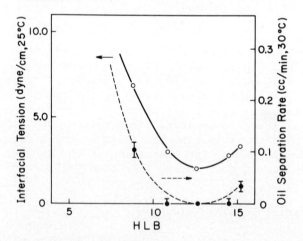

Fig. 3.   *Interfacial tension and oil separation rate of ester/water system*

aration rate are plotted against HLB for the paraffin/water(
single emulsifier) system and for the ester/water(single emulsifier) system.   Maximum, minimum, and median are shown by
stem and circle.   The data fluctuation of the oil separation
rate is very small with the ester emulsion, but is ± 10% with
paraffin emulsion.

B.   Stability and Temperature

In Fig. 4, the curves of the oil separation rate vs HLB

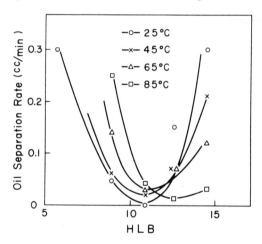

*Fig. 4.   Oil separation rate at different
temperatures of paraffin/water emulsion*

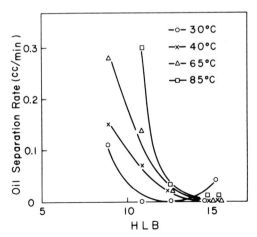

*Fig. 5.   Oil separation rate at different
temperature of ester/water emulsion*

are shown at different temperatures for the paraffin/water (
single emulsifier) system.   Fig. 5 is the same for the ester
/water emulsion.

C.   Mixed Emulsifier

In Fig. 6, the oil separation rate and the interfacial

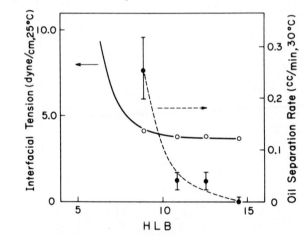

*Fig. 6.   Interfacial tension and oil
separation rate of ester/water(mixed) system*

TABLE 3
Stability of Butyl Stearate/Water Emulsion

Temp.		Oil Separation Rate (cc/min)		
(°C)	HLB	Single	Mixture	
30	5.7	1.0(2)		
	8.9	0.1(4)	0.26(2+13)	
	10.9	0(6)	0.04(2+13)	∿0(4+8.5)
	12.6	0(8.5)	0.04(2+13)	
	14.5	0(13)		
85	5.7	>0.5(2)		
	8.9	0.5(4)	0.03(2+13)	
	10.9	0.3(6)	0.01(2+13)	0.15(4+8.5)
	12.6	0.02(8.5)	0.01(2+13)	
	14.5	0(13)		

tension are plotted for the ester/water system.   The emulsifiers are composed of those of oxyethylene number 2 and 13.

In Table 3, the oil separation rates are list up for single and mixture of (2 + 13) and (4 + 8.5) at two temperatures of 30 85°C.

IV. DISCUSSIONS

A. Stability and Interfacial Tension

It is shown from Figs 2 and 3 that both the oil separation rate curve and the interfacial tension curve have a minimum at the same value of HLB. Furthermore, in Fig. 6 for blended emulsifiers, no marked minimum is observed and, at the same time, the rate curve also shows no minimum. The behaviors of the two curves are quit similar.

Of course the importance of interfacial tension in emulsion is well-known. However, as for the role of interfacial tension in emulsion technology, the data have been regarded as to provide information relating to the efficiency with which the discontinuous phase is dispersed during emulsification (1).

The emulsion stability is indeed ruled by many factors, but the interfacial tension should be one of them, and seems to be a useful measure to get a stable emulsion, if considering that the low tension means energetically stable interface and this, in turn, may suggest a stable structure of the adsorbed emulsifier at the interface.

B. Temperature Shift of Optimum HLB

From Fig. 4, especially, and Fig. 5, it is obvious that the emulsifiers of lower HLB lose the emulsification ability with increasing temperature, and on the contrary, those of higher HLB get the ability. This may be realized by shifting the curve to the higher HLB with increasing temperature.

HLB is expressed by

$$HLB - 7 = \frac{RT}{1680} \ln\left(\frac{C(w)}{C(o)}\right)$$

where $C(w)$ and $C(o)$ are the partition concentration of the emulsifier in water and oil phase, respectively, and T is absolute temperature, R gas constant(2). It may be assumed that the most stable emulsion is given when the emulsifier takes a stable and fitted orientation or configuration at the interface, and this occurs at the definite ratio of $C(w)/C(o)$, because C represents affinity or pulling tendency of emulsifier into the given phase. This assumption is a kind of application of the principle of corresponding state. Then HLB equation becomes

$$HLB - 7 = KT \quad K: \text{constant}$$

Taking for example the case of Fig. 4, the optimum HLB is roughly 11 at 25°C. It is easy by this equation to estimate the optimum HLB at 85°C, that is, 12. For Fig. 5, the optimum HLB at 30°C is 12.5, and estimated is 14. These estimated values are close to the observed HLB at which the oil separation rate is minimum. However, this is just an apploximation and should be refined further.

C. Predominance of the Higher HLB Component

Many researches have been conducted on the predominance of blended emulsifier. Shinoda has reported that the blend of widely different emulsifier is less effective for emulsification (3). Also the algebraical averaging of HLB for the blend has been examined by Shinoda from the measurement of phase inversion temperature (4) and by Ohba from the measurement of optimum HLB and the predominance of the higher HLB component is concluded.

In the present study, the predominancy of the higher component is obvious from Table 3 both in the mixtures of (2 + 13) and of (4 + 8.5) at 30°C and 85°C. However, as is seen from the comparison of three emulsifiers of HLB 10.9, any simple relation is difficult to find whether the wider blend is less effective or not.

REFERENCES

1. Sherman, P., in "Emulsion Science" (P. Sherman, Ed.) p. 178, Academic Press, New York, 1968
2. Davies, J. T., Proc. 2nd Intern. Congr. Surface Activity, 1, 426 (1957)
3. Shinoda, K., Saito, H., and Arai, H., J. Colloid Interface Sci. 35, 624 (1971)
4. Shinoda, k. and Saito, H., J. Colloid Interface Sci. 30, 258 (1968)
5. Ohba, N., Bull. Chem. Soc. Japan 35, 1016 (1962)

THE STABILITY OF FLUOROCARBON EMULSIONS
THE EFFECT OF THE CHEMICAL NATURE OF THE OIL PHASE

S.S. Davis, T.S. Purewal, R. Buscall,
A. Smith and K. Choudhury
University of Nottingham

I. ABSTRACT

Perfluorochemical liquids have a high capacity to dissolve gases such as oxygen and carbon dioxide and many research groups are exploring the possibilities of using emulsified perfluorocarbons as red-blood cell substitutes. Surprisingly, little attention seems to have been paid, as yet, to the relevant colloid science aspects. For example, an important problem concerns the judicious choice of oil and surfactant to give minimum toxicity, maximum stability and optimum characteristics in vivo. As a consequence we have studied the effect of the chemical nature of the oil phase on the stability of o/w emulsion systems using three different approaches; coalescence of single droplets at the o/w interface, bulk emulsion stability, and the energetics of surfact-surfactant adsorption to interfaces.

The data obtained by these methods are compared and contrasted and attempts are made to correlate selected variables with fundamental physicochemical quantities. Perfluorocarbon and hydrocarbon oils have been employed as disperse phase together with anionic and non-ionic emulsifiers. The chemical nature of the oil phase can have a profound effect on emulsion stability.

Methods for producing satisfactory perfluorocarbon emulsions are discussed. The possibility of employing a dual emulsifier system (low molecular weight and polymeric species) is also being considered.

II.   INTRODUCTION

Perfluorochemical liquids used as coolants have long been known to have the capacity to dissolve large quantities of gases such as oxygen and carbon dioxide. Various workers have further shown that such fluids can be used for 'liquid breathing' where a test animal is immersed in the liquid and can effect satisfactory gas transport (1). It was a natural progression to attempt to use perfluorochemical emulsions to perfuse isolated organs of the body in place of whole blood or tyrode solution (2,3). Perfluorochemical oils that have been used included perfluorotributylamine and perfluoro-tetrahydrofuran, with perfluorinated surfactants, non-ionic Pluronics (polyoxyethylene polyoxypropylene block polymers) or egg lecithin as emulsifying agents.

The next step was to use emulsified perfluorochemicals as blood substitutes for administration to whole animals (and eventually man). Some of the advantages of such an artificial blood substitute include good shelf life, no blood group compatibility problems, ready accessibility and so on. Geyer (4) has proposed that a suitable replacement system should be of low toxicity, with little adverse effect on the behaviour of normal blood. The oxygen and carbon dioxide exchange properties should be satisfactory and the emulsion should be cleared from the body in a 'suitable' period of time. Ideally the artificial blood should take over from the natural material for a period long enough for new red blood cells to be generated. That is, it should have a lifetime of about 1-2 weeks in the circulation before being cleared.

Various groups throughout the world are now actually engaged in perfecting a suitable system and the total replacement of blood in experimental animals by means of perfluorochemical emulsions has been achieved (5). A review of the subject has been given by Maugh (6) and the proceedings of a workshop on artificial blood have been published recently (7).

Surprisingly the literature indicates that the colloid science aspects of formulating stable biocompatible perfluorochemical emulsions seem to have been neglected. For example a leading expert has contended that the research effort would be greatly speeded if there was a centralized source of high quality materials and emulsions so that experimenters could simply test the emulsions for physiological effect rather than waste time learning how to produce them. As a colloid science group we take the opposite view. We consider that the success of perfluorochemicals as red blood cell substitutes will depend to a great extent on the application of the relevant areas of colloid science. We need to know far more about the adsorption of surfactants onto perfluorochemical oils and thus be able to decide *a priori* the optimum combination of oil and surfactant that will give maximum stability and minimum toxicity.

It has been reported that the chemical nature of the oil can affect the dwell time of an emulsion in the blood of an animal and its subsequent fate (8). For example, perfluoro-tributylamine and perfluorotetrahydrofuran emulsions are cleared slowly from the blood and are deposited in the liver, and spleen, apparently for the lifetime of the animal. It is thought likely that these oils bond physically to some iron-containing proteins in these organs via the hetero-atoms in the C-O-C and C-N-C moieties (9). In contrast perfluoro-decalin and perfluoromethyldecalin are cleared more rapidly from the blood and are excreted. Relationships between elimination rates and vapor pressures of the oils have been presented by Yokoyama and others (10).

The nature of the oil phase also affects stability. Emulsions made using oils containing a hetero-atom (e.g. perfluorotributylamine) are generally stable whereas perfluorocarbon oils tend to give somewhat unstable emulsions. Unfortunately the oils with the most promising characteristics *in vivo* seem to have poor stability *in vitro*. Perfluoro-decalin leaves the liver and spleen rapidly but it tends to form short-lived, somewhat unstable emulsions, whereas perfluorodimethyladamantane is excreted more slowly but forming much more stable emulsions.

No doubt in the future other oils will be investigated since there are great possibilities for chemical modification among the perfluoroalkanes resulting in several types of red blood cell substitute. For example, the substitution of one or more of the fluorine atoms with another halogen would lower the vapor pressure without necessarily conferring possible binding properties via a hetero atom (11).

III. EXPERIMENTAL PROGRAM AND CHOICE OF MATERIALS

In order to provide data on the relevant areas of colloid science we have initiated a program to study the stability of oil in water emulsions and the effect of the nature of the oil phase.

The following studies have been performed:-

1. Life-time of single droplets at the plane oil/water interface.

2. Adsorption of surface active agent to various oil/water interfaces.

3. Assesment of the stability of perfluorochemical o/w emulsions using particle size analysis (electron microscope) and viscometry.

Oils and surfactants were chosen on the following grounds:

Sodium dodecyl sulfate (SDDS) has been used in our earlier studies on the stability of hydrocarbon emulsions (12,13). Thus it provides a link between past and present work. The adsorption of this surfactant to various oil/water interfaces is well documented and it is available in a highly purified form.

Pluronic F68 is a non-ionic surface active agent that is generally accepted as safe for intravenous administration. It has been used widely as an emulsifier for perfluorochemicals as well as for vegetable oils used in intravenous nutrition (14). Egg lecithin (purified) was chosen for the same reasons.

Intuitively the most suitable emulsifying agent for a perfluorochemical oil should be a perfluorochemical surfactant. A range of such materials is available, unfortunately they are of unknown composition and toxicity. Two were selected for the present study, a low molecular weight anionic material and a higher molecular weight non-ionic material. They were used as a mixed system since it has been recognised that a surfactant combination may provide optimum conditions for an initially small particle size plus long term stability (15, 16).

Three perfluorochemical oils were chosen. Two of these were reported to have poor stability when emulsified (perfluoromethylcyclohexane, perfluorodecalin) but good *in vivo* characteristics. The other, perfluorotributylamine has been reported to have good stability when emulsified but poor *in vivo* characteristics.

IV. EXPERIMENTAL

A. Materials

1. Surfactants

Sodium dodecyl sulfate (SDDS) from British Drug Houses, purified in the manner described by Davis and Smith (12) to give greater than 99% purity by GLC and surface tension measurements. Egg lecithin from British Drug Houses (90% purity) further purified by column chromatography to greater than 99% purity by TLC. Pluronic F68 (polyoxyethylene-polyoxypropylene copolymer) from Ugine Kuhlmann Chemicals Limited. The fluorinated surfactants Forafac 1111 (Ugine Kuhlmann) (molecular weight 887) and FC 126 (3M Company) (molecular weight 393) were non-ionic and anionic respectively. Full details regarding their composition and purity were not available from the respective manufacturers, however, both were said to contain a fluorocarbon 'tail'.

2. Oils

Perfluordecalin (95% purity by GLC) from Aldrich Chemicals, Perfluorotributylamine (97% purity by GLC) from Koch-Light Laboratories. Perfluoromethylcyclohexane (98% purity GLC) from P.C.R. Incorporated, Florida.

3. Water

Double distilled from an all glass still

B. Methods

1. Emulsification

The perfluorochemical emulsions were prepared using an ultrasonic probe (Dawe Soniprobe - 7532A). The contents of each emulsion are given in Table 2. The oil phase (10% phase volume) was placed in the 'rosette' flask and then the aqueous phase containing the surfactant (4% w/v) was added. The tip of the sonic probe was placed at the o/w interface and the mixture sonicated for 5 minutes. The 'rosette' flask was cooled in an ice-bath during sonication.

2. Particle Size Analysis

The mean size on a number basis and particle size distribution of the fluorochemical emulsions were determined using an electron microscope (AEI, Corinth). A carbon replica (N.G.N. Coating Unit Model 12SG2) palladium-gold shadowing technique was used. The magnification of the electron micro-scope was checked regularly and a control analysis of latex particles was conducted every time the microscope was used.

A Zeiss Particle Size Analyser TGZ3 was used to measure the particle size distribution of each emulsion direct from electronmicrographs.

3. *Viscosity Measurement*
A standard U-tube viscometer (A) was used to measure the relative viscosities of the emulsion systems. This viscometer gave a flow time of 316 ± 0.5 sec. for triple distilled water at 25 ± 0.05^{o}C. The viscometer was cleaned with chromic acid and rinsed thoroughly before use. Each flow time determination was repeated twice. The emulsion density was determined by density bottle method.

4. *Coalescence of Droplets*
The coalescence cell was based on the design of Nielsen Wall and Adams (17). Uniform sized droplets were formed using an Agla Syringe and detached after a period for aging Full details of the construction of the apparatus and the standardization of the method have been given elsewhere (12). In the present work the system was inverted since the oils being studied were denser than water.

The rest-times of droplets at the o/w interface were determined by stop-watch and a minimum of 80 drops was measured in each experiment. A distribution of rest-times was obtained and where possible the data were analysed on the basis of a first order kinetics model (18)

$$\log N = \frac{-kt}{2.303} + \text{Const.} \qquad (1)$$

where N is the number of drops remaining at time t and k is the first order rate constant for coalescence calculated from the gradient of the linear portion of the log N versus t plot.

The first order half life ($T\frac{1}{2} = 0.693/k$) is related to the time for half the droplets to coalesce ($t\frac{1}{2}$) by

$$t\frac{1}{2} = T\frac{1}{2} + t_d \qquad (2)$$

where t_d is the drainage time.

5. *Measurement of Interfacial Tension*
The interfacial tensions between oil and surfactant solutions were measured using the Wilhelmy plate method (19). The apparatus was based on a Beckman LM.800 microforce balance.

It was calibrated in the usual way using pure oils of known surface tension and interfacial tensions against water. Interfacial tension data were collected over a range of surfactant concentration. The area occupied by each surfactant molecule at the o/w interface at a given concentration was calculated using the Gibbs' surface excess equation

$$\frac{d\gamma}{RT} = -\Gamma \, d\ln a \tag{3}$$

where γ is the interfacial tension,
$\quad \Gamma$ is the surface excess,
$\quad$ R is the gas constant,
$\quad$ T is the absolute temperature,
and a the activity of the surface active species. At low concentrations it can be assumed that a $\simeq$ C, where C is the surfactant concentration

V. RESULTS

A. The Effect of Oil Phase and Surfactant on Droplet Stability

The droplet life-time data for some perfluorochemical oils stabilized with SDDS and Pluronic F68 are shown in figure 1. Some of the curves correspond well to a first order kinetic process whereas others do not. Consequently droplet stability was characterised by the time for 50% of drops to coalesce ($t\frac{1}{2}$) (table 1).

It is clear that the nature of the oil phase and the nature of the surfactant both play an important role in droplet stability. For a given surfactant the lifetimes of the droplets are in the order Pf. decalin < Pf. methylcyclohexane < Pf. tributylamine and for a given oil the droplets stabilized by the non-ionic surfactant (Pluronic F68) are much more stable than those stabilized by the anionic surfactant (SDDS), even though the latter was at tenfold higher concentration.

A tenfold reduction in the concentration of SDDS leads to a corresponding decrease in the stability of Pf. tributylamine droplets. Data for a hydrocarbon oil (hexadecane) (12) have been included for comparison.

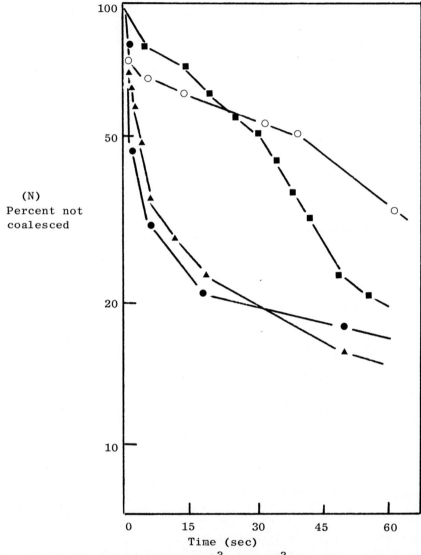

(N)
Percent not
coalesced

Time (sec)

■ Pf. tributylamine - SDDS 10^{-3} mol dm^{-3}

▲ Pf. methylcyclohexane - SDDS 10^{-3} mol dm^{-3}

● Pf. decalin - SDDS 10^{-3} mol dm^{-3}

○ Pf. decalin - Pluronic F68 10^{-4} dm^{-3}

Fig. 1. The coalescence of perfluorochemical oil droplets at the surfactant solution/oil interface.

TABLE 1

The Stability of Perfluorochemical Droplets at the Oil-Aqueous Surfactant Solution Interface (25°C)

Surfactant	Oil	Time for 50% of drops to coalesce (sec)
SDDS-10^{-3} mol dm^{-3}	Pf. decalin	2.0
SDDS-10^{-3} mol dm^{-3}	Pf. methyl-cyclohexane	3.2
SDDS-10^{-3} mol dm^{-3}	Pf. tributylamine	30
SDDS-10^{-4} mol dm^{-3}	Pf. tributylamine	2.8
Pluronic F68 10^{-4} mol dm^{-3}	Pf. decalin	35
Pluronic F68 10^{-4} mol dm^{-3}	Pf. methyl cyclohexane	260
Pluronic F68 10^{-4} mol dm^{-3}	Pf. tributylamine	>500
SDDS-10^{-3} mol dm^{-3}	Hexadecane	14.9
SDDS-10^{-4} mol dm^{-3}	Hexadecane	6.7

B. Interfacial Tension

The change of surface pressure (π) with surfactant concentration is shown in figure 2 for the three perfluorochemical oils and SDDS. Similar data for hexadecane (12) are also given. It is clear that SDDS is not well absorbed to the fluorochemical oil/water interface. Compare for instance the surface pressures at 1×10^{-3} mol dm^{-3} where the areas occupied by each surfactant molecule are Pf. decalin 14 nm^2, Pf. methylcyclohexane 5 nm^2, Pf. tributylamine 7 nm^2 and hexadecane 1 nm^2 (12).

C. Emulsion Stability

1. *Particle Size Analysis*

The mean particle sizes of the fluorocarbon emulsions at different storage times are summarized in table 2.

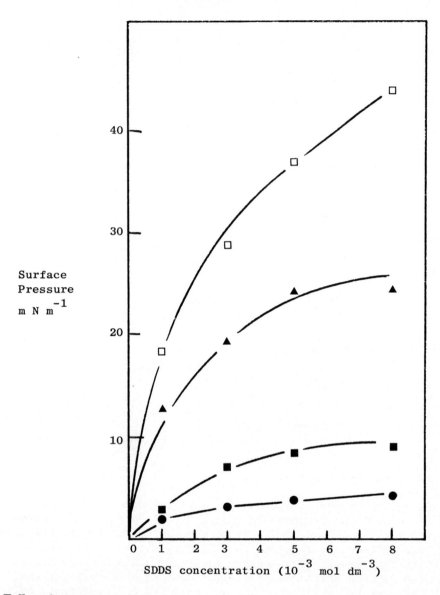

Fig. 2. Surface pressures of various oils in the presence of added SDDS.

TABLE 2

Mean Particle Size (Number Basis) of Perfluorochemical Emulsions and the Effect of Storage

		Diameter (μm)		
Oil (10% w/v)	Surfactant (4% w/v)	Day/0	32	57
Pf. methyl-Cyclohexane	Pluronic F68	0.41	0.44	0.51
	SDDS	1.62	1.94	2.25
	Fluorinated surfactants*	0.63	0.51	1.38
	Lecithin	0.43	0.56	0.57
Pf. decalin	Pluronic F68	0.40	0.41	0.48
	SDDS	0.48	0.70	1.77
	Fluorinated surfactants*	0.31	0.52	2.28
	Lecithin	0.23	–	0.86
Pf. tributylamine	Pluronic F68	0.45	0.48	0.59
	SDDS	0.60	0.67	0.74
	Fluorinated surfactants*	0.57	0.40	1.14
	Lecithin	0.41	0.46	0.73

* 2% each of FC126 and Forafac 1111

A distribution of particle sizes was obtained for each system. These distributions did not follow any particular distribution function (e.g. normal, log normal) and in some cases (particularly oils emulsified using SDDS and the fluorinated surfactant mixture) the particle size distribution was bimodal in nature. An example is given in figure 3. Some representative electron micrographs are shown in figure 4 for different oil-surfactant combinations.

It is apparent from the data shown in table 2 that the nature of the oil phase and surfactant has a profound effect on the initial particle size and stability for the fluoro-chemical emulsion systems. The mean particle size data have been plotted in figures 5 and 6 using the normalizing function D_t/D_o; where D_t is the mean diameter at storage time t and D_o is the initial particle size.

Considering first the initial size of the droplets we find that for a given oil the mean sizes are affected by the emulsifier, lecithin < pluronic < fluorinated surfactants < SDDS. And for a given surfactant the mean sizes are affected by the oil phases; Pf. decalin < Pf. methylcyclohexane and

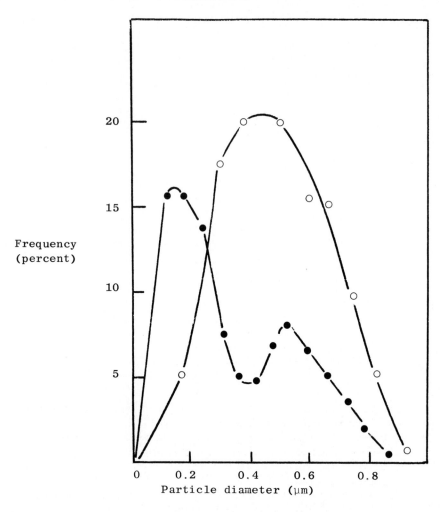

O Initial
● Stored 57 days

Fig. 3. The effect of storage on the particle size distribution of a perfluoromethylcyclohexane emulsion stabilized by Pluronic F68.

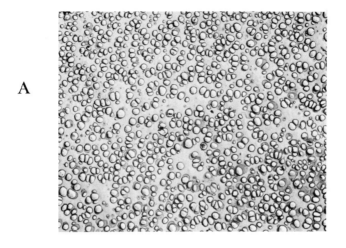

A

⊢————⊣ 1 μm

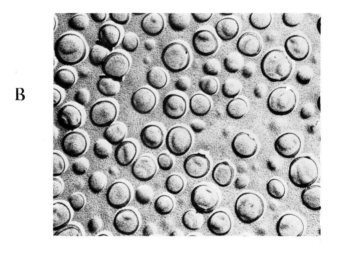

B

Fig. 4. Electronmicrographs of perfluorochemical emulsions. A. Perfluorodecalin emulsified with 4% egg lecithin. B. Perfluorotributylamine emulsified with 4% Pluronic F68.

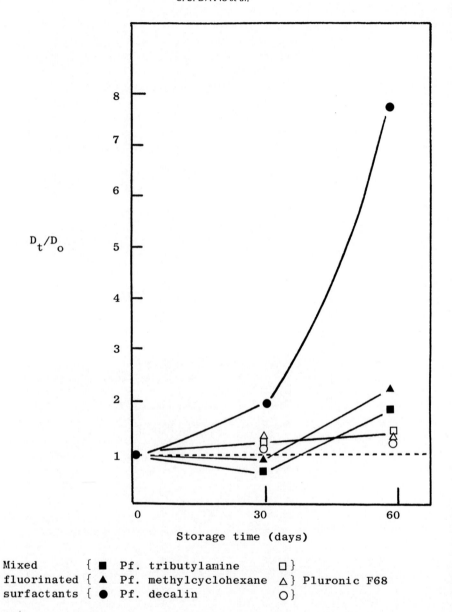

Fig. 5. The effect of storage on the mean particle size of perfluorochemical oils stabilized by Pluronic F68 and a mixed fluorinated surfactant system.

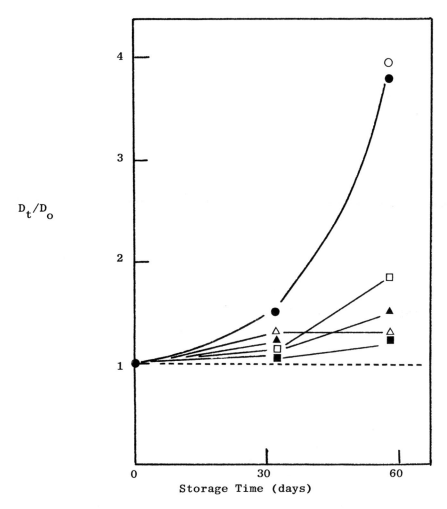

Fig. 6. The effect of storage on the mean particle size of perfluorochemical oils stabilized by SDDS and egg lecithin.

Pf. tributylamine, except for SDDS as emulsifier where the order is reversed. Upon storage for a given oil we find that the mean size is affected by the surfactant; pluronic < lecithin = SDDS < fluorocarbon surfactants. And for a given surfactant the mean sizes are affected by the oil phase; Pf. tributylamine < Pf. methylcyclohexane << Pf. decalin. Emulsions produced using Pluronic F68 and any of the three oils were the most stable, while Pf. decalin and lecithin gave very fine particle sized emulsions initially.

The data for the emulsions emulsified using the fluorocarbon surfactant mixture demonstrates an interesting effect; the mean size decreases on storage and then increases rapidly.

2. Viscosity
The relative viscosities of the emulsions changed little on storage. The time span without significant change in viscosity was taken as an index of stability (table 3)

TABLE 3

The Viscosity of Perfluorochemical Emulsions

Surfactant	Time (days) without viscosity change		
	Oil		
	Pf. tributylamine	Pf. decalin	Pf. methyl-cyclohexane
SDDS	7	0	0
lecithin	6	0	0
Pluronic F68	13	6	13
Fluorinated surfactants	13	7	13

The results suggest that Pf. tributylamine gives reasonably stable emulsions with all the surfactants studied but Pf. decalin and Pf. methylcyclohexane are stable only when emulsified using Pluronic F68 or the fluorinated surfactant mixture.

VI. DISCUSSION

A. Droplet Stability

The stability of emulsions of perfluorochemicals is dependent on the nature of the surfactant and the oil phase. In single droplet stability studies Pf. tributylamine (a fluorochemical containing a hetero-atom) gave the most stable systems. The non-ionic polymeric agent (F68) was much more effective a stabilizer than SDDS.

The data for SDDS may be considered in detail and can be compared with similar data derived by Davis and Smith (12) for hydrocarbon oil droplets. These authors attempted to correlate droplet stability data with the physicochemical properties of the oil (solubility parameter, interfacial tension, density difference, oil viscosity) and they concluded that coalescence would be dependent on one of two different rate determining steps; the draining of the film of continuous phase between the oil droplet and the o/w inter-face; the ease of displacement of surfactant molecules from the o/w interface.

1. Film Drainage Models

Following the suggestions of various authors (20) we have examined the roles of phase viscosity ratio, density difference, and interfacial tension. Generally coalescence times are expected to decrease with increased interfacial tension (20) but our data for the perfluorinated oils do not support this view.

The relation between density difference ($\Delta\rho$) and stability is not a simple one. An increase in $\Delta\rho$ results in greater buoyancy leading to more rapid film drainage. However, large differences in density can also cause deformation of the droplet and/or the interface which extends the area of the draining film and leads to increased stability. Provided any deformation is small and neglecting the effects due to double layer repulsion, electroviscous effects or the physical presence of an adsorbed film, the time for the film to drain evenly (t_h) to a given thickness can be related to interfacial tension and density difference by (21)

$$t_h \propto \Delta\rho^{-1} \text{ (spherical-planar model)}$$

$$t_h \propto \Delta\rho/\gamma ow^2 \text{ (parallel plate model)}$$

Smith and others (22) have found that the parallel plate model

gave good correlation of droplet stability of various hydrocarbon oils stabilized by SDDS and that the critical rupture
thickness for the alkanes was about 150 nm. A similar
correlation has been attempted here (figure 7). It is clear
that the data for the perfluorochemical oils do not follow
the same pattern as the hydrocarbon oils. A full analysis of
droplet coalescence data for a wide variety of oil phases
stabilized by different concentrations of SDDS will be
presented elsewhere (22).

We note that the easiest way to increase the stability of
perfluorochemical droplets in SDDS is to increase the
surfactant concentration. Nielsen and others (17) derived an
expression

$$t_{\frac{1}{2}} = c^n \tag{4}$$

The index n varied from 0.45 to 3. The data for Pf. tributylamine suggest a value of n close to unity. An increased
surfactant concentration will affect adsorption of the
stabilizer to the o/w interface as well as the interfacial
tension. At high surfactant concentration the second of these
two will be the dominant factor.

2. *Adsorption and Displacement of Surfactants from Interfaces*
The adsorption of SDDS to the o/w interface will be the
resultant of two processes (23); separation of oil molecules
as the molecules of surface active agents penetrate the oil
phase (cohesive forces); interaction between hydrocarbon
chains of the surfactant and molecules of the oil phase
(adhesive forces). When the adhesive forces are greater than
the cohesive forces the hydrocarbon chain of SDDS will
penetrate the oil phase readily. An estimate of the cohesive
nature of the oil phase can be gained from its solubility
parameter (which is related to the heat of vaporization and
solvent molar volume) (12). Perfluorocarbon oils have
low solubility parameters (in the region of 6 to 7) as
compared to hydrocarbons which have higher values (7 to 9)
(24). Thus the cohesive forces between perfluorocarbon
molecules are low. This is reflected likewise in their low
surface tensions (20 m N m^{-1} or less). The estimation of
adhesive forces between the surfactant chain and oils is more
difficult. Davis and Smith (12) have proposed that values for
the activity coefficient at infinite dilution ($f\infty$) of the
corresponding binary hydrocarbon mixtures (e.g. dodecane and
perfluorochemical oil) provide a qualitative estimate. Such
values are limited in the literature but Scott and his
colleagues (25) have studied a number of mixtures of aliphatic
hydrocarbons.

282

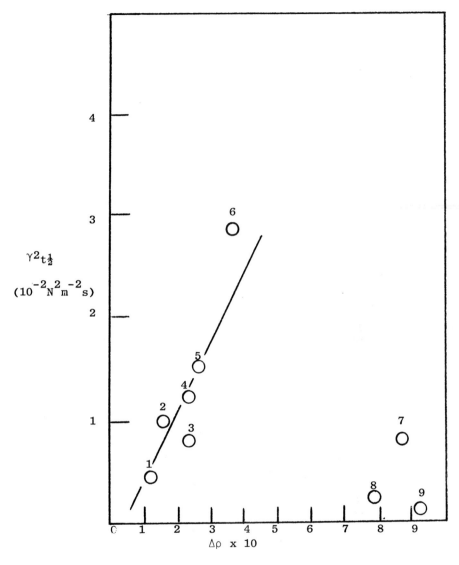

1. Benzene
2. Toluene
3. Hexadecane
4. Cyclohexane
5. Dodecane

6. Hexane
7. Pf. tributylamine
8. Pf. methylcyclohexane
9. Pf. decalin

Fig. 7. Analysis of droplet coalescence data by the uniform film model for film drainage.

Such mixtures have positive excess free energies. For perfluorohexane-hexane and perfluoroheptane-heptane mixtures f∞ is in the region of 2.00. An ideal system (e.g. dodecane in hexadecane) would have f∞ = 1.00.

Thus, for perfluorochemical oils both cohesive and adhesive forces are relatively low. The poor adhesion between surfactant chain and oil will result in poor penetration of SDDS and hence poor stability. The hetero-atom in Pf.tributylamine may provide some increased adhesional force in comparison to the almost inert perfluorocarbons.

The non-ionic polymeric material, Pluronic F68 provides a different picture. The adsorption of polymers to interfaces is very different to that for the adsorption of ionic low molecular weight surfactants. Effective penetration of the oil phase is not a prerequisite of adsorption to the interface and hence good droplet stability can be achieved even with inert oils such as the perfluorocarbons. The mechanics of stabilizing emulsions by non-ionic emulsifiers have been reviewed by Florence and Rogers (26) who have pointed out that one must consider not only electrostatic forces of repulsion and London, van der Waals forces of attraction but also the presence of the polymer molecules at the interface. A free energy of interaction force can arise due to steric hindrance, involving the free energy of mixing solvated adsorbed layers. The repulsion between two particles originating in this way has been termed entropic repulsion (26) because of the loss of configurational entropy of the adsorbed molecules on mixing. The electron micrographs in figure 4 indicate that Pluronic F68 may form quite thick adsorbed layers such that the stability against coalescence may be related to the mechanical properties of the interfacial film (27). This does not mean that the behavior of the oil droplet is independent of the nature of the oil for once again we see that stability increases as we pass from Pf. decalin to Pf. tributylamine. The adsorbed polymeric film is the most single important stabilizing factor, but secondary variables such as density difference, and interfacial tension will also play a part. Perfluorotributylamine stabilized with 10^{-4} mol dm^{-3} Pluronic F68 gives very stable droplets, the majority of which have life-times in excess of 500 seconds.

B. Emulsion Stability

The particle size analysis and viscosity results show that the stability of perfluorochemical emulsions depends on both the emulsifier and the surfactant. For SDDS there is good correlation between emulsion stability data and single droplet coalescence life-times (Pf. decalin < Pf. methylcyclohexane << Pf. tributylamine). To find such a correlation is pleasing for in our previous studies on hydrocarbon oils (13) we were unable to find a positive correlation, no doubt due to different mechanisms of instability. The major route for 'breakdown' with hydrocarbon emulsions stabilized by SDDS was Ostwald ripening (molecular diffusion), and was thus almost totally dependent on the solubility of the oil in SDDS solution (13). Perfluorochemical oils have very low solubility in water and instability due to Ostwald ripening will probably be negligible.

We conclude that coalescence is the major route for breakdown. The oils with poor affinity for the hydrocarbon chain of SDDS (Pf. decalin and Pf. methylcyclohexane) are less stable when emulisified than Pf. tributylamine.

Pluronic F68 gives good stability as would be expected from a polymeric material. The different values for the rest times in table 1 are not reflected in the stability data.

The mixed fluorinated surfactant combination was disappointing. We expected this system to give good stability since the forces of adhesion between the fluorocarbon tail of the surfactant and the various oils should have been significant. We plan to undertake further studies using these materials as single stabilizers and in combination, provided that greater detail about composition and purity can be obtained. Droplet stability studies for various oils should also be instructive. We note, however, that the order of stability is as for SDDS (Pf. decalin << Pf. methylcyclohexane < Pf. tributylamine). The minimum in stability found at 32 days storage for the two more stable emulsions is interesting and may reflect an Oswald ripening process. We have noted above that the particle size distributions of some of the systems became bimodal upon storage, as would be expected from Ostwald ripening (13).

Emulsions stabilized by egg lecithin are also unstable. The relative size changes with time are similar to those reported for the SDDS systems but the positions of Pf. tributylamine and Pf. methylcyclohexane are reversed at 57 days. Lecithin is a naturally occurring emulsifying agent. In structure it is similar to a fat, however, instead of being a simple triglyceride one position on the glycerine function is occupied by a substituted phosphoric acid (28). Egg lecithin is used as the emulsifier in intravenous fat emulsions for

285

parenteral nutrition. The oils in this case are of vegetable origin (soya, cotton seed) and emulsions of small particle size and long term storage stability can be formulated (14). The poor stability of lecithin stabilized perfluorochemical oils may be attributed to the poor affinity between the hydrophobic regions of the emulsifier (long chain fatty acid residues) and the oil.

The sizes of the oil droplets produced immediately upon emulsification show rather a different picture. SDDS produced the largest droplets whereas lecithin produced some of the smallest. Pluronic F68 was in between. The situation with the oils is also interesting. Perfluorodecalin gave the smallest droplet size for all emulsifiers yet this oil, with any emulsifier gave very poor stability. Thus, contrary to previous literature reports we can state that perfluorodecalin is easy to emulsify and produces very fine emulsions but these very fine particles are very unstable except when a polymeric emulsifier such as Pluronic F68 is used. The possibilities of employing a combined emulsifier will be discussed below.

C. Mixed Stabilizers

The rationale behind our choice of the mixed fluorinated surfactant system was based on the knowledge that high molecular weight polymeric and low molecular weight ionic surfactant mixtures can give enhanced stability as compared with either agent alone (16). The low molecular weight material will lower the interfacial tension so as to aid emulsification and will diffuse rapidly to the interface to form a relatively stable interface, preventing coalescence at short times. Thus, the emulsion produced will be of small particle size but may be unstable on prolonged storage. The higher molecular weight material will lower the interfacial tension to a lesser extent and will diffuse to the interface more slowly. Thus, the emulsions produced will be of large particle size but will be very stable on prolonged storage. A combination of the two types should give the best of both worlds provided they are compatible.

Our initial attempts to employ Pluronic F68 and the anionic fluorinated surfactant FC-126 were not successful because a precipitate was formed when the two were mixed. The mixed fluorinated surfactant system was compatible but gave emulsions with poor stability. The results described above suggest that small particle size and good stability could be achieved using a lecithin and Pluronic F68 combination and presently conducted stability studies with perfluorochemical oils bear this out. Other possible combinations are being considered, although with any system intended for intravenous administration the question of toxicity is ever present. Low molecular weight surfactants with good emulsifying characteristics are very often toxic and one could not contemplate their presence in the final emulsion system. Nevertheless, that does not preclude their use in the emulsification process *per se* since they could be removed by dialysis once the higher molecular weight material had time to diffuse to the various interfaces and form thick interfacial films. For example with the lecithin/Pluronic combination the lecithin can be dialysed out without affecting the stability of a perfluorochemical oil emulsion. This opens up the possibility of using a range of polymeric materials that are good stabilizers but poor emulsifiers. It is perhaps hoping too much to find one material that will do both jobs and also be acceptable for parenteral administration to man.

VII. CONCLUSIONS

The stability of perfluorochemical oils intended for intravenous administration as red blood cell substitutes is dependent on the nature of the oil and the emulsifier. Differences in stability of single droplets at plane interfaces and of bulk emulsions can be rationalized in terms of the intermolecular forces between oil molecules and between oil and surfactant molecules. The polymeric agent, Pluronic F68 provided droplets with good storage stability, whereas the natural material lecithin gave droplets of very small particle size initially but which coarsened on storage. A combined system of Pluronic and lecithin is proposed.

VIII. REFERENCES

1. Clark, L.C. and Gollan, F., Science 152, 1755 (1966).
2. Gollan, F. and Clark, L.C., Physiology 9, 191 (1966).
3. Triner, L., Verosky, M., Habif, D.B. and Nahas, G.G.,
    Fed. Proc. 29, 1778 (1970).
4. Geyer, R.P., Bull. Parent. Drug. Ass. 28, 88 (1974).
5. Geyer, R.P., Fed. Proc. 32, 927 (1973).
6. Maugh, T.H., Science 179, 669 (1973).
7. Workshop on Artificial Blood, Fed. Proc. 34, 1468 (1975).
8. Clark, L.C., Becattini, F., Kaplan, S., Obrock, V.,
    Cohen, D. and Becker, C., Science 181, 680 (1973).

9. Clark, L.C., Science 181, 680 (1973).
10. Vokoyama, K., Tamanouchi, K. and Murashima, R., Chem. Pharm. Bull. 23, 1368 (1975).
11. Clark, L.C., Wessler, E.P., Kaplan, S., Miller, M.L., Becker, C., Emory, C., Stanley, L., Becattini, F. and Obrock, V., Fed. Proc. 34, 1468 (1975).
12. Davis, S.S. and Smith, A., Kolloid Z. 251, 337 (1973).
13. Davis, S.S. and Smith, A. in "Theory and Practice of Emulsion Technology" (A.L. Smith, Ed.), p. 325. Academic Press, London, 1976.
14. Davis, S.S., J. Hosp. Pharm. 32, 149, 165 (1974).
15. Davis, S.S., J. Clin. Pharm. 1, 5 (1976).
16. Tadros, Th. F., J. colloid Interface Sci. 46, 3 (1974).
17. Nielsen, L.E., Wall, R. and Adams, G., J. colloid Sci. 13, 441 (1958).
18. Davis, S.S. and Smith, A., Kolloid Z. 254, 82 (1976).
19. Johnson, M.C.R. and Saunders, L., J. Pharm. Pharmac. 23, 895 (1971).
20. Jeffreys, G.V. and Davies, G.A., in "Recent Advances in Liquid-Liquid Extraction" (C. Hanson, Ed.), p. 495. Pergamon, Oxford, 1971.
21. Charles, G.E. and Mason, S.G., J. colloid Sci. 15, 235 (1960).
22. Smith, A., Davis, S.S. and Purewal, T.S., to be published.
23. Hutchinson, E., J. colloid Sci. 3, 235 (1948).
24. Hildebrand, J.H. and Scott, R.L., "Regular Solutions", Prentice Hall, Englewood, N.J., 1962.
25. Scott, R.L., J. Phys. Chem. 62, 136 (1958).
26. Florence, A.T. and Rogers, J.A., J. Pharm. Pharmac. 23, 153 (1971).
27. Kitchener, J.A. and Mussellwhite, P.R., in "Emulsion Science" (P. Sherman, Ed.), p. 77. Academic Press, London, 1968.
28. Becher, P., "Emulsions, Theory and Practice, p. 226. Reinhold, New York, 1965.

ELECTROCHEMICAL MEASUREMENTS IN
NONIONIC MICROEMULSIONS

R. A. Mackay, C. Hermansky and
R. Agarwal

Drexel University

Specific ion electrode, conductivity and some limited
polarographic measurements have been performed on oil
in water microemulsions stabilized by various tween surfactants and pentanol. The phase volumes studied were
generally in the range 0.25-0.75. The sodium ion electrode gave the same results in all three systems examined, while the fluoride ion electrode data reflected
the composition of the continuous phase. The conductivity and polarographic data indicated that the effective
relative microviscosity of the continuous phase has
the same functional form as the relative bulk viscosity
of a suspension of random spheres.

I. INTRODUCTION.
    A microemulsion is a clear, stable fluid consisting
of essentially monodisperse oil in water (o/w) or water
in oil (w/o) droplets with diameters normally in the range
10-60 nm(1,2). A microemulsion or micellar emulsion
generally consists of four components, water, oil, ionic
surfactant and alcohol, although a suitable nonionic surfactant may be effective (3,4). The formation of the
microemulsion is spontaneous, suggesting that it is thermodynamically stable (5,6). In any event, it is mechanically
stable. For o/w systems of the type examined here, the
internal structure may be described as a stable collection
of "oil" microdroplets in an aqueous continuous phase.
Each droplet consists of 6-60 nm diameter "bulk" oil drop
surrounded by a 2-3 nm thick surface phase consisting
mainly of surfactant (alcohol and/or detergent) molecules.
The "continuous" (aqueous) phase may contain some surfactant
-alcohol. The physical structure of the continuous phase
(solution or colloid) will depend upon a number of factors
such as the solubility of the surfactant and whether the
microemulsion is a thermodynamically or kinetically

stable system.

Electrochemical measurements are performed on solutions of aqueous ionic micelles for various purposes,
including determinations of the critical micelle concentration and the fraction of bond counterion. Two
of the principal types of measurements employed are
electrode potential and conductivity. In these systems,
both the volume fraction occupied by the micelles (phase
volume, ϕ) and the total electrolyte concentration are
usually small. In micellar emulsions ϕ is frequently
in the range 20-80%. When different types of electrochemical measurements are performed in these media the
question arises as to the quantities actually being
measured. For example, the potential measurements may
depend upon the nature of the continuous phase via its
effect on activity coefficient, a function of the electrolyte and its concentration, and on the electrode
potential, which depends upon the type of electrode and
its construction. The conductance measurements should
be affected not only by the composition of the continuous
phase, but also by the "obstruction effect" (7) due to
the presence of the droplets.

To begin these investigations, nonionic surfactants
have been used to avoid additional complications due to
coulombic ion binding. We report here the results of
conductance, specific-ion electrode, and some limited
polarographic studies in mineral oil or hexadecane in
water microemulsions stabilized by pentanol and various
tweens, and which contain added electrolyte.

II. EXPERIMENTAL.
     The potential and conductivity measurements were
carried out in a water bath at $25^{\circ}C$. The solutions were
stirred magnetically, care being taken to avoid the
creation of vortices. The sodium responsive glass
electrode (Fisher Scientific) and the solid membrane
chloride electrode (Orion) were used in conjunction with
a Ag/AgCl double junction reference electrode (Sensorex),
the outer compartment containing gelled KNO_3. A solid
membrane combination electrode (Orion) was used for the
fluoride measurements. The potentials were read on either
a Beckman Expandomatic or a Coleman Model 38A Expanded
Scale pH meter. A dip cell containing platinized platinum
electrodes (cell constant ≈ 0.9) in conjunction with a
Serfass conductance bridge operating at 1KHz was employed
for the conductivity measurements. All values of the
equivalent conductance are corrected for the conductivity

of the solvent. A Beckman Electroscan 30 and a three compartment cell was used for the d.c. polarographic measurements. The indicator, auxillary and reference electrodes were dropping mercury (DME), platinum and saturated calomel (SCE), respectively. The solutions were degassed by purging with nitrogen for at least 15 minutes. A bubbler containing microemulsion was used to prevent composition changes. The tween 40,60 and 81 were supplied courtesy of ICI United States Inc.. The Hexadecane (99%) is from Aldrich and the Mineral Oil (U.S.P.) from Fisher Scientific. Distilled water of specific conductivity 6μmhos/cm was used throughout. The standard procedure for the potential and conductivity studies was to mix the surfactant, oil, alcohol, and usually the solid salt (NaCl or NaF), and then add water or aqueous salt solution to obtain successive dilutions as the measurements were being made. For the polarographic measurements, individual dilutions were made separately.

III. MICROEMULSION SYSTEMS

The nonionic surfactants employed for the systems in this study are polyoxyethylene (20) sorbitan monopalmitate (tween 40), polyoxyethylene (20) sorbitan monostearate (tween 60), and polyoxyethylene (5) sorbitan monooleate (tween 81). The microemulsion compositions are described in Table 1.

Table 1

Microemulsion Systems

HLB^a	15.6	14.9	13.4
surfactant	tween 40	tween 60	tween 60/
wt. $\%^b$			tween 81
surfactant	72.3	48.2	38.6/17.1
pentanol	21.0	25.2	30.7
oil	6.7^c	26.6^d	13.6^c
% water (approx. range)	25-85	25-80	30-45

a. Atlas hydrophile-lipophile balance(HLB).

b. initial composition to which varying amounts of water are added.

c.   mineral oil
d.   hexadecane

The lower HLB numbers generally reflect a decreased water solubility of the surfactant and an increasing tendency to form w/o rather than o/w emulsions. This is also reflected in a decreased range of clear solutions. Decreasing the oil content of the tween 60/81 system does not effect an increase in (% water) range. On this basis, one would expect that the composition of the (aqueous) continuous phase would more closely approach that of water with decreasing HLB. This is consistent with the spectrum of 1-methyl-4-cyanoformylpyridinium oximate (CPO) in the three microemulsions. This molecule possesses intramolecular charge transfer bands which have very solvent sensitive band maxima (8). In addition, CPO is expected to be soluble only in the aqueous phase, not adsorbed by the drop nor affected by the presence of micelles[1].

The short wavelength band maxima in the tween 40, tween 60, and tween 60/81 systems are 364, 360, and 346 nm, respectively. These may be compared with values of 363 nm in methanol and 341 nm in water. It may be noted that the specific ion electrode data (vide infra) are also consistent with this interpretation of the effect of surfactant HLB on the composition of the continuous phase.

The two salts employed as electrolytes were sodium chloride and sodium fluoride. In general, NaCl concentrations in excess of 1M were required to break the microemulsions, while NaF concentrations of about 0.1M were sufficient to break the microemulsions of high phase volume. It may, therefore, be anticipated that sufficiently high salt concentrations will alter the (nonionic) composition of the aqueous phase, NaF producing a much larger effect than NaCl at the same effective concentration.

The phase volume may be estimated from the composition and density of the microemulsion. Assuming that all of the water forms the continuous phase, all of the other components form the drops, and that the density of water remains unchanged, the phase volume based on composition, ϕ_{comp}, is given by

$$\phi_{comp} = 1 - \omega\rho/\rho_o \qquad (i)$$

Here, ω is the weight fraction of water, ρ and ρ_o are the densities of the microemulsion and water, respectively.

---

[1] R. Mackay, unpublished results.

IV. POTENTIAL MEASUREMENTS

    A. Sodium Ion

    The behavior of the electrodes in both water and the microemulsions are essentially Nernstian, as shown in Fig. 1.

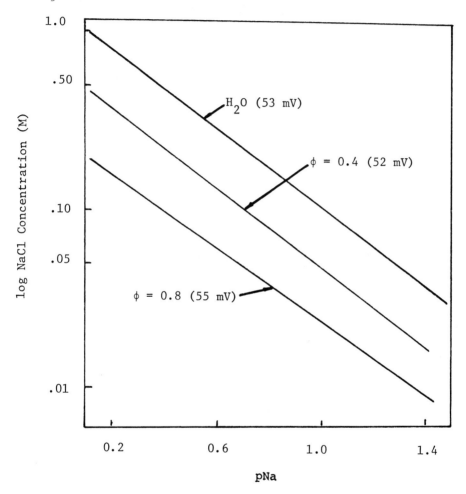

    Fig. 1. Sodium ion electrode response for NaCl in water and in the tween 40 system at two different phase volumes (ϕ). The numbers in parentheses are the potential changes in mV per unit change of log (NaCl).

293

The slopes are a bit less than 59mV, but are all comparable. We have used concentrations in place of activities since we are generally only working over a range (0.1M - 0.5M) where the activity coefficient is not changing too rapidly, and since concentration is the quality we wish to measure. The results shown in Fig. 1 are qualitatively the same for all of the electrodes (Na^+, Cl^-, F^-) in the three systems examined; namely, that the "effective" concentration measured by the electrode by using the aqueous calibration curve is higher than the stoichiometric concentration in the microemulsion. Also, this effective concentration increases with increasing phase volume. Four factors which can contribute to this shift are the liquid junction potential, the activity coefficient, an assymetry potential and the effect of phase volume on the concentration in the aqueous phase. This is shown in equation (ii).

$$\varepsilon = \varepsilon^0 + \varepsilon_{lj} + \varepsilon_{ap} - .059 \log \gamma - .059 \log C_{eff} \quad \text{(ii)}$$

Here, ε, ε^0, and ε_{lj} are the measured, standard and liquid junction potentials, respectively. The quantity ε_{ap} is a specific solvent effect on the electrode, γ is the activity coefficient resulting from the solvent effect on the electrolyte, and C_{eff} is the effective concentration in the continuous phase. Since the salt should be located only in the continuous phase, $C_{eff} = C/(1 - \phi)$ where C is the stoichiometric concentration. We will, therefore, define a measured phase volume (ϕ_m) by

$$\phi_m = 1 - C/C_m, \quad \text{(iii)}$$

where C_m is obtained by converting the electrode reading in the microemulsion to a concentration using the aqueous calibration curve. Thus, all of the factors mentioned above can contribute to ϕ_m. It is expected that the effect of the liquid junction potential will be minor since the medium is aqueous and both KCl and KNO_3 junctions give the same results.

It should be noted that the nonionic tween surfactants contain some ionic impurities which contribute to the electrode readings. All of the reported results have been "corrected" for this background, again using the aqueous calibration curve. In the case of sodium ion, this background was equivalent to an effective sodium chloride concentration of about 0.02M. The background was much less for chloride and entirely negligible for fluoride.

The effect of varying the stoichiometric salt concentration on the value of ϕ_m is given in Table 2. In general ϕ_m

Table 2.
Phase volumes from Sodium Ion Specific Electrode Measurements in the tween 60 System as a Function of the Stoichiometric Salt Concentration.

ϕ_{comp} \ ϕ_m	NaCl Concentration (M)			
	0.03	0.10	0.30	0.70
.76	.83	.85	.86	.89
.72	.79	.82	.83	.84
.65	.76	.77	.77	.78
.56	.70	.70	.68	.76
.49	.63	.65	.59	.63
.42	.57	.59	.51	.56
.36	.50	.52	.42	.48
.30	.44	.39	.35	.44
.21	.38	.35	.23	.27

is independent of the salt concentrations used to obtain it, indicating that the presence of NaCl in this concentration range does not have a significant effect on the microemulsion, and confirms the paralled behavior of the curves in both water and microemulsion as depicted in Fig. 1 for the tween 40 system.

The values of ϕ_m obtained from the pNa data are all higher than the corresponding values of ϕ_{comp}. The actual ϕ values will be lowered by the presence of surfactant and alcohol in the continuous phase since it is clearly not pure water (vide supra). However, the inclusion of some water as part of the droplet would tend to increase ϕ, and it easily might be expected that this effect would dominate. However, if some NaCl is also included with the droplet water, then ϕ_m would be nonetheless lower than the actual ϕ. While it is clear that the value of ϕ can depend on the method used to measure it, and while it is likely that $\phi_{comp} \leq \phi$, the pNa values of ϕ_m seem excessively high. Finally, it may be noted that there is more nearly a constant difference between the values of $(1 - \phi_{comp})$ and $(1 - \phi_m)$ rather than a constant ratio. We will return to this point below.

B. Chloride and Fluoride Ion.

The results for the chloride ion measurements using NaCl and for the floride ion using NaF are qualitatively the same as for the sodium ion. The principal difference however, is that the ϕ_m values from pCl data are higher than from pNa, while the pF values are higher still. The results are summarized in Table 3.

Table 3.

Comparison of Phase Volumes Obtained from Various Specific Ion Electrodes in each Microemulsion system.

Ion System	$\left(\phi_M - \phi_{comp}\right)^a$		
	Na^+ (.1M NaCl)	Cl^- (.1M NaCl)	F^- (.03M NaF)
Tween 40[b]	.12 ± .02	.19 ± .03	.40 ± .03
Tween 60[c]	.13 ± .03	.20 ± .01	————
Tween 60/81[d]	.14 ± .01	.14 ± .01	.28 ± .01

a.   average over range of ϕ values examined.

b.   number of points (n) = 13 for Na^+, Cl^- (ϕ_{comp} = .15 - .79); n = 6 for F^- (ϕ_{comp} = .18 - .56).

c.   n = 9 for Na^+, Cl^- (ϕ_{comp} = .21 - .76).

d.   n = 3 for Na^+, Cl^-; n = 8 for F^- (ϕ_{comp} = .47 - .68).

While the results are about the same in the tween 40 and tween 60 systems, in the tween 60/81 system the ϕ_m values from pNa and pCl are the same, and those from pF are closer. Thus, as the composition of the continuous phase becomes more aqueous (vide supra), the ϕ_m values converge. How much of the difference in ϕ_m values for the various ions can be attributed to a change in γ and how

much to ε_{ap} is yet to be determined. Both effects have
been reported for the fluoride electrode in 60% aqueous
ethanol (9), and judging from these data it would seem
that a solvent effect on γ is of the right order of
magnitude. This is also consistent with the effect being
in the order $F^- > Cl^- \geq Na^+$. In fact, the pNa results are
the same in all three systems. However, one puzzling
feature is the roughly constant difference between ϕ_{comp}
and ϕ_m from pNa. It is tempting to attribute this to a
solvent effect on the electrodes but there are some ob-
stacles to this interpretation. First, it has been reported
that no specific solvent effect was observed on sodium
ion glass electrodes in methanolic NaCl solutions (10).
Second, as long as the composition of the continuous phase
is not strongly dependent on ϕ, a constant ratio of $(1 - \phi_{comp})$ to $(1 - \phi_m)$ is expected. Of course, this assumes
that ϕ_{comp} is proportional to the actual ϕ, but this seems
reasonable. Third, the nature of the continuous phase is
clearly different in the tween 60/81 system. Experiments
are currently in progress to investigate this effect.

V. CONDUCTIVITY MEASUREMENTS.

It might be anticipated that the electrical conductivity of o/w microemulsions may be quite similar to the
conductivity of o/w emulsions. There have been numerous
investigations of the dielectric properties of emulsions,
and there is a good review of this subject by Hanai (11).
The results may be summarized by saying that the conductivity
of polydisperse o/w emulsions best follow Bruggeman's
equation (12), while for monodisperse systems the data lie
between the Bruggeman and Maxwell (13) equations, which are
shown in Fig. 2.

There has been a recent report of measurements of the
dielectric constant of a benzene in water microemulsion
stabilized by a mixture of tween 20 and Span 20 with no
alcohol (14). Dilutions were made with a water-surfactant
mixture, and the system is apparently a kinetically stable
one since the results depend upon the initial volume of
benzene. However, at the highest initial ϕ examined
(.633), the results followed the Bruggeman equation. For
dilutions from lower initial values of ϕ, the deviations
were in the direction of the Maxwell equation, and were
attributed to the formation of micelles by excess emulsifier
in the continuous phase.

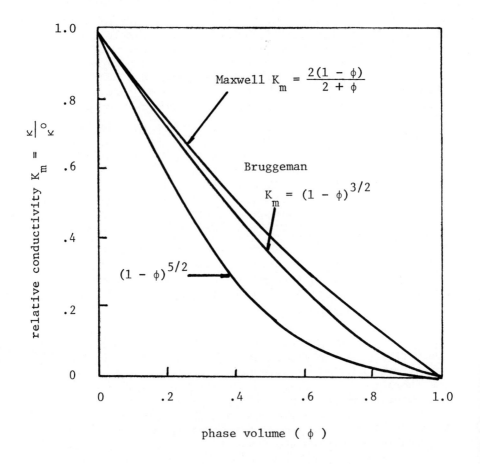

Fig. 2 Theoretical equations for the electrical conductivity of o/w emulsions (vide text, where κ is the specific conductivity of the emulsion and κ^0 is the specific conductivity of the continuous phase.

To put our results on a uniform basis, we have used the equivalent conductance (Λ) rather than the specific conductivity κ). If we begin with the empirical equation $\kappa/\kappa^0 = (1 - \phi)^n$, then $\Lambda/\Lambda^0 = (1 - \phi)^{n-1}$ since Λ is calculated from the measured conductivity and the stoichiometric concentration (C), while Λ^0 is based on the concentration in the continuous phase, $C/(1 - \phi)$. If the Bruggeman equation is obeyed, then $n-1 = 1/2$. In order to compare our results with theory, we must choose values for the phase volumes. The two values at hand are ϕ_{comp} and ϕ_m from pNa. In plotting the conductance date for the tween 40 and tween 60 systems, two principal features emerged. First, that the coefficient n-1 was definitely greater than one, and second, that the best value of the phase volume which fit the data was between ϕ_{comp} and ϕ_m. The data was, therefore, fit to the equation $\Lambda=\Lambda^0(1-a\phi)^m$, equation (iv), where Λ^0, **a** and **m** were adjustable parameters. The procedure employed was to fit Λ^0 and a by least squares for a range of m values from 0.3 to 2.5 using the values of ϕ_{comp}. For the tween 40 data, the best fit was obtained for m = 1.5. A plot of data are shown in Fig. 3 along with the values of Λ^0 and **a**.

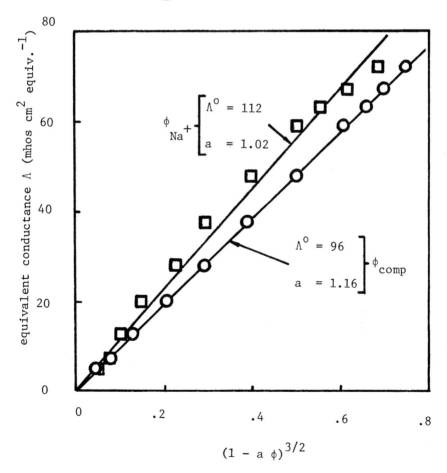

Fig. 3. Equivalent conductance vs.$(1-a\phi)^{3/2}$ for ϕ_{comp} and ϕ_m from pNa measurements (vide text.) The values

299

of Λ^O and $\underline{a}$ shown are the best values from least squares fit.

Also shown is a plot using ϕ_m from pNa. The values of $\underline{a}$ and Λ^O are for the value m = 3/2. The best fit for the tween 60 data are for m = 1.1, a = 1.23 and Λ^O = 98. However, for m = 1.5 the data also fit well with a = 1.10 and Λ^O = 109. All of these measurements were carried out at a stoichiometric concentration of 0.1 M NaCl, and Λ^O may be compared with our measured value of 109 for 0.1M NaCl in water.[2] For m = 3/2, n = 5/2, and plot for K_m = $(1 - \phi)^{5/2}$ is also shown in Fig. 2. While it is not established that n = 5/2 for nonionic microemulsions it is clear that n > 2 for the tween 40 and tween 60 systems examined here. The $\underline{a}$ values indicate that the effective phase volume is 10 – 20% greater than ϕ_{comp} The values of Λ^O in the microemulsion are, as expected, $\leq \Lambda$ (0.1M NaCl) and Λ^O (tween 40) $\leq \Lambda^O$ (tween 60). This, the conductivity measurements are in agreement with the expectations of the HLB system and the results of the CPO and potential measurements with respect to the nature of the continuous phase. There is at present no explanation of why the experimental dependence of K_m on ϕ lies below that of the Bruggeman equation.

---

2. It may be noted that the value of Λ was independent of stoichiometric concentration (within ± 5% at low ϕ up to ± 10% at high ϕ) over the range 0.01 – 0.30M.

VI. POLAROGRAPHIC MEASUREMENTS.

Walden's Rule is, in principle, only valid in the limit of infinite dilution, and the salt concentrations employed in this study are far from this limit. However, since Λ appears to be relatively independent of concentration in the microemulsions, we shall examine the results when Walden's rule (Stokes' Law) is applied with respect to the cintinuous phase. Proceeding on thus basis, $\Lambda^O \eta^O = \Lambda_{eff} \eta_{eff}$, where η is the viscosity and the "effective" subscript denotes that quantity with respect to the continuous phase. Then it follows that $\Lambda_{eff} \eta_{eff}$

$$= \frac{\kappa}{C_{eff}} \eta_{eff} = \frac{\kappa}{C} (1 - \phi) \eta_{eff}$$

$$= \Lambda (1 - \phi) \eta_{eff} = \Lambda^O \eta^O ,$$

where the other symbols have their previously defined mean-
ings. If $\Lambda = \Lambda^o$ $(1 - \phi)^{3/2}$, then $\dfrac{\eta_{eff}}{\eta^o} = \eta_{rel}$

$= (1 - \phi)^{-2.5}$ (V)

An equation with the form of (V) has been derived for the
bulk viscosity of a concentrated suspension of polydisperse
spheres (15). The bulk viscosity of the tween 40 system
does not follow any equation of this form above a phase
volume of 0.3.[1] However, if equation (V) is accepted as
holding in the continuous phase, then the diffusion coefficient (D) of a species in that phase is inversely
proportional to the viscosity, and it follows from equation
(V) that

$$D = D^o \ (1 - \phi)^{2.5}$$ (vi)

We have obtained some limited values of D by means
of d.c. polarography on Cd(II). These values involve
measurement of the limiting diffusion current and the mercury
drop mass, and are quite sensitive to temperature. We
estimate that our preliminary measurements of D are only
good to ± 20%. Nonetheless, the measured values of the
diffusion coefficient in the tween 40 system are consistent
with equation (vi) as shown in Fig. 4. The value of D^o
$= 6 \times 10^{-6}$ cm^2/sec may be compared with our measured

value of 6.5×10^{-6} cm^2/sec for Cd(II) in water. In this
case, ϕ_{comp} provides a much better fit than ϕ_m the pNa
measurements. Since the $CdCl_2$ concentration was only $10^{-3}M$,
it is likely that the presence of $LiClO_4$ as supporting
electrolyte caused a decrease in ϕ because of the increased
hydration of the lithium ion. In any event, the polarographic data also supports a value of n > 2.

VII. SUMMARY

The sodium ion electrode values of ϕ_m are the same in
all systems examined. They are higher than ϕ_{comp}, and seem
to be higher than the actual phase volumes. The chloride
ion electrode, and especially the fluoride ion electrode,
can be used in conjunction with the sodium ion glass
electrode to yield information on the composition of the
continuous phase in o/w microemulsions. The conductivity
data appear to obey an equation with a $(1 - \phi)^m$ dependence
of m = 5/2, and in any event greater than m = 3/2 dependence
of the Bruggeman equation which seems to hold for coarse o/w
emulsions. Limited polarographic diffusion coefficients

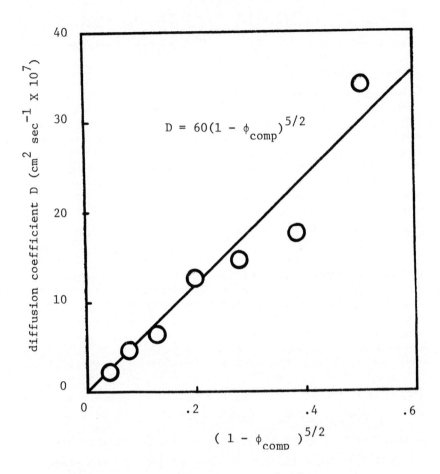

Fig. 4.   Polarographic diffusion coefficient of
$CdCl_2$ in the tween 40 microemulsion, 0.05M in $LiClO_4$.

are in accord with the conductance data, yielding an equation
for the effective microviscosity of the continuous phase of
the same form as for the bulk viscosity of a concentrated
emulsion.

VIII. REFERENCES

1. Schulman, J.H. Stoeckenius, W. and Prince L.M., J. Phys.
    Chem., 63, 1677 (1959).

2. Stoeckenius, W., Shulman, J. H., and Prince, L.M.,
    Kolloid Z., 169, 170 (1960).

3. Shulman, J.H., Matalon, R., and Cohen, M., Discuss. Faraday Soc., 117, (1952).
4. Shinoda, K. and Kunieda, H., J. Colloid Interface Sci, 42, 381 (1973).
5. Tosch, W. C., Jones, S.C. and Adamson, A. W., J. Colloid Interface Sci., 31, 297 (1969).
6. Levine S. and Robinson, K., J. Phys. Chem., 76, 876 (1972).
7. Robinson, R. A. and Stokes R. H., "Electrolyte Solutions", Butterworths, London, 1970, pp 310-313.
8. Mackay, R. A., Poziomek, E. J., J. Am. Chem. Soc, 94, 6107 (1972).
9. Lingane, J. J. Anal. Chem., 40, 935 (1968)
10. Ivanovskaya, I.S., Gavrilova, V.I. and Shul'ts, M.M., Soviet Electrochemistry, 6, 975, (1970).
11. Hanai, T., in "Emulsion Science", P. Sherman Ed., Academic Press, New York, N.Y., 1968, chapter 5.
12. Bruggeman, D.A.G. Ann. Phys. 24, 636 (1935).
13. Maxwell, J. C., "Electricity and Magnetism", Vol. 1., 3rd ed., p 440, Oxford (1892).
14. Clausse, M. and Sherman, P., C. R. Hebd. Seances Acad. Sci., Ser. C, 279, 919 (1974).
15. Cooke, C. E., Jr. and Schulman, J. H, in "Surface Chemistry", Ekwall, P., Groth, K., and Runnstrom - Reio, V., eds., Academic Press, New York, N. Y., 1965, pp. 231-251.

ACKNOWLEDGMENT

This work was partially supported by a grant from the Army Research Office.

ROLE OF IONS IN HETEROMOLECULAR NUCLEATION

A. W. Castleman, Jr.
Cooperative Institute for Research in Environmental Sciences
University of Colorado, Boulder

ABSTRACT

Heterogeneous nucleation leads to the formation of clouds, as well as aerosols from both natural and pollutant molecules in the atmosphere. Unfortunately, at the present time the basic theories pertaining to the various subclasses are not far advanced, and there is an urgent need for experimental data on the molecular aspects of the phenomenon to guide further theoretical developments. Heteromolecular nucleation is the subclass responsible for new particle formation and proceeds via the development of small clusters involving bonding between two or more chemically dissimilar species. Since ions provide an attractive central force field for polar molecules, ion induced nucleation is theoretically tractable and a good prototype for further developments in the theory of the general heterogeneous phenomenon.

The results of our research established that small ion clusters are a segment of the overall nucleation size spectrum and that the phase transition proceeds via the formation of small clusters which grow to a critical size and spontaneously nucleate. The thermodynamic properties of molecular clusters formed by water and ammonia clustered about ions of atmospheric importance were derived from laboratory experiments employing high pressure mass spectrometry. This paper presents data on the bond energies and entropies of successive clusters formed about ions of both open (e.g., Bi^+, Pb^+, Sr^+, and NH_4^+) and closed (e.g., alkali metal ions) electronic configurations. The importance of chemical bonding in effecting the formation of atmospheric ions and the stabilization of prenucleation embryos is discussed.

This research was sponsored in part by the National Science Foundation under grant AG423.

ELECTRICAL DISPERSION OF LIQUIDS FROM CAPILLARY TIPS -
A REVIEW

Raghupathy Bollini, *Southern Illinois University
at Edwardsville*
and
Steven B. Sample, *University of Nebraska, Lincoln*

ABSTRACT

*Numerous studies have been conducted on the atomization
or dispersion of liquids from capillary tips with the aid of
applied electric fields. The process makes possible the
production of monodisperse charged droplets having charge-to-
mass ratios that can be controlled over a very wide range.
The process lends itself to a broad spectrum of applications, and as a result, investigations of this process have
been carried out by researchers from many different areas.
Scientific applications of electrical dispersion include:
preparations of thin films of radioactive salts for nuclear
cross sectional studies; synthesis of hypersonic rain ero-
sion; and drop coalescence studies. In more applied fields,
electrical dispersion or spraying processes find application
in such areas as printing, painting, and deep space
thrustors.*
*This paper will summarize experimental and theoretical
results in the field of electrical dispersion, and will attempt to establish a number of general principles and
simple models. The paper will also discuss the effects of
various physical parameters on the electrical dispersion
process in terms of current experimental data.*

This paper appears in full in Colloid and Interface Science
Vol. III, page 291-300.

307

THE CONDENSATION OF METALLIC VAPORS*

David J. Frurip and S. H. Bauer
Cornell University

ABSTRACT

A kinetic model which is fully consistent with statistical-thermodynamic principles, has been developed for the condensation of metallic vapors. This is based on the solution of the master equation for n-mer growth:

$$A_{n-i} + A_i + M \rightleftharpoons A_n + M$$

$$\frac{1}{[M]} \frac{[d\,A_n]}{dt} = \sum_{i=1}^{5} k_{n-i}^{n} \left[A_{n-1} \right] \left[A_i \right] - \sum_{j=1}^{5} k_n^{n+j} \left[A_j \right] \left[A_n \right]$$

$$+ \sum_{j=1}^{5} k_{n+j}^{n} \left[A_{n+j} \right] - \sum_{i=1}^{5} k_n^{n-i} \left[A_n \right]$$

In these calculations the forward and reverse rate constants were related <u>via</u>: $-RT \ln K_{eq}[n;n-i;i] = \Delta H^{\circ} - T\Delta S^{\circ}$, for the corresponding standard enthalpy and entropy changes. Of course, the principal unknown term is the configurational entropy. We demonstrated that for the model to represent our observations adequately it is essential that ΔS_{conf} be set at its upper statistical limit.

Concurrently, the growth rates of Pb and Bi droplets were estimated from laser light scattering and turbidity data in supersaturated vapors. These measurements covered the range $20 < r < 90$ (Å). The data are expressed in terms of a time dependent parameter which characterizes the size distribution. Its evolution with time, in turn, is limited by the diffusion of monomers (and $2 \rightarrow 5$-mers) to the condensation centers. We also have preliminary data on rates of vaporization when a suspension of clusters $[10^3 < n < 10^6]$ under supersaturated conditions is subjected to a sudden rise in temperature, such as is imposed by a reflected shock.

Supported by the Material Science Center of Cornell University.

A STUDY OF THE GROWTH OF AMMONIUM BISULFATE AEROSOLS BY WATER-VAPOR CONDENSATION*

I. N. Tang and H. R. Munkelwitz
Brookhaven National Laboratory

ABSTRACT

The importance of understanding the behavior of sulfate aerosols under changing relative humidities in the atmosphere is well recognized. Substantial size changes of these submicron droplets can affect not only their transport and ultimate removal from the environment but visibility as well. The presence of ammonium bisulfate as a component of particulate sulfate in air under certain atmospheric conditions, although frequently inferred in the past, has been verified only recently by infrared spectral analyses of aerosol samples collected in urban areas. In the present work the phase transformation and growth of NH_4HSO_4 aerosols were investigated as a function of relative humidity. Experiments were carried out in a unique flow apparatus developed for studying size-distribution changes exhibited by deliquescent and hygroscopic salt aerosols. Briefly, a carrier gas stream containing a polydisperse aerosol is passed through an eletrostatic particle size separator, and a small fraction of the entering aerosol having a well-defined size distribution is extracted. This size selected fraction is subsequently injected into a growth chamber where the humidity can be varied and precisely controlled. The changes in particle size during growth are continuously monitored using a light-scattering photometer coupled with a pulse height analyzer and a multichannel data storage and retrieval system. Experimental data of phase transformation at the onset of growth and the subsequent droplet growth will be presented. Theoretical growth calculations based on requisite thermodynamic properties, which were determined for the system in separate studies, will also be given for comparison.

---

*This research performed under the auspices of U.S. Energy Research & Development Administration.

311

A STUDY OF DROPLET EVAPORATION RETARDATION

H. T. DelliColli, R. E. Shaffer and C. J. Cante
Edgewood Arsenal

ABSTRACT

Droplet evaporation has been retarded by the use of insoluble film-forming materials such as n-alkanols with chain lengths of from 14 to 20 carbon atoms. Droplet lifetime was found to be directly proportional to the chain length of the monolayer-forming molecules. However, the poor spreadability of the longer chain alcohols such as octadecanol and eicosanol, C_{18} and C_{20}, respectively, at the air/water interface required the use of a shorter chain material. Optimum results were obtained with 1-hexadecanol, cetyl alcohol, and mixtures of the C_{16} and C_{18} or C_{20} alcohols. The droplets employed were made of triply distilled water, electrolyte solutions, artificial and natural sea waters, and naturally occurring fresh waters.

Studies were made on the evaporation characteristics of single droplets suspended in a two-field hyperboloidal chamber which permitted the determination of the mass, charge, and surface tension of the evaporating droplet at any given point in time. The results obtained for monolayer-covered water droplets point out the inadequacy of older conventional means of droplet evaporation studies such as suspension on a fiber or the use of Stokes' law.

313

BROWNIAN COAGULATION IN THE TRANSITION REGIME

Marek Sitarski[*] and John H. Seinfeld
California Institute of Technology

ABSTRACT

The rate of Brownian coagulation of aerosols is studied theoretically. The coagulation process is represented by the steady state Fokker-Planck equation, which is solved analytically by the 13-moment method of Grad (1). For generality it is assumed also, that a given fraction α of collisions do not lead to a coagulation event but to specular reflection. The solution gives the steady state flux of particles of radius a_j to a stationary particle over the entire range of a_i and a_j. In the same way the flux of particles of radius a_i to the stationary particles of radius a_j is obtained. Assuming independence of the two processes, the coagulation rate is determined as the sum of the two processes. The expression obtained for the coagulation constant $\beta(a_i,a_j)$ is

$$\beta(a_i,a_j) = 4\pi(a_i+a_j)$$

$$\left[D_i \frac{1-\alpha^2+0.2(1-\alpha)^2\kappa_i}{0.06\pi(1-\alpha^2)\kappa_i^2+(0.68(1-\alpha)^2+0.1\pi(1+\alpha)^2)\kappa_i+ 1-\alpha^2} \right.$$

$$\left. + D_j \frac{1-\alpha^2+0.2(1-\alpha)^2\kappa_j}{0.06\pi(1-\alpha^2)\kappa_j^2+(0.68(1-\alpha)^2+0.1\pi(1+\alpha)^2)\kappa_j+1-\alpha^2} \right] \quad [1]$$

Where D_i and D_j are the diffusion coefficients of the particles of radii a_i and a_j, respectively, which are determined by the Stokes-Einstein law with the slip correction of Phillips (2). κ_i and κ_j are effective Knudsen numbers, which are defined as the ratio of the relaxation times of the particles in the background gas to the times during which the particles move the distance (a_i+a_j) with the average Maxwellian velocities $\bar{c}_i$ and $\bar{c}_j$. The effective Knudsen number κ_i is expressed by the following relation

$$\kappa_i = 10\sqrt{\frac{m_i}{2\pi kT}} \; D_i/(a_i + a_j) \quad [2]$$

where m_i is the mass of the particle of radius a_i and k is the Boltzmann constant. (Analogous expression for κ_j).

---

[*] On leave from the Institute of Physical Chemistry of the Polish Academy of Sciences.

The coagulation rates predicted by the above theory are close to those predicted by Fuchs' interpolation formula (3) and to experimental rates recently reported by Chatterjee et al. (4). (See Table 1.)

Table 1. The Ratio of Coagulation Constants Predicted by Various Theories and Measured by Chatterjee et al. (4) to the Smoluchowski Continuum Coagulation Constant for Coagulation of Equal-sized Droplets of Diethylhexylsebacate in Helium at 298°K.

$Kn = \lambda_{He}/a_i$	κ_i, Eq.[2]	$\dfrac{\beta_{EXP}^{\dagger}}{\beta_c}$	$\dfrac{\beta_F^{\ne}}{\beta_c}$	$\dfrac{\beta(a_i,a_i)}{\beta_c}$ Eq.[1] with $\alpha=0$
0.83	0.084	1.59±0.14	2.04	1.80
1.10	0.102	2.06±0.24	2.43	2.16
1.66	0.141	2.78±0.29	3.19	2.92

†Ratio measured experimentally by Chatterjee et al. (4)
$\ne\beta_F$ refers to the Fuchs constant (3).

References

1. Grad, H., Comm. Pure Appl. Math., 2, 331 (1949).

2. Phillips, W. F., Phys. Fluids, 18, 1089 (1975)

3. Fuchs, N. A., "Mechanics of Aerosols," Pergamon Press, N.Y., 1964.

4. Chatterjee, A., Kerker, M., and Cooke, D. D., J. Coll. Interface Sci., 53, 71 (1975).

SURFACE CHEMICAL ASPECTS OF ELECTROSTATIC PRECIPITATION OF FLYASH

Robert W. Coughlin and Pavel Ditl
Lehigh University

ABSTRACT

The efficiency of electrostatic precipitation is closely related to the electrical resistivity of the collected fly-ash which in turn is strongly influenced by the chemical composition of the associated flue gas. Experimental measurements and a theoretical development are presented to show that the lowered resistivity of flyash brought about in practice by the addition of so called "conditioning agents" to the flue gas may be understood and explained in terms of capillary condensation of liquid at the points of contact in a precipitated ash layer, thereby providing additional pathways for the flow of electrical current through the ash. A quantitative physicochemical description of such phenomena provides a framework for interpreting and correlating such observations often reported and frequently exploited practically but heretofore only poorly understood.

DYNAMIC ASPECTS OF W/O MICROEMULSIONS

Hans-Friedrich Eicke

Institute of Physical Chemistry
University of Basel
CH-4056 BASEL, Switzerland

Abstract:

In contrast with the considerable amount of information already available with respect to (static) properties of microemulsions, derived mostly from equilibrium thermodynamical experiments and calculations there appear to be few considerations as to possible exchange processes of solubilized material within these microemulsions.

In relation to a recent investigation (1) on the stability of micelles containing solubilized polar liquids in apolar media, it was thought worthwhile to study the exchange of polar solubilizate between micelles in an apolar environment. By means of fluorescent measurements it is shown that there is a rapid exchange of electrolyte solutions and water between AOT micelles in isooctane occurring during collisions (2). A theoretical approach supported by experiments is offered with regard to a possible exchange mechanism. The various conclusions drawn from this model describing different aspects of the exchange are consistent with each other and with data taken from the literature.

REFERENCES

(1.) Eicke, H.F., J. Colloid Interface Sci. 52, 65 (1975).
(2.) Eicke, H.F., Shepherd, J.C.W., and Steinemann, A., J. Colloid Interface Sci. (1976) in press.

A LABORATORY STUDY ON THE CORRELATION OF INTERFACIAL CHARGE
WITH VARIOUS INTERFACIAL PROPERTIES IN RELATION TO OIL
RECOVERY EFFICIENCY DURING WATER FLOODING

M. Chiang, K. S. Chan, and D. O. Shah
University of Florida

ABSTRACT

The objective of the proposed research was to correlate
the interfacial charge with other interfacial properties of
crude oil/brine interface. In the present study, the crude
oil (Seeligson Oil Field, Texas) was dispersed ultrasonical-
ly in aqueous solutions of different NaCl concentrations.
The crude oil droplets exhibited the maximum electrophoretic
mobility at 3.5% NaCl concentration. At the same NaCl concentration, the drop-volume of the crude oil in brine was
minimum which suggested that the interfacial tension at the
crude oil/brine interface was also minimum. Moreover, the
coalescence rate was slowest at 3.5% NaCl concentration. We
propose that these changes observed at 3.5% NaCl concentration are due to the surface charge density at the crude oil/
bring interface. Furthermore, we observed the maximum oil
recovery efficiency (i.e. the minimum residual oil) in sand-
packs when the crude oil was displaced by 3.5% NaCl solution.
A possible explanation for the maximum oil recovery efficiency is proposed in terms of the interfacial charge at the
crude oil/brine interface and that at the sand/brine interface.

THE STRUCTURE OF MICROEMULSIONS
CONTAINING NONIONIC SURFACTANTS

Stig Friberg
The Swedish Institute for Surface Chemistry
Stockholm, Sweden

The structure of microemulsions has been intensely discussed ever since their introduction three decades ago. Suggestions about their structure and the reason for their stability have been offered, and in earlier contributions a difference from micellar solutions was implicated. This has later been shown not to be correct if the condition on thermodynamic stability of the system is retained.

Microemulsions containing nonionic surfactants constitute simple model systems since the influence from the diffuse electric double layer is eliminated.

The results of the investigations show the existence of a) normal micelles b) combination micelles which required a minimum amount of hydrocarbon for stability c) inverse micelles, and d) isotropic solutions containing concurrently solubilized water and hydrocarbon in equilibrium with an aqueous _and_ a hydrocarbon solution.

^{13}C NUCLEAR MAGNETIC RESONANCE STUDIES ON MOTION AND STRUCTURE OF NONIONIC SURFACTANT MICELLES AND MIXED MICELLES WITH PHOSPHOLIPIDS

Anthony A. Ribeiro and Edward A. Dennis[1]
University of California at San Diego

Natural abundance ^{13}*C nuclear magnetic resonance studies on nonionic surfactants in organic solvents and as micelles in aqueous media are presented and spectral assignments given. Specifically, p,tert-octylphenylpolyoxyethylene ethers (such as Triton X-100) with a polydisperse distribution of oxyethylene chains and an* n-*alkyl polyoxyethylene ether of defined oxyethylene content were studied. NMR parameters were found to be similar for both the polydisperse and homogeneous surfactants. Spin-lattice relaxation times* (T_1) *for these surfactants in* CD_3OD *and* $CDCl_3$, *suggest that in these organic solvents free tumbling of these molecules occurs with some degree of internal or segmental motions in the hydrophobic alkyl and polar polyoxyethylene chains. Upon micelle formation in aqueous media, the surfactant molecules become restricted from free tumbling. Interpretations of the relaxation times in terms of a classical micellar structure with hydrophobic alkyl chains in an interior region and oxyethylene chains in a polar palisade layer suggest that a maximal restriction of motion occurs at the hydrophobic-hydrophilic interface and progressively increasing segmental motions occur in both chains outward from the interface.*
Additionally, we have used ^{13}*C-NMR to study mixed micelles of these surfactants with phospholipids. These mixed micelles serve as useful membrane models and are an excellent form of phospholipid substrate for lipolytic enzymes. The* ^{13}*C-NMR studies show that the presence of phospholipids in mixed micelles does not alter the structural characteristics of the nonionic micelle. The phospholipid in mixed micelles, however, exhibits narrower lines and longer* T_2^* *relaxation times than it does in bilayer structures and this is consistent with a model in which less motional restrictions occur for phospholipids in mixed micellar structures than in bilayer structures.*

---

[1] To whom correspondence should be addressed.

I. INTRODUCTION

The *p,tert*-octylphenylpolyoxyethylene ethers (OPE) (I) and the alkylpolyoxyethylene ethers (APE) (II) constitute

$$(CH_3)_3CCH_2C(CH_3)_2 \text{—} \langle \text{ring} \rangle \text{—} (OCH_2CH_2)_nOH \qquad (I)$$

$$CH_3(CH_2)_X(OCH_2CH_2)_nOH \qquad (II)$$

major classes of nonionic surfactants which are widely used as detergents, solubilizers, and emulsifiers. Many of these ethers are currently used to solubilize biological membranes and are a major means of purifying membrane-bound proteins and enzymes with retention of their biological activity (1). These amphipathic nonionic surfactants form micelles in aqueous media (2-6), and an understanding of the detailed structure of these micelles is important as these surfactants form mixed micelles with phospholipids and other amphipaths as a product of membrane solubilization (1,7,8). Furthermore, mixed micelles of these surfactants and phospholipids serve as an excellent form of phospholipid substrate for enzymes of phospholipid metabolism, while phospholipids in bilayer structures are by comparison poor substrates (9,10).

II. ^{1}H-NMR RELAXATION STUDIES ON NONIONIC MICELLES

Several ^{1}H-NMR relaxation studies on nonionic surfactants in solution have been reported (11-14). They are consistent with the classical picture of micelles in which the nonionic surfactant molecules aggregate with their hydrophobic alkyl chains in an interior core and their more polar polyoxyethylene chains forming an external palisade layer (15). Water molecules presumably hydrate the polyoxyethylene chains (4). Recent 220 MHz ^{1}H chemical shift and T_1 relaxation time studies on the OPE system suggest that little or no water penetrates to the hydrophobic core (13,14). Unfortunately, ^{1}H-NMR spectra for these surfactants yield overlapping, envelope lines for alkyl chains and a broad envelope of partially resolvable closely-spaced, but chemically-shifted lines for the oxyethylene protons. Even at 220 MHz, resolution of all of the polyoxyethylene groups is not possible and spectra at much higher fields are required for more detailed ^{1}H-NMR work. Fortunately, ^{13}C-NMR allows resolution of some individual carbon atoms in both the alkyl and polyoxyethylene chains of these ethers and a further assessment of mobility in both regions of these surfactants is possible (16).

III. ^{13}C-NMR STUDIES ON NONIONIC MICELLES

A.   Spectral Assignments in Various Solvents

We have initiated ^{13}C-NMR studies on a number of poly-oxyethylene containing surfactants. Fig. 1 shows ^{13}C-NMR spectra for p,tert-octylphenol, tetraethylene glycol, and Triton X-100 (I, n = 9-10), a commercial surfactant containing a polydisperse distribution of oxyethylene chains, along with their suggested spectral assignments. Peaks corresponding to alkyl, phenyl, and oxyethylene regions are readily disting-uishable. Several OPE surfactants (16), ethylene glycols of varying oxyethylene content, and a few mixed alkyl glycol monoethers (manuscript in preparation) were studied in order to distinguish individual alkyl and oxyethylene carbons.

We found that the polyoxyethylene carbon signals at 70.5 ppm (in $CDCl_3$) increase with oxyethylene content, suggesting that this peak corresponds to most of the interior oxymethyl-ene carbons. Peaks (f, 1) were found to be present for all surfactants and glycols studied, clearly suggesting that these lines correspond to carbons in the terminal oxyethylene unit. Peaks (g,h) were present only for the OPE surfactants clearly suggesting that these lines correspond to carbons in the oxy-ethylene unit adjacent to the phenol ring (16). It should be noted, however, that from all of the available data, the two pairs of methyl carbons and non-protonated carbons in the p,tert-octyl group and the two carbons within the terminal unit (peaks f, 1) are not unequivocally assigned at this time, although a probable assignment is suggested (16). With these caveats in mind, parentheses have been placed around certain peak pairs in Fig. 1. Specifically-labelled ^{13}C-enriched or deuterated surfactants will be required to make definitive assignments.

Fig. 2 shows ^{13}C-NMR spectra for a polydisperse surfact-ant Triton X-100 (I, n = 9-10) and a homogeneous surfactant n-dodecyl octaethylene glycol monoether (II, x = 11, n = 8) in a micellar state in aqueous media. Clear structural differ-ences are reflected in the ^{13}C spectra of the OPE and APE sys-tems as can be seen for the two different kinds of alkyl chains and the presence of the phenyl ring in the OPE system. Moreover, the APE exhibits a new resonance (peak j), which is assigned to the C-1 carbon of the alkyl chain as it is pre-sumably shifted to lower field by the polar oxygen heteroatom.

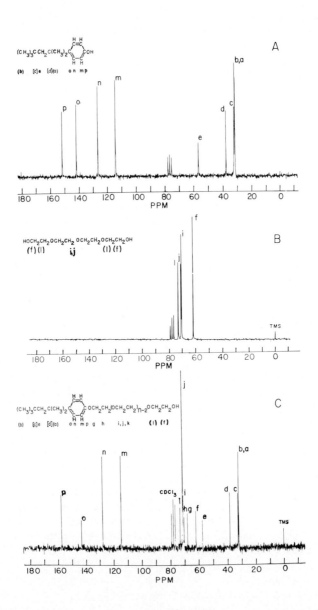

Fig. 1. Natural abundance ^{13}C-NMR spectra at 25 MHz and 40°C of (A) 1.5 M p,tert-octylphenol (B) 1.4 M tetraethylene glycol and (C) 0.4 M Triton X-100 in CDCl$_3$-1% TMS. Parentheses indicate pairs of peaks that are not unequivocably assigned.

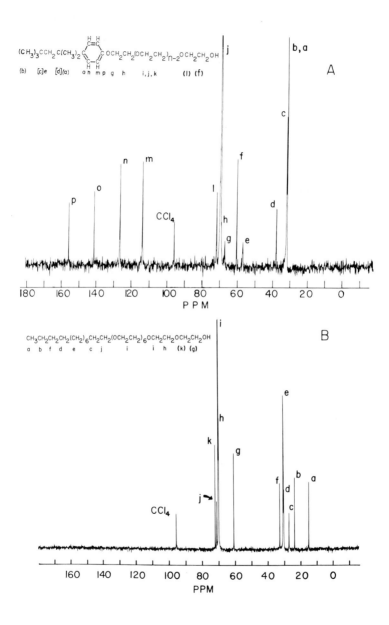

Fig. 2. Natural abundance ^{13}C-NMR spectra of nonionic surfactant micelles in D_2O at 25 MHz and 40°C of (A) 0.4 M Triton X-100 and (B) 0.4 M n-dodecyl octaethylene glycol mono-ether.

It is clear that both surfactants give similar ^{13}C-NMR spectra for the oxyethylene chains, indicating that the NMR parameters obtained for polydisperse samples are characteristic of their homogeneous analogues as is the case for several other physical properties (17).

We have found that the alkyl and phenol peaks for OPE and the alkyl peaks of APE have similar chemical shifts in organic solvents and in micellar form in D_2O. This is consistent with a similar microenvironment for these hydrophobic groups in micelles and in organic solution. By contrast, all of the hydrophilic oxyethylene carbons for both surfactants in micelles shift upfield about 1 ppm from their resonance position in $CDCl_3$ and CD_3OD (16).

B. T_1 Relaxation Times in Various Solvents

The NMR spectra of these nonionic micelles show narrow NMR lines [see ref. (16)] indicating that considerable fluidity exists within these aggregates. In an attempt to quantitate the mobility, ^{13}C T_1 values were determined for the protonated carbons of both the OPE and APE surfactants and are shown as plots of NT_1 versus the molecular axis in Fig. 3.

Ideal NMR T_1 studies on micellar systems require measurements for the surfactant molecule as a monomer below the critical micelle concentration (CMC) in a given solvent and then as a micellar aggregate, in order to determine changes in molecular motion upon micelle formation. Measurements of this type have been accomplished for some ionic micellar systems (18). Unfortunately, the nonionic surfactants studied here have CMC $<10^{-4}$ M (2-6), making natural abundance ^{13}C-NMR measurements on the monomer prohibitively tedious. Such measurements will require ^{13}C-enriched surfactants.

For the reasons above, it was necessary to use solvents in which the aggregation of nonionic amphipaths is minimized for comparison with results for micelles in aqueous media. The choice of a non-aqueous solvent is critical as solvents with two or more potential hydrogen bonding centers enhance micelle formation similar to that in water (19). Thus, micelle formation by these surfactants has been reported in ethylene glycol and formamide (20-23). By contrast, in hydrocarbon solvents like decane or cyclohexane inverted micelles may form (24). Linear hydrogen bonding solvents with a single hydrogen bonding center, however, do not support micelle formation (21) and several reports have appeared to show the destruction of aqueous micelles in the presence of high amounts of the lower alcohols like ethanol and methanol (25-28). With this background in mind, we have made ^{13}C T_1 determinations for the nonionic surfactants in methanol and

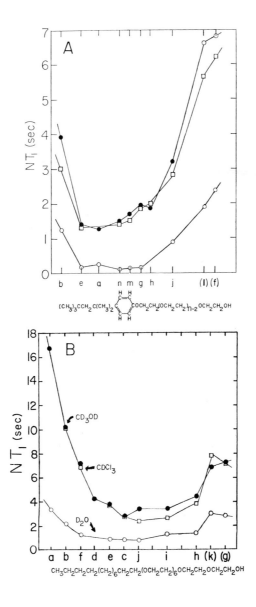

Fig. 3. Plots of NT_1 values of protonated carbons of nonionic surfactants against the long axis of the molecule for (A) 0.4 M Triton X-100 and (B) 0.4 M n-dodecyl octaethylene glycol monoether in CD_3OD (●), $CDCl_3$ (□), and D_2O (○).

chloroform for comparison with determinations in aqueous media.
The relaxation data yield the following points:

1. *In all solvents, the relaxation times appear shortest for the OPE in the vicinity of the rigid phenyl ring and for the APE at the confluence of the alkyl and oxyethylene chains. The T_1 values progress to higher values at each carbon successively further away from these regions.*

2. *T_1 is clearly shorter for all peaks when the surfactant forms micelles in D_2O than when it is in a non-micelle promoting solvent.*

3. *In both CD_3OD and $CDCl_3$, the T_1 values for the alkyl chain are virtually identical but the T_1 values for the oxyethylene chain are slightly higher in CD_3OD than in $CDCl_3$.*

C.   Interpretation of Relaxation Times

Interpretations of ^{13}C T_1 values for protonated carbon
nuclei of large molecules are usually discussed in terms of
dipole-dipole interactions between the carbons and their
directly attached protons (29). In terms of these dipolar
mechanisms, motions can be discussed in terms of 1.) an overall tumbling of a given molecule, which implies that every
carbon atom of a spherical molecule should have an equal NT_1
value, and 2.) independent rotation about single bonds in the
molecule, which in principle can be different for every carbon
of the molecule (29,30). Obviously, for long chain molecules,
local viscosity effects might influence internal motions.
Several attempts to quantitatively treat multiple internal
motions of alkyl chains have appeared (31-33).

A simple interpretation in terms of 1.) and 2.) above
suggests that for large molecules all carbon atoms which cannot undergo internal rotation and have directly-bonded protons
will usually show the same NT_1 values, while values of NT_1 for
other carbons could increase progressively along a chain from
rigid parts of the molecule to the termini. Observations of
these phenomena have usually been made for alkyl chains
attached to a rigid group and thus this kind of segmental
motion has been observed in 1-decanol (34), phospholipids (35),
large alkanes (33) and ionic micelles (23, 36-38).

Our present results suggest that these phenomena may also
be observed in the nonionic polyoxyethylene surfactants. In
CD_3OD and $CDCl_3$, the NT_1 values for the alkyl methylene,

dimethyl and protonated phenyl carbons (peaks e, a, n and m) are approximately equal for the OPE system. Moreover, the NT_1 values in the intersecting region of the alkyl and oxyethylene chains of APE (peaks c-h) are also about equal. Presumably, the NT_1 for these groups reflect only molecular tumbling, and using standard formulae, suggest tumbling correlation times of $\sim 10^{-11}$ sec for these surfactants. Longer NT_1 values suggest the presence of internal motions about single bonds for the tert-butyl methyl and the oxyethylene carbons of OPE and for the other alkyl and oxyethylene carbons of APE. Interestingly, both the alkyl and polyoxyethylene chains exhibit segmental motions, with gradients of mobility to the ends of the chains. It is also interesting to note the alkyl chain shows greater internal mobility than the polyoxyethylene chain.

Upon micelle formation in D_2O, all NT_1 values are reduced, consistent with restriction of these molecules from free tumbling when in a micellar environment. The largest reduction in NT_1 is observed for peaks n-g of OPE and peaks c-j of APE, respectively, and their NT_1 values now indicate molecular reorientations of $\sim 10^{-10}$ sec. These peaks correspond to those groups that would constitute the hydrophobic/hydrophilic interface within these micelles.

Significant internal motions about single bonds, however, must remain in other groups. For OPE, the NT_1 value of the tert-butyl methyl carbon (peak b) is now 8.1 NT_1 of the protonated phenyl ring carbons; this situation approaches the theoretical limit of 9 for rotating methyl groups on the reorienting axis of a molecule (29). For the resolved alkyl carbons of APE, the NT_1 values of chain-terminal carbons (peaks a-f) remain longer than for peaks e-j. These findings imply significant mobility in the hydrophobic core of these micelles, with increasing segmental motion away from the interface.

Similarly, for the resolved oxyethylene carbons the NT_1 values of the terminal unit remain longer than the unit near the interface. The evidence suggests increasing segmental motions may occur in this polar chain from the interface through the palisade layer to the terminal unit, presumably at the outer periphery of the nonionic micelle. Previous ^{1}H-NMR T_1 studies have implied that such a mobility gradient could exist in this polyoxyethylene layer; however, overlapping lines prevented more definitive conclusions (13,14).

IV. ^{13}C-NMR STUDIES ON MIXED MICELLES

In the presence of an excess of a nonionic surfactant, phospholipid bilayer structures are converted to mixed micellar structures, as shown by ^{1}H-NMR (7,8), gel chromatographic and ultracentrifugation techniques (39). A ^{13}C-NMR spectrum of mixed micelles of Triton X-100 and dimyristoyl phosphatidylcholine is shown in Fig. 4. Several resonance

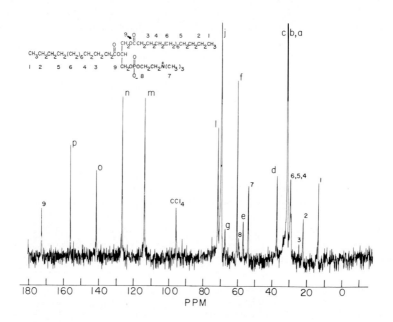

Fig. 4. Natural abundance ^{13}C-NMR spectrum at 25 MHz and 40°C of 0.4 M Triton X-100 plus 0.133 M dimyristoyl phosphatidylcholine in D_2O. Triton X-100 resonance lines are assigned as in Fig. 2A and only the resolvable phospholipid lines are assigned here.

lines in both the hydrophobic and hydrophilic regions of the phospholipid could be resolved. By contrast, ^{13}C-NMR spectra for phospholipids in bilayer structures generally yield spectra with broader resonance lines (see below).

^{13}C T_1 values for phospholipids in these mixed micelles are compared with reported values of phospholipids in bilayer-structures in Table I (40-42). Despite differences in field strengths, temperatures and phospholipids used by various investigators, it appears that to a first approximation, the T_1 values for phospholipids in the three aqueous structures are similar and may measure the same motional process. On the other hand, a comparison of T_2^* values from line widths in Table II, clearly suggest that the T_2^* values of phospholipids

TABLE I

^{13}C-NMR T_1 (sec) for Phosphatidylcholine in Mixed Micelles

Peak	Unsonicated Multibilayers[a] 15.1 MHz, 23°C Egg PC	Sonicated Vesicles[b] 25.1 MHz, 34°C Egg PC	Mixed Micelles[c] 25.1 MHz, 40°C Dimyristoyl PC
1	3.31	2.8	1.8
2	1.36	1.4	1.1
6	0.45	0.40	0.33
7	0.51	0.62	0.61
8		0.41	0.27
9		1.8	1.7

a. See ref. (40)
b. See ref. (41)
c. See ref. (42)

TABLE II

T_2^* (sec) for Phosphatidylcholine in Mixed Micelles[a]

Peak	Unsonicated Multibilayers[b] 15.1 MHz, 23°C Egg PC	Sonicated Vesicles[b] 15.1 MHz, 23°C Egg PC	Mixed Micelles[c] 25.1 MHz, 40°C Dimyristoyl PC
1	0.0267	0.0376	0.13
2	0.011	0.0282	0.12
4-6	(0.0035)	(0.0135)	(0.039)
9	(0.0077)	(0.0307)	(0.050)

a. Parentheses are placed around T_2^* values obtained from peaks with overlapping resonance lines or chemical shift non-equivalences.
b. See ref. (40)
c. See ref. (42)

in mixed micelles are considerably longer than in either unsonicated multibilayers or sonicated vesicles. Indeed, for the two singlet peaks which are clearly resolvable in the fatty acid chains of the mixed micelles (terminal methyl carbon and adjacent methylene), the T_2^* values increase dramatically from the multibilayer to vesicle to mixed micellar structures. Unfortunately, clear interpretations for other peaks is not possible due to overlapping lines, spin-coupling to heteronuclei, or chemical shift non-equivalence (42).

Relaxation phenomena of phospholipids have also been discussed in terms of intramolecular dipole-dipole relaxation processes, although the interpretation of relaxation times in phospholipids is quite complicated. Presumably, the similar T_1 relaxation times for phospholipids in bilayers, vesicles and mixed micelles can be attributed to similar internal, segmental motions parallel to the long axis of the phospholipid molecule. The dissimilar T_2^* values may reflect off-axis motions which may be a result of the phospholipid packing. As a structure becomes more ordered, large off-axis motions could become more restricted and overall motions more anisotropic. Thus, although the phospholipid in mixed micelles shows anisotropic motions, the narrow lines suggest less restriction in the anisotropic motions when compared with bilayer structures. This is presumably due to a less ordered packing of phospholipids in mixed micelles than in vesicles or multibilayers.

Finally, the chemical shift, relaxation time, and line width of all resonance lines of the surfactant micelles are unchanged within experimental error when phospholipid is present (16). These data and previous work clearly suggest that in the mixed micelle the phospholipid is intercalated into a nonionic surfactant micelle with the hydrophobic chains in the core and the polar groups in the palisade layer (14,42). The presence of phospholipid in the mixed micelle does not appear to greatly affect the microenvironment of the surfactant micelles, nor the motional behavior of individual groups in the micelle.

V. ACKNOWLEDGEMENT

We thank Dr. John Wright and Mr. Rick Freisen for aid in the operation of the JEOL-PFT-100 Fourier transform spectrometer system used in these studies. This work was supported by grant NSF BMS 75-03560. A.A.R. was a NIH predoctoral fellow GM-1045. The NMR facilities were supported by NIH grant RR-00-708.

VI.   REFERENCES

1. Helenius, A., and Simons, K., Biochim. Biophys. Acta 415,
    29 (1975).
2. Crook, E. H., Fordyce, D. B., and Trebbi, G. F., J. Phys.
    Chem. 67, 1987 (1963).
3. Ray, A., and Némethy, G., J. Am. Chem. Soc. 93, 6787
    (1971).
4. Kushner, L. M., and Hubbard, W. D., J. Phys. Chem. 58,
    1163 (1954).
5. Deguchi, K., and Meguro, K., J. Coll. Int. Sci. 48, 474
    (1972).
6. Nishikido, N., Moroi, Y., and Matuura, R., Bull. Chem.
    Soc. Japan 48, 1387 (1975).
7. Dennis, E. A., and Owens, J. M., J. Supramol. Struct. 1,
    165 (1973).
8. Ribeiro, A. A., and Dennis, E. A., Biochim. Biophys. Acta
    (Biomembranes) 332, 26 (1974).
9. Deems, R. A., Eaton, B. R., and Dennis, E. A., J. Biol.
    Chem. 250, 9013 (1975).
10. Warner, T. G., and Dennis, E. A., J. Biol. Chem. 250,
    8004 (1975).
11. Corkill, J. M., Goodman, J. G., and Wyer, J., Trans.
    Faraday Soc. 65, 9 (1969).
12. Clemett, C. J., J. Chem. Soc. (A), 2251 (1970).
13. Podo, F., Ray, A., and Némethy, G., J. Am. Chem. Soc. 95,
    6164 (1973).
14. Ribeiro, A. A., and Dennis, E. A., Biochemistry 14, 3746
    (1975).
15. Nakagawa, T., in "Nonionic Surfactants, Surfactant
    Science Series" (M. J. Schick, Ed.) Vol. 1, p. 558.
    Marcel Dekker, New York, 1976.
16. Ribeiro, A. A., and Dennis, E. A., J. Phys. Chem. 80
    (1976). In press, July 29 issue.
17. Becher, P., in "Nonionic Surfactants, Surfactant Science
    Series" (M. J. Schick, Ed.), Vol. 1, pp. 478-515.
    Marcel Dekker, New York, 1967.
18 Williams, E., Sears, B., Allerhand, A., and Cordes, E. H.,
    J. Am. Chem. Soc. 95, 4871 (1973).
19. Kresheck, G. C., in "Water: A Comprehensive Treatise,
    Aqueous Solutions of Amphiphiles and Macromolecules"
    (F. Franks, Ed.), Vol. 4, p. 141. Plenum Press, New
    York, N. Y., 1975.
20. Ray, A., and Némethy, G., J. Phys. Chem. 76, 809 (1971).
21. Ray, A., Nature 231, 313 (1971).
22. McDonald, C., J. Pharm. Pharmac. 22, 774 (1970).
23. McDonald, C., J. Pharm. Pharmac. 22, 148 (1970).

24. Fendler,J. H., and Fendler, E. J., "Catalysis in Micellar and Macromolecular Systems", p. 314, Academic Press, New York, N.Y., 1975.

25. Deguchi, K., Mizuno, T., and Meguro, K., J. Coll. Int. Sci. 43, 485 (1973).

26. Becher, P., and Trifeletti, S. E., J. Coll. Int. Sci. 43, 485 (1973).

27. Becher, P., J. Colloid Sci. 20, 728 (1965).

28. Sasaki, H., and Sata, N., Koll. Z., 199, 49 (1964).

29. Doddrell, D., Glushko, V., and Allerhand, A., J. Chem. Phys. 56, 3683 (1972).

30. Lyerla, J. R., Jr., and Levy, G. C., in "Topics in Carbon-13 NMR Spectroscopy" (G. C. Levy, Ed.), Vol. 1, pp. 79-149. Wiley-Interscience, New York, 1974.

31. Levine, Y. K., Partington, P., and Roberts, G. C. K., Mol. Phys. 25, 497 (1973).

32. Levine, Y. K., Birdsall, N. J. M., Lee, A. G., Metcalfe, J. C., Partington, P., and Roberts, G. C. K., J. Chem. Phys. 60, 2890 (1974).

33. Lyerla, J. R., Jr., McIntyre, H. M., and Torchia, D. A., Macromolecules 7, 11 (1974).

34. Doddrell, D., and Allerhand, A., J. Am. Chem. Soc. 93, 1558 (1971).

35. Levine, Y. K., Birdsall, N. J. M., Lee, A. G., and Metcalfe, J. C., Biochemistry 11, 1416 (1972).

36. Levy, G. C., Komoroski, R. A., and Halstead, J. A., J. Am. Chem. Soc. 96, 5456 (1974).

37. Roberts, R. T., and Chachaty, C., Chem. Phys. Lett. 22, 348 (1973).

38. Brown, J. M., and Schofield, J. D., J. Chem. Soc. Chem. Comm. 434 (1975).

39. Dennis, E. A., Arch. Biochem. Biophys. 165, 764 (1974).

40. Sears, B., J. Membrane Biol. 20, 59 (1975).

41. Godici, P. E., and Landsberger, F. R., Biochemistry 13, 362 (1974).

42. Ribeiro, A. A., and Dennis, E. A., J. Coll. Int. Sci. 55, 94 (1976).

NUCLEAR MAGNETIC RESONANCE STUDIES ON HYDRATION AND INTERACTIONS OF COUNTERIONS IN SOME COLLOID SYSTEMS

Hans Gustavsson and Björn Lindman
Physical Chemistry 2, Chemical Center, Lund, Sweden

Nuclear magnetic resonance, NMR, of alkali and halide ions has been shown to be sensitive to interactions and microdynamics of surfactant systems. In this work the shielding of $^{23}Na^+$ and $^{133}Cs^+$ nuclei has been studied, as well as the quadrupole relaxation of $^{23}Na^+$ and $^{37}Cl^-$. By the introduction of Fourier transform techniques much lower concentrations could be studied than previously feasible; for example with $^{23}Na^+$, measurements were made down to $0.5 \cdot 10^{-3}$ m. This makes many more problems accessible to study. Systematic studies of especially the sodium ion binding was performed as a function of surfactant concentration, alkyl chain length and polar end-group. The mechanism of counterion binding was inferred to change markedly with polar end-group, while alterations in the other factors have smaller effects on the ionic interactions. The water isotope effect in shielding was found to provide interesting novel information on counterion hydration. As a complement to the study of surfactant aggregates, the $^{23}Na^+$ and $^{37}Cl^-$ relaxation and shielding were also studied in the presence of different polyanions, polyampholytes and polycations at different acid dissociation degrees. An anomalous variation of counterion binding with the polyion acid dissociation degree is observed for certain systems.

I. INTRODUCTION

One important application of various NMR techniques concerns the study of ionic interactions in systems of biophysical interest. Much interest has been devoted to the use of $^{35}Cl^-$ NMR in the biological field (1), and $^{23}Na^+$ NMR has also been found to be of great utility (2-6). By the introduction of pulse and Fourier transform techniques, ^{23}Na has become one of the most conveniently studied nuclei in NMR. Thus we have determined sodium relaxation rates and chemical shifts in aqueous solutions down to $5 \cdot 10^{-4}$ molal concentrations.

There are essentially two kinds of electrically charged colloid systems occuring in biological systems: charged macromolecules (as e.g. proteins, nucleic acids, and polysaccharides) and aggregates of amphiphile molecules as in for example cell membranes. To have the basic knowledge, necessary to understand the mechanisms of ionic interactions in complex biological systems, we have for some years been working on model systems, like association colloids of simple surfactants (7-11). Recently we have also started NMR studies on the binding of small ions to synthetic polyelectrolytes (12).

The present report concerns $^{23}Na^+$ NMR relaxation rate and chemical shift studies in aqueous solutions of some simple surfactants as a function of polar end-group and alkyl chain length. Furthermore, $^{23}Na^+$ and $^{133}Cs^+$ chemical shift studies as well as studies of the water isotope effect in $^{133}Cs^+$ shielding were performed for the systems alkali octanoate-octanoic acid-water. Included are also $^{23}Na^+$ and $^{37}Cl^-$ NMR investigations of ionic interactions in aqueous solutions of poly-(methacrylic acid) (PMA), of poly(N,N-dimethylaminoethylmethacrylate) (PDMEM), of an ampholytic co-polymer of methacrylic acid and N,N-dimethylaminoethylmethacrylate (P(MA-DMEM)) and finally of a terpolymer of the two above mentioned monomers and ethylmethacrylate (P(MA-DMEM-EM)). In all these cases the ion binding was studied as a function of the degree of neutralization.

II. EXPERIMENTAL

The sodium octyl, nonyl and dodecyl sulfates were bought from Merck AG., sodium decyl sulfate and sodium octyl sulfonate from Eastman Kodak Co., and sodium octanoate and octanoic acid from BDH Ltd. Sodium octylbenzene sulfonate was a kind gift from Dr. J. Rouvière, Facultè des Sciences, Montpellier. These substances were used without any further purification. Cesium octanoate was obtained by neutralization of dry octanoic acid in ether with ethanolic cesium hydroxide solution. The obtained substance was recrystallized twice from ethanol and dried in vacuum. In order to exchange the acid proton in octanoic acid for ^2H, the acid was shaken four times with D_2O and then dried over molecular sieve. The polyelectrolytes were kind gifts from Prof. B. Törnell, Chemical Center, Lund.

For the $^{23}Na^+$ and $^{37}Cl^-$ experiments, a modified Varian XL-100 pulsed Fourier transform NMR spectrometer was utilized. The spectrometer modification, which is similar to that described by Traficante et al. (13), enables us to study most of the NMR active nuclei in the frequency range 0 to 100 MHz at

a magnetic field of 2.35 T. The $^{23}Na^+$ resonances were recorded at the frequency 26.47 MHz and those of $^{37}Cl^-$ at 8.16 MHz.
The sample temperatures were 30 ± 1°C. For ^{23}Na a pulse width
of 35 μsec. was used as a 90° pulse with an aquisition time of
0.4 sec. At least 100 transients were accumulated and each
reported value is the average of at least three separate determinations. An external reference was used, as described in
Ref. 12, to provide both the ^2H lock and the sodium reference
signal. The ^{37}Cl signals were recorded with the pulse width
80 μsec., an aquisition time of 0.1 to 0.4 sec., and external
^1H frequency lock. The experiments on ^{133}Cs were carried out
on a Varian Wide Line spectrometer at 7.80 MHz and at 26 ± 2°C.
The measuring technique is described elsewhere (8).

The chemical shifts were measured as the distance between
the peaks of the reference and the sample signals and were
converted by extrapolation to shifts relative to the respective ion at infinite dilution in water. A positive chemical
shift corresponds to a shift to higher magnetic field.

The $^{23}Na^+$ and $^{37}Cl^-$ transverse relaxation rates, R_2, were
calculated according to the formula $R_2 = \Delta\nu.\pi$, where $\Delta\nu$
symbolizes the line width at half height of a signal.

Solutions of the alkyl sulfates were studied freshly prepared in order to avoid hydrolysis. All the surfactant studies were performed on individually prepared samples, while
some of the polymer experiments were done by titrating a
sample with small volumes of HCl or NaOH solutions. The definition of α characterizing the degree of neutralization will
throughout be such that at α = 0 the solution contains equivalent concentrations of acid and polymer. α = 1 corresponds
to equivalent concentrations of base and polymer. Sodium
chloride was used to adjust the sodium or chloride concentration.

III. RESULTS AND DISCUSSION

A. Surfactant Systems

Previous studies (7,8,14) have demonstrated that so-
called rapid exchange conditions apply for counterions in
surfactant systems. Therefore, the observed NMR chemical
shift, δ, is given by the expression

$$\delta = \sum_i p_i \, \delta_i \qquad (1)$$

where δ_i is the intrinsic chemical shift for the counterion at
site i and p_i is the fraction of ions in site i.

For micellar systems it is usually assumed that only two sites have to be considered (10), namely free (f) and micelle bound (m) ions. We also assume that the pseudo-phase separation model of micelle formation (15) is valid with sufficient precision, and finally that the degree of counterion association, β, to the micelles is independent of the soap concentration. (β is given by the molar ratio of counterions and amphiphile in the micelle.) With these assumptions one easily derives that

$$\delta = \delta_f \qquad (2a)$$

at concentrations below the critical micelle concentration, cmc, while above the cmc

$$\delta = \delta_f + \beta(\delta_m - \delta_f) - \frac{\beta \cdot c_m}{c_t}(\delta_m - \delta_f) \qquad (2b)$$

Here c_m is the cmc, c_t the total surfactant concentration, δ_m is the intrinsic chemical shift for bound ions and δ_f is the chemical shift for the free aqueous ions. Since δ_f is expected to be approximately concentration independent, a plot of δ against the inverse surfactant concentration should consist of two straight lines intersecting at the cmc.

The experimental data from the ^{23}Na chemical shift studies of octyl, nonyl, decyl and dodecyl sulfates are shown in Fig. 1, where the chemical shifts are plotted against the inverse soap concentration. Fig. 1 demonstrates that the assumed model is reasonably obeyed for the four alkyl sulfates under consideration. The cmc values obtained (see Table 1) are in good agreement with those obtained with other methods (16). According to Eq. (2b), one can from the slope of the high concentration range or from the intercept with the chemical shift axis, evaluate the product $\beta \cdot \delta_m$ when δ_f is negligible compared to δ_m. We find for all the alkyl sulfates studied that $\beta \cdot \delta_m = 0.37 \pm 0.01$, i.e. the product is independent of the alkyl chain length. As cancelling of opposite effects is improbable this finding strongly suggests that both β and δ_m are constant in the series. Unfortunately, we have not yet been able from the NMR experiments to obtain either β or δ_m independently, but from diffusion studies (17), β has been found to be close to 0.6 for alkali ions in micellar systems of octanoate (18) and dodecyl sulfate (19) in a wide concentration range. Thus with $\beta = 0.6$, δ_m is obtained to 0.62 for the sodium alkyl sulfate micelles studied.

From these findings we conclude that the sodium ion binding mechanism is very similar for the four alkyl sulfates

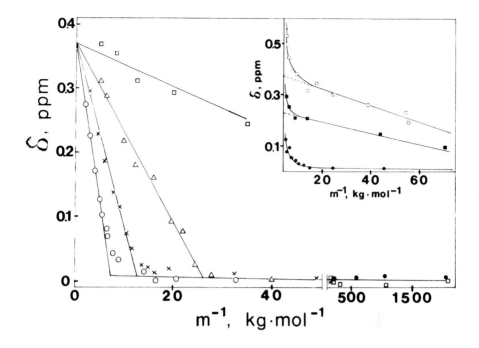

Fig. 1. $^{23}Na^+$ chemical shifts, δ (ppm), at 30^oC for aqueous solutions of sodium octyl (0), nonyl (X), decyl (Δ) and dodecyl ($\square$) sulfate as a function of inverse soap molality. The insert gives the high concentration region for sodium dodecyl sulfate as well as chemical shift data for sodium octylbenzene sulfonate ($\blacksquare$) and octyl sulfonate ($\bullet$).

studied. One would expect some reduction of the area per polar group with increasing alkyl chain length as a result of increasing hydrophobic interactions. Apparently this effect is small enough not to appreciably affect either the counterion association degree or the shielding of micellarly bound Na^+ ions.

In the insert of Fig. 1 is shown the sodium ion chemical shifts for octylbenzene sulfonate and octyl sulfonate in comparison with the dodecyl sulfate shifts. For octylbenzene sulfonate we obtain a cmc value of 0.010 and $\beta \cdot \delta_m$ = 0.23 ppm, which gives (also here with β = 0.6) an intrinsic micelle shift δ_m = 0.38 ppm. For sodium octyl sulfonate the shifts are too small to enable us to evaluate any cmc value as well as to locate the intersection point with the shift axis, but δ_m can be estimated to 0.1-0.2 ppm. For sodium octanoate we obtain as shown in Table 1 a δ_m value of -0.72 ppm.

TABLE 1

Values of CMC, $\beta \cdot \delta_m$ and δ_m Obtained from the Concentration Dependence of the $^{23}Na^+$ Chemical Shifts

Amphiphile	CMC	$\beta \cdot \delta_m$	δ_m
Sodium octyl sulfate	0.14	0.37	0.62
Sodium nonyl sulfate	0.077	0.37	0.62
Sodium decyl sulfate	0.038	0.37	0.62
Sodium dodecyl sulfate	0.0086	0.36	0.60
Sodium octylbenzene sulfonate	0.010	0.23	0.38
Sodium octyl sulfonate	-	-	ca. 0.1
Sodium octanoate	0.37	-0.43	-0.72

The considerable dependence of δ_m on polar end-group reflects a markedly different sodium ion interaction with micelles with different polar groups. As has already been discussed, in e.g. Ref. 10, this can be understood if the water of sodium ion hydration is considered to play a role in the ionic interactions. Thus differences in hydrogen bonding strength between sodium hydration water and the charged micelles may strongly affect the chemical shifts observed. According to the Kondo-Yamashita orbital overlap model (1) for the chemical shifts, an increased overlap should result in a downfield shift (paramagnetic shift). As increased water – sodium ion overlap integrals can be expected with increasing polarization of the water molecule it may be possible to relate an enforced hydrogen bonding with a paramagnetic chemical shift. Thus a hydrogen bond strength of H_2O in the order $-COO^- >$ $H_2O > -SO_3$ (alkyl-$SO_3 >$ aryl-SO_3) $>-SO_4$ is suggested.

From the insert of Fig. 1 may be inferred that for dodecyl sulfate a faster increase than expected from the simple two-site model occurs in the high concentration range. Analogous deviations are observed also with the other surfactants. These observations will be discussed in connection with the relaxation rate studies below.

The $^{23}Na^+$ and $^{37}Cl^-$ NMR relaxation rates, R, are generally dominated by the time dependent interaction between the quadrupole moment and the electric field gradient at the nuclei. The relaxation rate can, under extreme narrowing conditions, be written (20)

$$R = \frac{3\pi^2}{10} \left(\frac{e^2 qQ}{h}\right)^2 \cdot \frac{2I + 3}{I^2(2I - 1)} \cdot \tau_c \qquad (3)$$

Here eq is the largest component of the electric field gradient tensor, eQ is the electric quadrupole moment of the nuclei and τ_c is the time constant of the variation of the field gradient. The spin quantum number, I, equals for both nuclei 3/2. While previously we have presented pulsed NMR studies of both longitudinal and transverse relaxation (10) we will in this report be concerned only with the transverse relaxation rate, $R_2 = 1/T_2$. Using the two-site model, the pseudo-phase separation model and the assumption of a constant association degree of counterions to the micelles, we obtain, in the same way and with the same notations as for the chemical shifts above

$$R_2 = R_{2f} \qquad (4a)$$

for concentrations below the cmc, where R_{2f} is expected to be nearly concentration independent. Above the cmc we obtain

$$R_2 = R_{2f} + \beta \ (R_{2m} - R_{2f}) - \frac{\beta \cdot c_m}{c_t} \ (R_{2m} - R_{2f}) \qquad (4b)$$

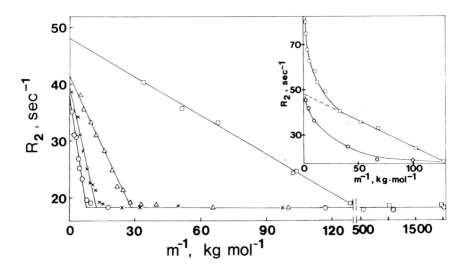

Fig. 2. The concentration dependence of the ^{23}Na$^+$ transverse relaxation rate (sec^{-1}) for the sodium octyl (0), nonyl (X), decyl (Δ) and dodecyl (□) sulfates.

The insert shows the relaxation rates for sodium dodecyl sulfate in the high concentration region and also $^{23}Na^+$ relaxation rates for sodium octylbenzene sulfonate (0).

In Fig. 2, the observed $^{23}Na^+$ relaxation rates are plotted against the inverse surfactant concentration. As can be seen from the figure, the model used is also for the relaxation rate studies reasonably obeyed and the cmc values obtained from the intersection points are in good agreement with those earlier obtained (Table 2). From the data given in the figure it is also possible to evaluate values of $\beta \cdot (R_{2m}-R_{2f})$ from the slopes of the high concentration ranges or from the intercepts with the relaxation rate axis. These values are given in Table 2. As can be seen from the table, these values are not constant but increases slowly in the series octyl sulfate to dodecyl sulfate. With a constant β (=0.6) this results in a set of increasing R_{2m} values also included in Table 2. For the octyl, nonyl and decyl sulfates this increase is just on the limit of significance, whereas the dodecyl sulfate value is significantly higher.

TABLE 2

Values of CMC, $\beta(R_{2m}-R_{2f})$ and R_{2m} Obtained from the Concentration Dependence of the ^{23}Na Relaxation Rates[a]

Amphiphile	CMC	$\beta \cdot (R_{2m}-R_{2f})$ sec^{-1}	R_{2m} sec^{-1}
Sodium octyl sulfate	0.14	20	52
" nonyl sulfate	0.085	22	54
" decyl sulfate	0.038	23	57
" dodecyl sulfate	0.0077	30	68
" octylbenzene sulfonate	0.010	–	–
Sodium octanoate	0.37	28	66

a. The value of R_{2f} used is 18.2 sec^{-1}. The data for sodium octanoate are from Ref. 10.

From Eq. (3) we see that the relaxation rate depends on the product of the correlation time, τ_c, and the square of the field gradient, eq. It is unfortunately difficult to quantitatively separate the effects of these two parameters except

for the so-called long correlation time case. The field gradient is expected to be affected by the alkyl chain length
primarily via the stronger hydrophobic interactions with longer chains leading to smaller areas per polar group.

It has been demonstrated earlier that the molecular motion determining τ_c is not directly connected with the overall
tumbling time of the micelles but instead related to a local
motion of water molecules (11,14,21). This motion can be e.g.
water molecule rotation or translation or an exchange of hydrated counterions between being bound to the micelle and free
in the intermicellar solution.

The insert of Fig. 2 shows that deviations from linearity
occur at higher concentrations for dodecyl sulfate. Such
increases are also observed for the other surfactants but are
not explicitly shown here.

For octylbenzene sulfonate a smoothly increasing R_2 is
obtained. This can be due to the space demanding benzene
groups giving much more loosely packed micelles than the alkyl
sulfate micelles. This effect can also be combined with a
weaker binding of the sodium ion to the sulfonate end-group.
This correlates well with the findings in the chemical shift
study.

For sodium dodecyl sulfate the additional increase in R_2
starts at about 0.03 m. This increase in the relaxation rate
can be interpreted as an increase in the intrinsic micelle
relaxation rate, R_{2m}, in β or in both R_{2m} and β. A simple
calculation shows that an increase in β alone is not sufficient to explain the observed data. With a β of about 0.6 for
the first formed micelles ($0.008 < c_t < 0.030$) we obtained $R_{2m} =$
68 sec^{-1}. If, for higher concentrations, we assume the maximal β value possible (i.e. $\beta = 1$), the highest obtainable
value of the observed relaxation rate would be 68 sec^{-1}. This
value is exceeded at a concentration of about 0.13 m. Thus,
we conclude, the increased relaxation rate must include an
increase in R_{2m} and thus some change in the binding state of
the bound counterions.

The phase equilibria of the system sodium octanoate –
octanoic acid – water have been characterized in detail by
Ekwall and co-workers (22,7), who also studied complex formation in the solution phase rich in octanoic acid (22). To
provide further insight into the ionic interactions of complexes formed, we have studied the $^{23}Na^+$ and $^{133}Cs^+$ shieldings
as well as the water isotope effect in $^{133}Cs^+$ shielding, of
solutions of water and octanoate in octanoic acid. For the
$^{23}Na^+$ chemical shifts given in Fig. 3a, two important features
are noted: 1) The chemical shift is upfield and 2) a rapid
decrease is observed for water concentrations below 4 moles of
water per octanoate. (The extensive broadening of the sodium
signal in this region (7) made the shift determinations more

difficult. Approximative experimental errors are shown in the figure.)

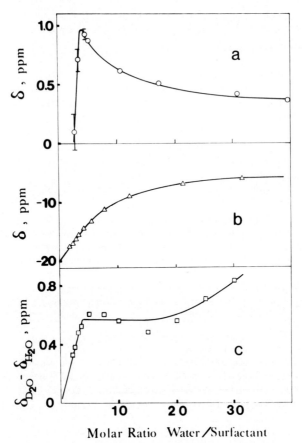

Fig. 3. a) The $^{23}Na^+$ chemical shift, δ (ppm), in the octanoic acid-sodium octanoate-water system as a function of the water concentration. The molar ratio octanoate to octanoic acid is held constant at 0.350. In b) the $^{133}Cs^+$ chemical shift and in (c) the water isotope effect in $^{133}Cs^+$ shielding are given as functions of the water concentration in the system octanoic acid-cesium octanoate-water. The molar ratio octanoate to acid is for the solutions in b) and c) 0.307.

The interaction between the sodium and carboxylate ions is expected to result in a negative shift (10), while, as has been shown by Bloor and Kidd (23), the interaction between Na^+

and carboxylic acid should give a positive shift. As a preliminary interpretation of the positive chemical shifts, we
thus propose that they are due to the interaction of the hydrated sodium ion both with octanoate and octanoic acid molecules. A specific complex between aqueous Na^+, $-COO^-$ and
$-\underline{C}OOH$ mentioned above seems thus to be of importance in the
reversed micellar region in this system. When the water concentration is below about four moles per mole octanoate the
decrease in sodium hydration results in the formation of direct contact ion pairs (COO^-Na^+). These are expected to give
a downfield shift.

It is most interesting to note, that in contrast to sodium, the $^{133}Cs^+$ chemical shifts, given in Fig. 3b, have
"normal" downfield values over the entire studied concentration range. Thus, for the cesium ion in this system, specific
complexes seem to be of much less importance than for sodium.
The water isotope effect in shielding has not yet been theoretically explained in detail, and our investigations are
still on a preliminary stage, but it may be proposed that an
essentially unchanged isotope effect on attachment of a counterion to an aggregate means an unchanged hydration of the ion.
A decrease in the effect should consequently be due to a decrease in the degree of counterion hydration. The water isotope effect in $^{133}Cs^+$ shielding in dilute aqueous solution
(11,24) is 1.25 ppm. In aqueous solutions of cesium octanoate
the effect remains at about this value up to high surfactant
concentrations (11). For the solutions studied here, the
isotope effect varies markedly with the water concentration
(Fig. 3c). In the concentration range studied, there is first
a decrease from about 0.8 ppm at higher water concentrations
down to about 0.5 ppm at around 4 moles of water per mole of
octanoate. Thereafter a more rapid and roughly linear decrease is observed. The straight line is pointing to the expected value zero, when no water is present. These observations indicate a slow decrease in the cesium hydration over a
wide concentration region and a faster final dehydration at
water concentrations below 4 molecules of water per cesium
molecule.

Since the hydration properties of cesium and sodium ions
are expected to be rather different due to the different sizes
of the outer orbitals of the ions, it is difficult to transfer information on cesium hydration to conclusions on sodium
hydration. It is, however, quite safe to assume that if the
cesium ion is extensively hydrated down to low water concentrations in a system, then is the sodium ion hydrated to at
least the same extent.

The preliminary conclusions of the data in Figure 3 are
thus:

349

i)   The sodium ion is extensively hydrated down to very low water contents.
ii) Specific complex formation seems to be of importance for the sodium ion in the reversed micellar system studied.
iii) The fast final dehydration of both sodium and cesium ions starts at ca. four moles of water per mole of counterion.

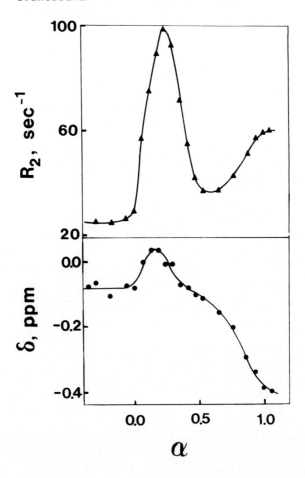

Fig. 4. The $^{23}Na^+$ relaxation rate in sec^{-1} and chemical shift in ppm for a 4% (by weight) aqueous solution of PMA as a function of the degree of neutralization, α. The sodium ion concentration was 0.481 m.

B. Polyelectrolyte Systems

We have recently started investigations of ion binding to
various synthetic polyelectrolytes (12). Even if such systems
are simple compared to natural biological polyelectrolytes,
they show in some cases rather complex ionic interactions. As
an example, Fig. 4 displaces the ^{23}Na$^+$ transverse relaxation
rate and chemical shift as a function of the stoichiometric
degree of neutralization, α, for poly(methacrylic acid), PMA.
Instead of giving regular titration curves, both shift and
relaxation deviate anomalously around α = 0.2. Such an anomaly in ^{23}Na$^+$ relaxation is not observed for poly(acrylic acid)
(25), but for PMA irregularities have been observed with several methods (26,27). It has been suggested that the anomal-
ies are due to a cooperative conformational transition from a
globular state at low α, to an expanded coil (28) for α higher
than about 0.2.

A correlation of these results with those obtained for
the system sodium octanoate-octanoic acid-water is interesting. For example, a tentative interpretation of the positive
chemical shift observed around α = 0.2, would be in terms of
complexes between Na$^+$, -COO$^-$ and -COOH.

By co-polymerization of acidic, basic and neutral monomers in various proportions, one obtains polyelectrolytes with
special characteristics. They have e.g. isoelectric point
with often pronounced solubility minima around this point, but
are easily soluble both on the basic and acidic sides of the
isolectric point. This kind of substances may serve as valuable model systems. In Fig. 5 are shown the observed ^{23}Na$^+$
relaxation rates and chemical shifts for two such polyampholytes as well as for PMA. The polyampholytes are a co-polymer
with the molar ratio methacrylic acid (MA) to dimethylamino-
ethylmethacrylate (DMEM) of 70/30, and a terpolymer composed
of equal parts of MA and DMEM and 30 mole-% ethylmethacrylate
(EM). We denote these substances P(MA-DMEM) and P(MA-DMEM-
EM), respectively. Their concentrations, expressed as concentrations of monomer, were in the experiments 0.124 and
0.078 molal and the sodium concentrations 0.150 and 0.098
molal, respectively. The PMA concentration was 0.113 molal
with a sodium ion concentration of 0.150 molal.

The relaxation rate enhancements of the different substances, when normalized as regards Na$^+$ and polyion concentration, are at high α-values approximately proportional to
the carboxylate content, and the same applies for the chemical
shifts. Thus, the presence of DMEM seems to have no marked
effect on the local sodium ion binding to the polyanion. We
have also measured the ^{23}Na$^+$ relaxation rate in PDMEM solutions and found no significant effects in any part of the pH

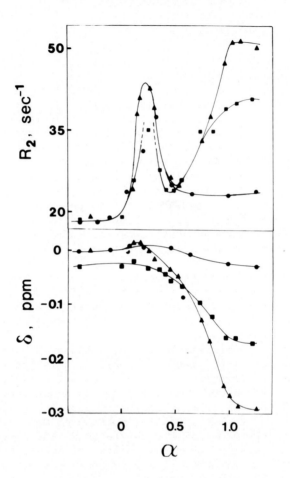

Fig. 5. The variation with the degree of neutraliz-
ation (α) of the $^{23}Na^+$ relaxation rate, R_2 (sec^{-1}), and
chemical shift, δ (ppm), for aqueous solutions of PMA
(▲), P(MA-DMEM) (■) and P(MA-DMEM-EM) (●). Concentrations are given in the text. The dashed part of the
curves indicates areas of insolubility for the polyampholytes.

range where the polymer is soluble.
    The ^{37}Cl transverse relaxation rate is shown as a funct-
ion of α in Fig. 6 for aqueous solutions of PMA and PDMEM.
Only very slight changes in the relaxation rate are observed
in the case of PMA. The main effect is a small step in the
relaxation rate at an α-value of about 0.2.
    For the PDMEM solutions .considerable effects are observed
in spite of the great chloride excess. It can be noted that

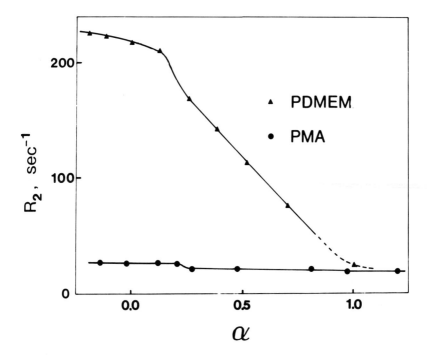

Fig. 6. The variation with the degree of neutralization (α) of the $^{37}Cl^-$ transverse relaxation rate, R_2 (sec^{-1}), for solutions of PDMEM (▲) and PMA (●). The concentrations, referred to monomer content, are 0.127 m for PDMEM and 0.156 m for PMA and the chloride to monomer molar ratios are 14.1 and 1.28, respectively.

under corresponding conditions, changes in $^{37}Cl^-$ relaxation are relatively much larger than for $^{23}Na^+$ relaxation. At low α-values, a very effective relaxation is observed. The linear decrease in relaxation rate for intermediate α-values, corresponds to a gradual titration of the dimethylammonium groups. (The substance is not completely soluble from about $\alpha = 0.8$.)

Thus here a similar conclusion can be drawn as for the sodium relaxation: The ^{37}Cl relaxation rate is expected to depend solely upon the chloride ion binding to cationic sites. As long as cooperative effects are not important one expects that the anionic binding sites should give very limited effects. Thus, it should be possible to characterize cooperative ion binding phenomena, in for example polyampholyte solutions, from deviations from the ideal behaviour.

Charged colloids containing the ionized groups considered above are of interest both in connection with synthetic

353

macromolecules and association colloids. As the local separated interactions can thus be expected to be similar in the two cases, comparative studies are useful for elucidating structural effects on counterion binding phenomena. For the ionic interactions in a number of biological systems, the modifying effects of adjacent groups and molecules are of great significance. Although our study demonstrates the utility of NMR shielding and relaxation investigations in this connection, the results hitherto obtained are much too limited to permit more than certain general conclusions. Also, in view of the limited space available more complete discussions of the results have to be deferred to future reports.

IV. REFERENCES

1. Lindman, B., and Forsén, S., "Chlorine, Bromine and Iodine NMR, Physico-Chemical and Biological Applications", Vol. 12 of "NMR, Basic Principles and Progress" (P. Diehl, E. Fluck, and R. Kosfeld, Eds.) Springer Verlag, Berlin, in press.
2. James, T.L., and Noggle, J.H., Proc. Nat. Acad. Sci. U.S. 62, 644 (1969).
3. Lindman, B., and Lindqvist, I., Acta Chem. Scand. 23, 2215 (1969).
4. Edzes, H.T., and Berendsen, H.J.C., Ann. Rev. Biophys. Bioeng. 4, 265 (1975).
5. Lindblom, G., Persson, N.-O., Lindman, B., and Arvidson, G., Ber. Brunsenges. Phys. Chem. 78, 955 (1974).
6. Andrasko, J., Lindqvist, I., and Bull, T.E., Chem. Scr. 2, 93 (1972).
7. Lindman, B., and Ekwall, P., Kolloid-Z.Z. Polym. 234, 1115 (1969).
8. Gustavsson, H., Lindblom, G., Lindman, B., Persson, N.-O., and Wennerström, H., in "Liquid Crystals and Ordered Fluids, Vol. 2", (J.F. Johnson, and R.S. Porter, Eds.) p. 161. Plenum Press, New York, 1974.
9. Gustavsson, H., and Lindman, B., J.S.C. Chem. Commun. 1973, 93.
10. Gustavsson, H., and Lindman, B., J. Amer. Chem. Soc. 97, 3923 (1975).
11. Gustavsson, H., and Lindman, B., in "Proc. Int. Conf. on Colloid and Surface Sci., Budapest 1975" (E. Wolfram, Ed.) p. 625. Akadémiai Kiadó, Budapest, 1975.
12. Gustavsson, H., Lindman, B., and Törnell, B., Chem. Scr. in press.
13. Traficante, D.D., Simms, J.A., and Mulcay, M., J. Magn. Resonance 15, 484 (1974).

14. Wennerström, H., Lindblom, G., and Lindman, B., Chem. Scr. 6, 97 (1974).
15. Shinoda, K., Nakagawa, T., Tamamushi, B.-I., and Isemura, T., "Colloidal Surfactants", p. 25. Academic Press, New York, 1963.
16. Mukerjee, P., and Mysels, K.J., Nat. Stand. Ref. Data Ser., Nat. Bur. Stand., No. 36 (1971).
17. Clifford, J., and Pethica, B.A., Trans. Faraday Soc. 60, 216 (1964).
18. Lindman, B., and Brun, B., J. Colloid Interface Sci. 42, 388 (1973).
19. Fabre, H., "Etude par self-diffusion des phases L_1 et L_2 des melanges ternaires", Thesis, Montpellier, 1976.
20. Abragam, A., "The Principles of Nuclear Magnetism", p. 314. Clarendon Press, Oxford, 1961.
21. Lindblom, G., Persson, N.-O., and Lindman, B., in "Chemie, physikalische Chemie und Anwendungstechnik der grenzflächenaktiven Stoffe", Vol. II, p. 925. Carl Hanser Verlag, München, 1973.
22. Ekwall, P., Mandell, L., Solyom, P., and Friberg, S., Kolloid-Z.Z. Polym. 233, 938 (1969); 233, 945 (1969); 233, 955 (1969).
23. Bloor, E.G., and Kidd, R.G., Can. J. Chem. 46, 3425 (1968).
24. Halliday, J., Hill, H.D.W., and Richards, R.E., Chem. Commun. 1969, 219.
25. Leyte, J.C., Zuiderweg, L.H., and van der Klink, J.J., in "Polyelectrolytes" (E. Sélégny, Ed.) p. 383. D. Reidel Publishing Company, Dordrecht, 1974.
26. Crescenzi, V., ibid, p. 115.
27. Van der Veen, G., and Prins, W., ibid, p. 483.
28. Leyte, J.C., and Mandel, M., J. Polymer Sci. A 2, 1879 (1964).

THE PROTON MAGNETIC RESONANCE SPECTRA AND MOLECULAR
CONFORMATIONS OF SODIUM N-ACYL SARCOSINATES
IN AQUEOUS SOLUTION (PART II)

Hirofumi Okabayashi, Koji Kihara
and Masataka Okuyama
*Department of Engineering Chemistry, Nagoya Institute
of Technology, Gokiso, Nagoya 466, Japan*

Proton magnetic resonance spectra were measured for sodium
N-acyl sarcosinates in deuterium oxide, and the concentration
dependence of molecular conformations of these molecules was
observed. The effect of temperature on the population of the
isomers was also investigated, and then the energy difference
between the two isomers was found to be extremely small.
The energy barrier to rotation about the carbonyl carbon-
nitrogen bond was estimated by line shape analysis in the monomolecular dispersion state and the micelle state.

I. INTRODUCTION

In a preceding paper (1), we described the concentration
dependence of the NMR spectra of sodium N-acyl sarcosinates
(SNAS) in deuterium oxide solution and showed that the
percentage of the trans form of the surfactant molecules
relative to that of the cis increases with an increase in the
concentration.
The conformational change of the hydrocarbon part in a
surfactant molecule has already been observed; our Raman
studies (2,3) on surfactants have indicated that the all-trans
form of the hydrocarbon chain increases when surfactant
molecules form micelles.
In this paper, a further PMR study of SNAS was made;
sodium N-butanoyl sarcosinate and sodium N-hexanoyl sarcosinate
are favorable for the investigation of physicochemical properties of the SNAS compounds at concentration below the critical
micelle concentration (CMC), while potassium N-hexadecanoyl
sarcosinate is most appropriate for the study of physicochemical properties in the micelle state as the CMC is extremely
small. Variable temperature NMR studies were made to provide

information on the energy difference between the two isomers
and the energy barrier to rotation about the C-N bond of SNAS.

II. EXPERIMENTAL METHODS

  a. Materials. The chemicals used are sodium N-butanoyl
sarcosinate (SNBS), sodium N-hexanoyl sarcosinate (SNHS),
sodium N-decanoyl sarcosinate (SNDS) and potassium N-hexadecanoyl sarcosinate (PNHDS). Details of the preparations of these
materials were described previously (1,4).
  b. PMR Measurements. PMR measurements were made with a
Hitachi-Perkin Elmer R-20B spectrometer (60 MHz), equipped with
a Takeda Riken TR-3824 electronic frequency counter.
The sample temperature could be controlled to within +1°C in
the region of 0 to 100°C.
  c. CMC Measurements. The CMC's of SNBS, SNHS and PNHDS
were obtained by dye method. The CMC values obtained were as
follows: SNBS, 480 mg/cm^3 (20°C); SNHS, 200 mg/cm^3 (20°C);
PNHDS, 0.3 mg/cm^3 (25°C) and 0.5 mg/cm^3 (95°C).

III. RESULTS AND DISCUSSION

1. Concentration Dependence of Populations of Isomers
  The proton magnetic resonance spectra for SNBS and SNHS
resemble that of sodium N-octanoyl sarcosinate (SNOS), which
were reported previously (1). In the spectra, two N-CH$_3$-group
peaks and split N-CH$_2$ peaks are observed in addition to the
(CH$_2$)$_n$- and CH$_3$-resonance peaks of the acyl group. For the
N-CH$_3$ peaks, the peak at higher magnetic field is assigned to
the cis form and that at lower magnetic field to the trans.
In the case of N-CH$_2$ resonance peaks, the peak at higher magnetic field is ascribed to the trans form and that at lower
magnetic field to the cis (Fig. 1).

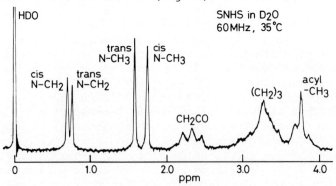

Fig. 1. *NMR spectrum of SNHS (100mg/cm^3).*

In the NMR spectra of SNBS and SNHS, the relative intensities of the two N-methyl peaks were found to be concentration-
dependent; for SNBS, below the concentration of 500 mg/cm^3 the
percentage of the trans form is about 50 % and constant,
whereas above this concentration it increases with an increase
in the concentration (Fig. 2a), and for SNHS the percentage of
the trans form also increases with increasing concentration
(Fig. 2b). In the case of SNHS, the trans-form percentage
slightly increases with an increase in the concentration below
the CMC. This observation may be due to the preaggregation of
the SNHS molecules. However, the change of the relative
intensity is marked at the CMC, as well as the case of SNOS
(1). Thus, the spectral change is considered to be caused by
the micelle formation; such behavior of SNBS and SNHS is
similar to that of SNOS and other long chain SNAS's.

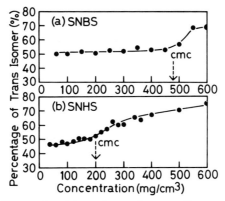

Fig. 2. Concentration dependence of percentage of the
trans isomer. The broken arrows show the CMC's of SNBS and
SNHS.

2. *Concentration Dependence of Internal Chemical shift*
For SNBS and SNHS, the internal chemical shift of the
N-CH$_3$ and N-CH$_2$ resonance peaks were also measured at different
concentrations. As is shown in Fig. 3, it was found that the
internal chemical shift ($\Delta\nu_{t-c}$) of the N-CH$_3$ peaks is independent of the concentration, while that of N-CH$_2$ peak decreases
with an increase in the concentration. This observation of the
N-CH$_3$ resonance shows that splitting of the N-CH$_3$ protons should
be ascribed to only the restricted rotation about the C-N bond,
and for the $\Delta\nu_{t-c}$ values of the N-CH$_2$ peaks the concentration
dependence is due to the conformation change about the N-CH$_2$
bond accompanying the micelle formation as discussed in our
previous paper (1).

(a)

(b)

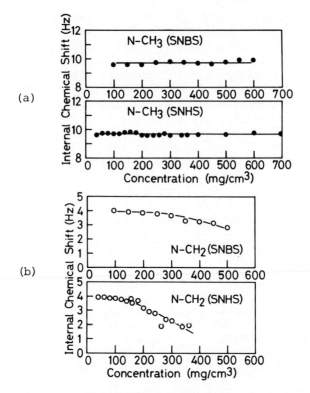

Fig. 3. *Concentration dependence of internal chemical shifts of N-CH$_3$ (a) and N-CH$_2$ (b).*

3. *Effect of Temperature on Isomer Population*

The intensities of two N-CH$_3$ peaks at different temperatures lead to the energy differences between the trans form and the cis of SNAS. In this report, variable-temperature studies of the N-CH$_3$ peaks were made for the SNBS-deuterium oxide solution (300 mg/cm^3), in which the molecules are in a monomolecular dispersion state, and for the micelle solutions of SNDS and PNHDS. Figure 4 shows the percentage of the trans form at various temperatures. When the temperature is raised in the range of about 35 to 95°C, the percentage of the trans form is almost unchanged. Observation of SNDS and PNHDS shows that the effect of the surfactant molecules in the monomolecular dispersion state on the isomer populations is negligible even at high temperature in these concentrated surfactant solutions; for the two surfactants, more than 99 % of these molecules in the investigated solutions are included in the micelle.

From the observations, it is concluded that the energy

differences between the trans and the cis forms must be very small in the micelle state of SNDS and PNHDS. For the SNBS molecules, such an energy difference in the monomolecular dispersion state is also expected to be extremely small.

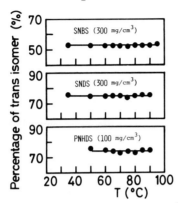

Fig. 4.   Effect of temperature on the trans-isomer percentage.

4.   *Concentration Dependence of Energy Barrier to Rotation about the C-N Bond*
The interconversion from the trans form to the cis is regarded as a rate process. Accordingly, the temperature dependence of the rate constant k for this interconversion is given by

$$k = k' \exp\ (-Ea/RT) \qquad (1)$$

where k' is the frequency factor and Ea the potential energy barrier hindering internal rotation about the C-N bond of a peptide group. Equation (1) can be reduced to equation (2).

$$\log_{10}\ (1/2\pi\tau\Delta\nu_{t-c}) = \log_{10}\ (k'/\pi\Delta\nu_{t-c}) - Ea/2.303RT \qquad (2)$$

$$\tau = \tau_t\ \tau_c/\ (\tau_t + \tau_c) \qquad (3)$$

where τ_t and τ_c are the mean life time in the trans form and the cis form, respectively. The energy barrier Ea to rotation about the C-N bond is derived from a linear plot of $\log_{10}\ (1/2\pi\tau\Delta\nu_{t-c})$ vs. $1/T$.
For the two N-CH$_3$ peaks of SNBS, SNDS and PNHDS measured at various temperatures, a line shape analysis was made according to Gutowsky and Holm's method (5), and the energy barrier to rotation about the carbonyl carbon-nitrogen bond was assumed.
In Fig. 5, the N-CH$_3$ resonance peaks of SNBS and SNDS at

361

different temperatures are shown together with the most fitting
calculated spectra for the corresponding τ values.

The Arrhenius plots for three SNAS's are shown in Fig. 6.
The energy barriers obtained from the slopes of the line are
11.4, 19.5, and 20.1 kcal/mol for SNBS, SNDS and PNHDS,
respectively. The relative error was within + 1.0 kcal/mol.

In the two different concentrations of $\overline{SNOS}$ (45 and 400
mg/cm^3), the energy barrier was also estimated. The activation
energy for the high concentration of SNOS was found to be
greater than that for the low concentration; for the samples of
400 and 45 mg/cm^3 the energy values of 20.0 and 13.3 kcal/mol
were obtained, respectively (Fig.7). In these calculations of
SNOS, the effect of preaggregation and the temperature-
dependence of the CMC were neglected. Accordingly, these
activation energies are regarded as the apparent values.
However, the difference between the activation energies of the
two sample solutions should be attributed to the effect of
micellar formations.

In the cases of SNBS, SNDS and PNHDS, the differences
between the energy barrier height of SNBS in monomeric state
and those of the other two surfactants in micelle state must
also be due to the effect of micellar formations.

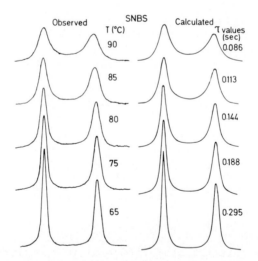

*Fig. 5a   The 60 MHz NMR-spectra of N-CH$_3$ protons and
matching theoretical spectra for SNBS in D$_2$O at various
temperatures. In calculations of these spectra, for the trans
and the cis N-CH$_3$ peaks T$_2$ values of 1.0 and 0.6 sec were used.*

In conclusion, it may be concluded that the activation
energy about the C-N bond of SNAS in the monomolecular
dispersion state differs from that in the micelle state.

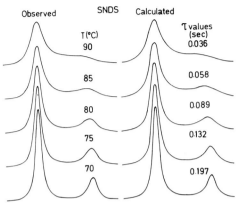

Fig. 5. The 60 MHz NMR-spectra of N-CH₃ protons and mathcing theoretical spectra for SNDS in D₂O at various temperatures. In calculations of these spectra, for both the trans and the cis N-CH₃ peaks T₂ of 0.4 sec was used.

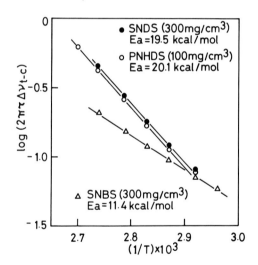

Fig. 6. The Arrhenius plot for SNBS, SNDS and PNHDS. In calculation of the PNHDS-line shape, for both the trans and the cis N-CH₃ peaks T₂ of 0.4 sec was used.

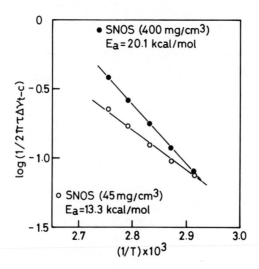

Fig. 7. The Arrhenius plot for SNOS. In a line shape analysis, for the 45 mg/cm³ solution T_2 values of the trans and the cis N-CH₃ peaks are 0.8 and 0.5 sec, respectively, and for the 400 mg/cm³ solution T_2 values of the trans and cis N-CH₃ peaks are 0.6 and 0.4 sec, respectively.

IV.   REFERENCE

1. H. Takahashi, Y. Nakayama, H.Hori, K. Kihara, H. Okabayashi and M. Okuyama, J. Colloid & Interface Sci., 54, 102 (1976).
2. H. Okabayashi, M. Okuyama, T. Kitagawa, and T. Miyazawa, Bull. Chem. Soc. Japan, 47, 1075 (1974).
3. H. Okabayashi, M. Okuyama, and T. Kitagawa, Bull. Chem. Soc. Japan, 48, 2264 (1975).
4. E. Jungerman, J. F. Gerecht and I. J. Krems, J. Amer. Chem. Soc., 78, 172 (1956)
5. H. S. Gutowsky and C. H. Holm, J. Chem. Phys., 25, 1228 (1956).

Surfactant Effectiveness in Concentrated Salt Solutions.

J.M. Sangster and H.P. Schreiber,
Department of Chemical Engineering,
Ecole Polytechnique, Montreal H3C 3A7
Canada.

A B S T R A C T

In research intended to provide guidelines to the interfacial properties of compositions which are becoming increasingly important in practical uses, four commercially available surfactants have been used as foamers and foam stabilizers in concentrated solutions of NH_4NO_3 and NH_4NO_3 in water, and in mixed solvents of propylene glycol/water. The performance of surfactants in these complex, concentrated salt solutions was assessed in terms of foam build, foam stability and mechanical strength. Physical properties and density were measured in attempts to correlate with foam characteristics. Surface tension lowering alone failed to correlate with the ability of the surfactants to build or to stabilize foams. Similar lack of correlation with viscosity and density effects was observed. On the other hand, sharp changes in the mechanical strengths of foams (brittle-stable transitions) were associated with a transition of the surfactants from solution to micellar states. In the case of one surfactant, detailed measurements of the critical micelle concentration (cmc) were made over a range of glycol/water compositions and at salt concentrations up to saturation. Major variations of cmc with solvent composition were observed; the sensitivity of cmc to salt concentration, while finite, is less pronounced.

THE NON-AQUEOUS MICELLAR CATALYSIS OF
THE DECOMPOSITION OF BENZYL CHLOROFORMATE

Frederick M. Fowkes and David Z. Becher[1]
(Department of Chemistry, Lehigh University)

The decomposition of benzyl chloroformate
in decane was chosen as a model reaction for the
study of micellar catalysis in non-polar solvents.
The surfactants chosen as the catalysts were the
sodium, zinc, chromium, aluminum, and cerium (III)
salts of di-(2-ethyl-hexyl) succinate sulfonic acid.
The reaction was strongly catalyzed (in one case more
than 5000 times) by the micelles. The plots of the
rate constants versus the catalyst concentrations
showed critical micelle concentrations and the characteristics of saturation kinetics which confirmed
that the catalysis was micellar. The catalytic
effectiveness of the different salts was not directly related to their Hammett acidities. The zinc
salt which was third most acidic salt was the most
powerful catalyst. The activation enthalpies of several of the catalyzed reactions were unusually large
(Zn(II), $\Delta H^{\neq}$ = 266 kilojoules/mole; Al(III), $\Delta H^{\neq}$ =
147 kilojoules/mole). These results were interpret-
ed as showing the effects of micellar geometry on
the effectivenesses of the catalysts.

I. INTRODUCTION

Catalysis by surfactant micelles in non-aqueous solvents
is an area which has not received much attention. This poten-
tially large area has a literature of less than twenty papers (1,2,3,4). In addition, most of the papers are studies of
very complicated reaction systems; including ones involving
reactants insoluble in the bulk solvent or so large that
they could not possibly be solubilized by a normal surfactant

---

[1] Present address, PPG Industries Inc. P.O. Box 9, Allison
Park, Pa. 15101

micelle. The completeness of these papers is further decreased by the fact that in only one case, O'Connor, Fendler, and
Fendler(5), were rate constants reported at more than one temperature, and the method used for the calculation of the activation parameters in that paper involved the use of a very
questionable assumption to separate the catalytic effects of
the monomeric and micellar forms of the surfactant. A study
of micellar catalysis in non-polar solvents was, therefore,
undertaken to investigate these systems.

For the study of micellar catalysis in non-polar solvents, a surfactant whose properties in non-aqueous solvents
have been relatively well studied was required. The available
data on non-aqueous micelles are not extensive (6), but the
properties of the salts of di-(2-ethyl-hexyl) succinate sulfonic acid are as well studied as those of any surfactant (6,
7,8). In addition, it is relatively simple to convert the
available sodium salt into other salts and there is some information on their properties (9). For these reasons, the aluminum (III), cerium (III), chromium (III), zinc (II), and
sodium (I) salts of di-(2-ethyl-hexyl) succinate sulfonic acid
were used as the catalysts.

A model reaction which possesses certain characteristics is necessary to study micellar catalysis. The reaction
to be used should have simple, easily observed kinetics and
the other characteristics which are desirable in a reaction to
be used in a kinetic's study. The reactants should be soluble in the hydrocarbon solvent and nevertheless still be polar
enough to be appreciably solubilized by the micelles. It is
desirable that the products have a lower polarity than the
reactants so that they will not tend to reduce the effective
catalyst concentration. In addition, since the effect of
varying the acidity of the counterion was to be studied, an
acid catalyzed reaction was desired. The thermal decomposition of benzyl chloroformate to benzyl chloride and carbon dioxide appeared to have all the desired characteristics. It
is a unimolecular reaction which can easily be followed by observing the evolution of the gaseous product. Benzyl chloroformate should be more strongly solubilized than benzyl chloride, so product interference would not be expected. The
decomposition rates of chloroformates have been shown to be
dependent on the dielectric constants of the solvents (10,11,
12). Preliminary work in this laboratory by C. Silebi (Master's Thesis 1974) and M. Marmo (Senior Thesis 1974) showed
that in solvents with strongly acidic or basic properties the
reaction rate was much greater than in neutral solvents with
the same dielectric constants. They, also, showed that the
surfactants would catalyze this reaction. It was, therefore,
chosen as a model reaction for the study of micellar catalysis.

II. Experimental

A. Materials

The methanol and benzene used in the catalyst preparation were obtained from either J. T. Baker Chemical Company or
the Lehigh Volley Chemical Company. The decane, dibutyl phosphate, and indicator grade phenylazodiphenylamine. N,N-
dimethyl-p-phenylazoaniline (Butter Yellow) and 1,9-di-
phenyl-1,3,6,8-nonatetraene-5-one were obtained from Eastman
Kodak. Purified 4-phenylazonapthylamine was generously provided by Dr. H. Benesi of Shell Development Company. Benzyl
chloroformate was obtained from the RSA Corporation. A sample
was analyzed by the method of Michels (13) and found to consist of approximately 70 percent benzyl chloroformate and 30
percent benzyl chloride. Di-(2-ethyl-hexyl) sodium sulfosuccinate was obtained from the Aldrich Chemical Company. Aluminum chloride, cerous nitrate, zinc nitrate, and chromium
nitrate were obtained from the Fisher Scientific Company. The
other salts of the di-(2-ethyl-hexyl) succinate sulfonic acid
were prepared by the method of Kitahara, Watanabe, Kon-no,
and Ishikawa (9). The product was dissolved in benzene and
freeze-dried to remove the solvent. Typical elemental analyses (Baron Consulting Co.) of the salts are presented in

Table 1.

The Elemental Analyses of the Surfactants
==

	Al(III) Salt		Ce(III) Salt	
	calculated[a]	observed	calculated[b]	observed
C	51.48	51.49	51.30	50.99
H	8.86	8.69	7.96	7.99
S	–	–	6.85	6.49
Al	1.93	1.72	–	–
Ce	–	–	9.97	9.35
Na	–	–	0.0	0.37

a. $Al(C_{20}H_{37}SO_7)_3$
b. $Ce(C_{20}H_{37}SO_7)_3$

These results indicate that the salts produced contained

small amounts of coprecipitated sodium salt. This impurity is believed to be the cause of much of the error in the measured rates.

B. Procedures

The reaction rates were determined by the following procedure. A weighed amount of catalyst was dissolved in 50 milliliters of decane and placed overnight in a constant temperature bath. The reaction flask was kept at a constant temperature to within $\pm 0.5°C$ by immersing it in an oil bath regulated by a Cole-Parmer magnetic "6x6" stirrer hot plate equipped with a temperature control probe. One milliliter of benzyl chloroformate was introduced into the flask with a syringe and the volume of carbon dioxide gas evolved was noted at constant time intervals. The volume was measured by observing the displacement of a mercury column in a graduated tube. The pressure within the apparatus was adjusted to atmospheric by means of a leveling bulb and a mercury manometer. Whenever possible the amount of gas evolved at infinite time was determined and the rate constant was calculated directly from the first order rate equation. This method gave values with an uncertainty of about five percent. In some cases a measurement of the volume of gas evolved at infinite time was not possible. In these cases, the rate constants were calculated by the use of the Guggenheim method (14). This method gave values with an uncertainty of about ten percent.

The Hammett acidities were measured spectrophotometrically (15,16). Dibutyl phosphate was used to convert the indicators to their acid forms so that their absorbances could be measured. The absorbances were determined using a Perkin-Elmer Model 402 ultraviolet-visible spectrophotometer.

The aggregation numbers of the surfactants at 75°C were determined by Mr. Thomas Harvey and Mr. R. L. Birkmeir of the Specialty Chemicals Research Department of ICI United States Inc. using a H-P Model 302B vapor pressure osmometer.

III. RESULTS

The rate of the uncatalyzed decompositions of benzylchloroformate in decane was too slow to measure accurately. The rate constant at 125°C was found to have a value of less than 10^{-6} sec^{-1}, which is insignificant in comparison with the catalyzed rates.

The variation of the pseudo-first order rate constants on surfactant concentration for the sodium and aluminum salts at

126°C are presented in Figure 1. The existence of critical

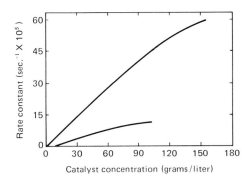

Fig. 1. The dependence of the pseudo-first order rate constant of the thermal decomposition of benzyl chloroformate in decane at 126°C on the surfactant concentrations.

concentrations for catalysis and the curvatures of the plots in Figure 1 are evidence for micellar catalysis.

The values of the critical concentrations, which can be identified with the critical micelle concentrations of the surfactants, were determined by extrapolation to be 1.5 grams/liter for the aluminum salt and 6.6 grams/liter for the sodium salt.

Plots of the reciprocal of the measured rate versus the reciprocal of the surfactant concentration minus the extrapolated critical micelle concentration (Figure 2) were linear. This suggests that the type of saturation kinetics which have been found in aqueous systems (1) are also obeyed here. The uncertainties in the critical micelle concentrations make the exact values of the intercepts of the reciprocal plots uncertain, and this makes a meaningful calculation of the quantities which compose the saturation kinetics expression impossible.

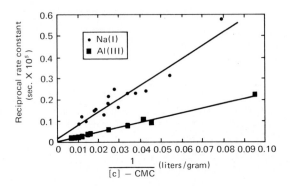

Fig. 2. The reciprocal of the rate constant versus the reciprocal of the surfactant concentration minus the critical micelle concentration.

The catalytic effectivenesses of the other salts were determined at 126°C at a catalyst concentration of 40 grams per liter. The results are presented in Table 2.

Table 2
The variation of the catalytic effectiveness of the surfactants with the cation

| Cation | Hammett Acidity | Rate Constant |
		$(sec^{-1} x10^5)$
Zn (II)	0.7	524
Cr (III)	-2.3	26.4
Al (III)	0.7	17.2
Na (I)	5.3	5.6
Ce (III)	3.1	4.2

The Hammett acidities are obviously not simply related
to the catalytic activity. There must be some other powerful
influence to account for these results.

The temperature dependencies of the rates were determined for the zinc, cerium, aluminum, and sodium salts. The
activation enthalpies and entropies were calculated by a
least square fit from the results (Table 3).

Table 3
The Activation Enthalpies and Entropies for the Reaction
Catalyzed by the Different Salts.

Cation	$\Delta H^{\ddagger}$ (kilojoules/mole)	$\Delta S^{\ddagger}$ (joules/mole-°K)
Zn (II)	266	374.6
Al (III)	147	50.3
Na (I)	124	-15.5
Ce (III)	83	-122.4

The high activation enthalpies and very large positive
activation energies are unusual. The values of the enthalpy
of activation are linearly related to the values of the entropy of activation (Figure 3) which obviously indicates
that both values are similarly dependent on some property of
the catalyst.

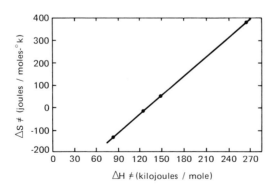

Fig. 3. The relation of the enthalpy of activation to
the entropy of activation for the salts of **di**-(2-ethyl hexyl)
succinate sulfonic acid.

In an attempt to confirm that the surfactants used formed micelles at elevated temperatures, their number average
aggregation numbers were measured at 75°C by vapor pressure
osmometry (Table 4).

Table 4
The Aggregation Numbers of the Surfactants in Decane at 75°C
by Vapor Pressure Osmometry.

Cation	Number of Sulfosuccinate Groups
Al (III)	6
Ce (III)	9
Na (I)	12

The values are rather small, but aggregation obviously does
occur. Since the number average values include a large contribution due to the monomer, the average micelle size should
be larger than the values presented. The vapor pressure
osmometry data show some curvature. This is consistent with
the expectation that the average micelle size would be noticeably concentration dependent for small aggregation numbers.

IV. DISCUSSION

It is apparent that except for the cerium salt the
activation enthalpies and entropies are usually large. The
values reported for the activation parameters of chloroformate decompositions in the literature (10,11,12) are of the
same magnitude as those observed for the cerium salt.
The much higher values observed for the other three
salts would seem to indicate a change in the mechanism.
Kevill and Weitl (12) have suggested two possible limiting mechanisms for chloroformate decomposition, a covalent
mechanism involving a four member cyclic intermediate and an
ionic mechanism involving the formation of chloride and
carbonium ions. An intermediate type of mechanism involving
a cyclic formation of an ion pair with only partial charge
separation can also be imagined. The evidence in the literature suggests an ionic mechanism, but in non-polar solvents
the degree of ionization is probably relatively small. In
the presence of the cerium salt, it seems likely, in view of
the similarities of the activation enthalpies, that the
reaction proceeds in a similar manner with the cerium salt
merely providing a more polar environment for the reaction.

The higher activation enthalpies of the other salts suggest that the transition state has a more ionic character. The higher activation enthalpies are caused by the increased amount of energy needed for the greater separation of the charges.

A mechanism which can explain the increased ionic character of the transition state is presented in Figure 4.

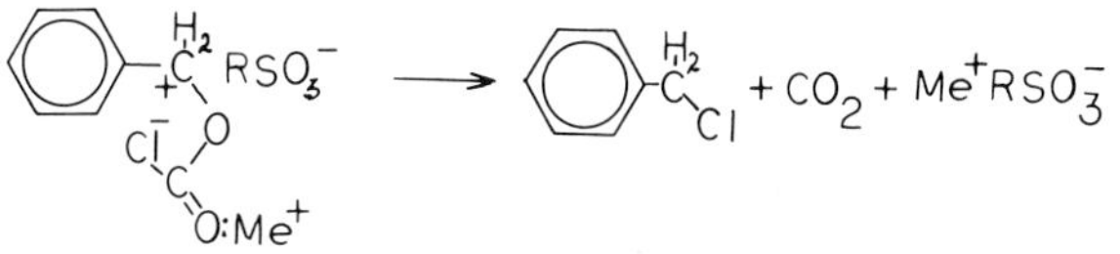

Fig. 4. Proposed Mechanism of Micellar Catalysis

The catalytic ability of the salts in this model would depend strongly on the size of the cation and the micellar geometry. The relatively poor catalytic ability of the cerium salt could be explained by the large size of the cation.

The high reaction rates observed despite the high activation enthalpies are due to the extremely large activation entropies. The catalysis of this reaction could reasonably be called entropic catalysis. The large positive activation entropies can be explained as being due to the large amount of freedom on the carbon dioxide group in the more ionic transition states. The relation of the entropies to the amount of ionic character and therefore to the enthalpies would explain the linear relation between them.

This mechanism would explain the non-linear dependence of the catalytic strength on the Hammett acidity in terms of a geometric factor. The effectiveness of the cation would depend not only on its acdity but also on its ability to form the cyclic intermediate (which would depend on the precise micellar geometry).

V. CONCLUSIONS

In this work catalysis by micelles in non-polar solvents has been demonstrated at elevated temperatures. The observation of critical concentrations for catalysis and saturation kinetics are evidence for micellar catalysis. The complicated dependence on acidity and the high activation enthalpies have interesting implications for both micellar catalysis and the study of the mechanism of the reaction. The proposed geometric factor may provide a method for investigating the micellar structure. The large rate enhancements indicate that non-aqueous micellar catalysis may have some useful applications.

VI. ACKNOWLEDGEMENTS

The support of the National Science Foundation in the form of an NSF Graduate Fellowship is gratefully acknowledged. Acknowledgment is also made to the Donors of The Petroleum Research Fund, administered by the American Chemical Society, for partial support of this research.

VII. REFERENCES

1. Fendler, J. H. and Fendler, E. J., "Catalysis in Micellar and Macromolecular Systems," Academic Press, New York, 1975.
2. Kon-no, K., Miyazawa, K., and Kitahara, A., Bull. Chem. Soc. Japan, 48, 2955 (1975).
3. Kon-no, K., Mitsuyama, T., Mizuno, H., and Kitahara, A., Nippon Kagaku Kaishi 1975, 1857 (1975).
4. Manabu, S., Shiraishi, S., Araki, K., Kise, H., Bull. Chem. Soc. Japan. 48, 3678-81 (1975).
5. O'Connor, C. J., Fendler, E. J., and Fendler, J. H., J. Org. Chem. 38, 3371 (1973).
6. Kertes, A. S., and Gutman, H., in "Surface and Colloid Science," (E. Matijevic, Ed.), Vol. 8, p. 193ff. John Wiley and Sons, New York, 1976.

7. Eicke, H. F. and Christen, H., J. Colloid Interface Sci. 46, 417 (1974).

8. Muto, S., and Meguro K., Bull. Chem. Soc. Japan 46, 1316 (1973).

9. Kitahara A., Watanabe, K., Kon-no, K., and Ishikawa, T., J. Colloid Interface Sci. 29, 48 (1969).

10. Wiberg, K. B., and Shryne, T. M., J. Am. Chem. Soc. 77, 2774 (1955).

11. Oliver, K. L., and Young, W. G., J. Am. Chem. Soc. 81, 5811 (1959).

12. Kevill, D. N., and Weitl, F. L., J. Am. Chem. Soc. 90, 6416 (1968).

13. Michels, J. G., Anal. Chem. 47, 1446 (1975).

14. Guggenheim, E. A., Phil. Mag. 2, 538 (1926).

15. Hammett, L. P. and Deyrup, A. J., J. Am. Chem. Soc. 54, 2721 (1932).

16. Fowkes, F. M., Benesi, H. A., Ryland, L. B., Sawyer, W. M., Detling, K. D., Loeffler, E. S., Folckemer, F. B., Johnson, M. R., and Sun, Y. F., Agricultral and Food Chemistry 8, 203, (1960).

SIZE, SHAPE AND SIZE DISTRIBUTION OF MICELLES
IN AQUEOUS SOLUTIONS OF SHORT-CHAIN LECITHIN HOMOLOGUES

R.J.M. Tausk[1] and J.Th.G. Overbeek

Van 't Hoff Laboratory, University Utrecht

Micelle formation in dilute aqueous solutions of three synthetic lecithin homologues with equal fatty acid ester chains of 6, 7 or 8 carbon atoms is described. The average micellar weight increases with lecithin concentration. This effect becomes more pronounced on increasing the lipid chain length and the concentration of the salting-out electrolyte NaCl. This phenomenon is in agreement with a stepwise open association equilibrium model in which the standard free energy of micellization per monomer is independent of the micellar size (above a certain minimum size), but increases with the chain length and salt content.

I.   INTRODUCTION

Studies on the association of lipid molecules are of great importance to come to understand the large numbers of problems encountered in research on biological membranes and lipid-protein interactions. The interactions between lipids and proteins can strongly depend on the structure of the individual lipid molecules. In some enzymatic reactions (1) it is the association structure of the lipid that plays a very important part. To obtain a better understanding of the factors governing lipid association, we studied lecithin homologues containing two hydrocarbon chains of equal length (2). To this end we synthesized lecithin molecules with fatty acid ester chains of 6, 7 and 8 carbon atoms. The structure of the diheptanoyl-lecithin (= di-C_7) is shown in Fig.1.

Fig.1.   Diheptanoyllecithin (= di-C_7)

1. Present address: Koninklijke/Shell-Laboratorium, Amsterdam (Shell Research B.V.), P.O. Box 3003, Amsterdam, the Netherlands.

The di-C_9 is the lowest homologue that associates into the familiar lamellar structures at very low concentrations. The lecithins with chains of 6 or 7 carbon atoms, on the other hand, show normal micelle formation (2b). Di-C_8 is a borderline case: depending on the electrolyte content, a phase separation occurs (2c), salting-out and salting-in being observed at high concentrations.

Electrolytes at high concentrations (C_s) also have an influence on the critical micelle concentrations (CMC) and their effect can be expressed by (2a)

$$\log(CMC)_{C_s} = \log(CMC)_{C_s=0} - k\,C_s \qquad [1]$$

in the following table (1) we give a few relevant data for the CMC's of the synthetic lecithins and the effect of NaCl.

TABLE 1
Critical Micelle Concentrations of the Lecithin Homologues and the Effect of NaCl Addition (see eq.[1])

	$(CMC)_{C_s=0}$ in g/l	k_{NaCl} in l/mol
di-C_6	6.9	0.26
di-C_7	0.71	0.21
di-C_8	0.14	
di-C_9	0.016	

From a purely physical chemical point of view, too, the micelle formation of these compounds is of importance, since the molecules (i) contain two hydrocarbon chains and (ii) have no net charge, which simplifies the interpretation of thermodynamic data. This last property is especially important in connection with the demonstration of micellar size distributions . With ionic surfactants, the large deviations from thermodynamic ideality caused by the repulsive forces originating from electrical charges obscure the size distributions of the micelles.

This paper will first deal with an association model giving rise to polydisperse micellar systems (2d). Specifically, we will discuss the dependence of the average micellar size on the concentration and chain length of the surfactants and on the salt concentration. We then deal with the interpretation of light scattering and sedimentation equilibrium experiments on di-C_6 and di-C_7 in terms of average micellar weights. Here thermodynamic nonideality is taken into account on the basis of the excluded volume of rigid particles and association is described in terms of the mass action law. Thereafter, we will describe micelle formation of the di-C_8 homologue. This lipid associates into very large micelles and a strong angular dependence of the light scattering was observed. The results for the different lecithins will then be discussed in terms of our theory for the formation of polydisperse micelles.

II.   MODEL FOR MICELLAR POLYDISPERSITY

Our model for micellar polydispersity is based on association equilibria between monomers and micelles (2d). The theory is an extension of the work of Mukerjee (3).

We define the equilibrium constants, k_i, with the following equation:

$$L_{i-1} + L_1 \rightleftharpoons L_i \quad ; \quad k_i = \frac{C_i}{C_{i-1} \times C_1} = \frac{C_i}{\prod_2^i k_i (C_1)^i} \qquad [2]$$

where L_1 represents a monomer molecule and L_i a micelle with association number i. C_i stands for the micellar concentration in amounts per unit volume. This equation holds at very low concentrations and also at higher concentrations if the chemical potential (μ_i) of each species can be described by eq.[3] , as is often a good approximation.

$$\mu_i = \mu_i^0(P,T,c') + RT \ln c_i + RTM_i (B_1 c_t + B_2 c_t^2 + \ldots) \qquad [3]$$

The standard condition is defined at constant pressure, temperature and concentration of all other solute components. c_t is the total lipid concentration, now in mass per unit volume. B_1, B_2, etc. are constants and independent of the molecular weight M_i.

A distribution in association constants will always lead to a micellar weight distribution and both distributions are intimately connected.

At low association numbers the hydrocarbon chains in the micelles are only partly in contact with each other and an appreciable hydrocarbon-water contact remains. As the decrease in this hydrocarbon-water contact is the main driving force for micellization these small micelles will have relatively small stability constants.

Increasing the association number will result in cooperative effects on the change in free energy of micellization, and k_i will increase. Addition of a monomer to an "incomplete" micelle will decrease the alkyl chain - water contact area of all the molecules already present in the associate. At a certain micellar size with association number i = n the most efficient packing of the molecules will occur. For the sake of simplicity one could visualize such a micelle as a sphere with radius equal to the length of the hydrocarbon tail of the molecule.

If the micelle grows beyond this size, its shape will have to change and prolate or oblate ellipsoids, cylinders or discs will be formed. For these large micelles the change in free energy of association becomes independent of the association number. This common association constant will be denoted by K. In Fig.2 (solid line) we give the expected general shape of the dependence of k_i on i.

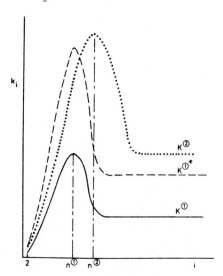

Fig.2. A schematic picture of the dependence of the association constant k_i on the association number i. The full line represents the dependence for homologue 1 in salt-free solutions. After addition of a salting-out agent the broken line is obtained. The dotted line represents the dependence in salt-free solutions for a higher homologue.

The concentration of micelles with i = n is given by

$$\frac{C_n}{C_1} = \Pi_2^n k_i (C_1)^{n-1} \equiv (K_1 C_1)^{n-1} \qquad [4]$$

in which we define K_1 as the geometrical average of all associ-

382

ation constants between 2 and n. At or above the critical micelle concentration the product K_1C_1 is quite near unity.

We now introduce two simplifications:

(i) Since the concentrations of micelles with i < n will be much smaller than C_n, we ignore them.

(ii) We assume the association constant K to apply to all micelles with i > n.

These assumptions give rise to a sharp transition between stability constants of "small" and "large" micelles and to a distribution as depicted in Fig.3.

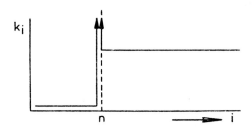

Fig.3. A simplified picture of the dependence of k_i on i.

The equivalent micellar concentration $(E_m = E_t - C_1)$, expressed in amount monomer per unit volume, can now be calculated as a function of K_1, K, n and C_1, and with straight-forward algebra we find:

$$E_t - C_1 = C_n \left(\frac{n}{1-X} + \frac{X}{(1-X)^2}\right) \qquad [5]$$

with $X \equiv kC_1$. Likewise we find for the weight average micellar association number:

$$N_w \equiv \frac{\Sigma\, i\, C_i}{\Sigma\, C_i} = n + \frac{X}{1-X} + \frac{X}{(1-X)^2\left(n + \frac{X}{1-X}\right)} \qquad [6]$$

For the micellar sizes to remain finite X must be smaller than unity.

For the dependence of N_w on C_t two important rather extreme situations can be distinguished:

(i) $1/n < 1 - X < 1$. This condition leads to micellar systems with narrow size distributions and $N_w \approx n$.

(ii) $1 - X < 1/n \ll 1$. In this case wide size distributions are obtained and N_w can be approximated by

$$N_w = 2 \ (K/K_1)^{\frac{n-1}{2}} \ (c_m/c_1)^{\frac{1}{2}} \qquad\qquad [7]$$

The concentrations are now expressed in mass per unit volume. c_m is the micellar and c_1 the monomer concentration ($\approx$ CMC). Wide distributions can only occur if K is somewhat larger than K_1; otherwise K_1C_1 approaches unity faster than $K C_1$ and C_n will strongly dominate all other micellar concentrations.

A.   Effect of Addition of Electrolytes

Electrolytes exert an influence on the critical micelle concentrations. For nonionic or zwitterionic surfactants these phenomena are often described with the general terms of salting-out or salting-in effects on the activity of the free monomer molecules. On the basis of this theory Mukerjee (4) derived eq.[1] .

Salt effects on uncharged micelles with result in a change in free energy of micellization per molecule independent of the micellar size, so that we expect all constants k_i in eq.[2] to change by the same factor. This means that K and K_1 will both change and that their ratio will remain approximately constant. As the value of n for these types of surfactants will mainly be determined by the geometry of the molecules, we furthermore expect n to be independent of the electrolyte content. The expected change in the dependence of k_i on i on addition of a salting-out agent is shown in Fig.2. As a consequence, the plot of N_w versus c_m/c_1 (see eq.[7]) will not be affected by the addition of salt. At constant micellar concentration, c_m, however, large effects on average micellar weights can occur due to changes in c_1.

B.   Effect of Chain Length

Geometrical factors will undoubtedly lead to an increase in n with increasing chain lengths. K and K_1 will also increase. The change in the ratio K/K_1, however, is more difficult to predict.

The area per molecule at the surface of the hydrocarbon core of densely packed spherical micelles with diameters proportional to the chain length is independent of the chain length. In larger elongated micelles of the same packing density and diameter, the area per molecule will be smaller than in the case of spherical micelles. The packing of the parts of the hydrocarbon chain close to the surface of the hydrocarbon core will be more constrained (smaller number of possible configurations) for large micelles than for small ones. On elongation of the carbon chains, this effect will become relatively less important, leading to an increase in K/K_1 (see dotted curve in Fig.2).

If repulsive forces between the polar groups, which extend away from the hydrocarbon core, are considerable, as in the case of ionic surfactants or strongly hydrated nonionics, another mechanism is also operating. The area per head group measured at some fixed distance outside the core of small densely packed spherical micelles decreases with elongation of the hydrocarbon chain. The repulsive forces will therefore increase with the chain length. The area per head group in strongly elongated micelles, however, will be less dependent on the chain length, and an increase in K/K_1 can be expected. For zwitterionic micelles the situation is more complicated, since depending on the orientation of the dipoles attractive or repulsive forces exist. The change in area per molecule will change the electrostatic interactions and the number of possible configurations of the polar groups.

The combined increase of n and K/K_1 will result in a strong increase in N_w as a function of c_m/c_1. The effect of N_w at fixed c_m will, of course, be even much greater, due to the decrease in the CMC with increasing chain length.

C.   Ionic Micelles

In principle, the association model outlined above is also applicable to ionic micelles, but the long range repulsive forces between the polar groups will strongly oppose the formation of larger micelles. Beyond $i = n$, k_i will therefore strongly decrease before approaching the constant value K and the micelles will be rather isodisperse. An increase of the salt content or of the surfactant concentration will be accompanied by a decrease in the electrostatic interactions. An increase in k_i and in polydispersity are expected.

III.   MICELLAR WEIGHT DETERMINATION OF $Di-C_6$ AND $Di-C_7$

A.   Method of Evaluation

Micelle formation of the synthetic lecithins $di-C_6$ and $di-C_7$ was studied with light scattering and ultracentrifugation. Details of the experiments and the procedures for calculating the micellar weights are described extensively in ref.2b. Here we will only briefly discuss the general approach.

Average total molecular weights, thus including monomer contributions, were calculated from equations for multicomponent systems (2b, 5, 6, 7):

$$\frac{K^1 c_t}{R_{90}} = \frac{1}{\sum_i f_i M_i n_i^2} + c_t \frac{\sum_i \sum_j f_i f_j n_i n_j A_{ij}}{(\sum_i f_i M_i n_i^2)^2} \qquad [8]$$

and

$$\left(\frac{RT}{\omega^2 r c_t} \frac{dc_t}{dr}\right)^{-1} = \frac{1}{\sum_i f_i M_i \rho_i} + c_t \frac{\sum_i \sum_j f_i f_j \rho_i \rho_j A_{ij}}{(\sum_i f_i M_i \rho_i)^2} \qquad [9]$$

$K^1 = 2 \pi n_0^2 \lambda_v^{-4} N_0^{-1}$, n_0 = refractive index of solvent, λ_v = wavelength in vacuum, N_0 = Avogadro's constant, R_{90} = excess Rayleigh ratio perpendicular to incident beam, c_t = total concentration in mass per unit volume, f_i = weight fraction of micellar species with association number i, n_i $(=\partial n/\partial c_i)$ = refractive index increment, R = gas constant, T = absolute temperature, ω = angular velocity, r = distance from centre of rotation and ρ_i $(=\partial \rho/\partial c_i)$ = density increment.

Experimentally, it was found that both the refractive and the density increments are smaller for micelles than for free monomers. We assumed the increments to be independent of the micellar size $(i \geq 2)$.

The interaction parameter A_{ij} is related to the change of the activity coefficient γ_i of component i with the molar concentration of component j at constant solvent chemical potential, temperature and molar concentration of the other solutes.

$$A_{ij} = (\partial \ln \gamma_i / \partial c_j)_{\mu_0, T, C'} \qquad [10]$$

To assess the weight average micellar weights we will have to subtract the contributions of the free monomers. Generally, light scattering data from micellar solutions have been interpreted by using the Debye approximation, in which the monomer activity $(\approx CMC)$ is taken constant. Using this method, however, some details are lost, especially in the case of rather small micelles and at micellar concentrations comparable with the monomer concentrations. We therefore used an association model in which the change of the monomer activity with the micellar concentration is allowed for. A procedure to do this has been extensively discussed by Adams and Filmer (8). Their theory is based on association equilibria where the chemical potential of each species participating in the association reaction can be described by eq.[3]. The monomer weight fraction

(f_1) can then be obtained from the dependence of the weight average total molecular weight $\langle M \rangle_w$ on the surfactant concentration. This association model also gives the possibility of calculating the number average molecular weights $\langle M \rangle_n$. Estimates of the weight average total molecular weights can be obtained when eqs.[8] and [9] are multiplied by the z-average increments $\langle (\partial n/\partial c)^2 \rangle_z$ and $\langle \partial \rho/\partial c \rangle_z$, respectively.

To calculate the second terms on the right-hand sides of eqs.[8] and [9] we further need the weight fractions of the different micellar species. The specific micellar distribution function used has only a small influence on the calculated second virial coefficient and we assumed the Schulz distribution (7) to apply. This two-parameter function is easy to integrate and one of these parameters can be expressed as $\langle M \rangle_w/\langle M \rangle_n$, which in turn can be assessed from the experiments with the method of Adams. From the experiments we found $\langle M \rangle_w/\langle M \rangle_n$ for di-C_7 micelles to be around 2 and for di-C_6 around 1.05 to 1.1.

The last problem in the calculation of the micellar weights is the estimation of the interaction parameter A_{ij}. As electrostatic dipole intermicellar interactions are probably quite small, one could visualize the micelles as rigid particles, which only interact because of their finite sizes. A_{ij} is related to the pair correlation function (5) and can be obtained from the mutual pair excluded volume. Isihara and Hayashida (9) solved this problem for particles of arbitrary size and shape and arrived at

$$A_{ij} = N_o \left(v_i + v_j + (1/4\,\pi)(x_i s_j + x_j s_i) \right) \qquad [11]$$

where v_i stands for the volume of molecule i with surface s_i. x_i equals 4π times the average distance of the centre of particle j to the tangent plane with particle i. We assumed the free monomer to be spherical and the micelles to be spherical or spherocylindrical. Several spherocylindrical micellar structures with different packing densities have been used in these calculations.

Although in general the excluded volume model with eqs.[10] and [11] is in conflict with the use of eq.[3], the assumption that this last equation is valid and that the method of Adams can be used to calculate the monomer concentrations does not lead to significant errors in micellar systems.

B.   Results

Some of our results (2b) on di-C_6-lecithin are shown in Fig.4.

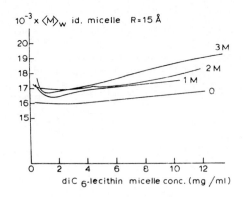

Fig.4. Weight average micellar weights, from light scattering, as a function of the di-C_6-lecithin micellar concentration, in aqueous solutions containing various NaCl concentrations. The micellar weights are corrected for non-ideality using the spherocylinder model with a radius of 15 Å.

The average micellar weights in the absence of added NaCl is around 16 000, which could correspond to spherical micelles with radii of 18 Å . This value is not unrealistic in view of the length of the whole monomer molecule. However, in this "compact" structure there is a substantial contact of the hydrocarbon part with the polar interface and elongated micelles are perhaps more probable.

One could also visualize the micelles in an alternative model, where the contact between the hydrocarbon part and the water and the polar groups is avoided (10). In this case we chose for a spherocylinder of which the radius of the hydro-carbon core is given by the length of the alkyl chains (≈ 7.8 Å). A sphere with this radius can only accommodate 6.2 monomers, where in fact an association number of around 35 was found. To accommodate the polar groups we assumed the outer radius of the cylinder to equal 15 Å. The results shown in Fig.4 were based on this last model. If the compact spherical structure is used in the evaluation of A_{ij} the calculated micellar weights increase somewhat less with the lecithin concentration (2b).

The average micellar weights of di-C_7-lecithin are shown in Fig.5. The different lines in this figure originate from different assumed micellar structures in the calculation of A_{ij}.

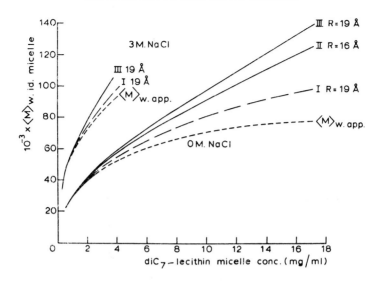

Fig.5. Weight average micellar weights of di-C₇-lecithin
as a function of the micellar concentration in 0 M and 3 M
NaCl (upper set of curves). The dotted lines represent the
apparent weights. The broken lines I were obtained from the
compact micelle model and the full lines II and III were
derived from the spherocylinder model with outer radii of 16 Å
and 19 Å respectively and with as little hydrocarbon-water
contact as possible.

In case I we assumed micelles with weights below 20 000 to
have compact spherical structures. Larger micelles then again
have spherocylindrical shapes. In cases II and III we used the
model in which the hydrocarbon-water contact was avoided and a
thickness of 7 or 10 Å respectively was allowed for the polar
groups.

From these experiments we conclude that:

(1) The micelles of di-C₆-lecithin have rather narrow size
distributions. The weight average micellar weight, in the
absence of added salt, is around 16 000 to 18 000
(association numbers ≈ 35) and increases only slightly with
concentration. Addition of NaCl leads to a very limited
increase of the average micellar weight and to a decrease
in the CMC.

(2) Di-C₇-lecithin associates into micelles with a broad size
distribution with average micellar weights between 20 000
and around 100 000. Accurate assessments of the micellar
weights at high lipid concentrations are complicated by the
relatively strong influence of the second virial coefficient.
The specific assumptions concerning the micellar structure

do have an influence on the details of the results, but the overall conclusions remain unaltered. The ratio $\langle M \rangle_w / \langle M \rangle_n$ for the micelles is close to 2.

Addition of 3 M NaCl leads to a strong increase in the micellar weight at fixed micellar concentrations, accompanied by a decrease in the CMC by a factor of 4. Within our experimental accuracy NaCl has, however, no effect on the dependence of N_W on $(c_m/c_1)^{\frac{1}{2}}$, as is shown in Fig.6.

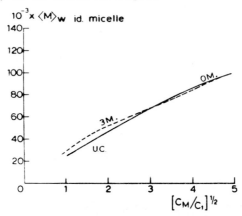

Fig.6. Ideal weight average micellar weights of di-C_7-lecithin as a function of the root of the ratio of micellar and monomer concentration. The molecular weights are idealized using the compact micelle model.
(———) = 0 M and (----) = 3 M NaCl.

In this graph the results from the "compact" micellar model are plotted. Use of the micellar structure with little hydrocarbon-water contact (especially curve III in Fig.5) lead to a slight upward curvature when N_W is plotted versus $(c_m/c_1)^{\frac{1}{2}}$. In this case the thermodynamic non-ideality was probably somewhat overestimated.

IV.   ASSOCIATION PHENOMENA OF Di-C_8-LECITHIN

Analysis of micelle formation of this lipid in electrolyte-free solutions is hampered by the appearance of a phase separation at concentrations slightly above the CMC. This demixing is sensitive to addition of electrolyte, and salting-out and salting-in phenomena can be observed at high salt concentrations. At lower concentrations (< 0.5 M) probably other electrostatic effects also play a role. In Fig.7 we have plotted the phase separation diagrams.

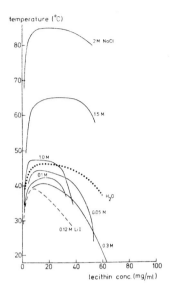

Fig.7. Phase separation diagrams of dioctanoyllecithin in aqueous solutions. The dotted line was obtained in electrolyte-free solutions. The full lines and the broken line refer to solutions containing NaCl and LiI, respectively.

The dependence of the critical temperatures on NaCl concentration is shown in Fig.8. LiI can suppress the demixing. Even in the presence of high concentrations of NaCl, addition of LiI leads to a decrease of the critical temperatures.

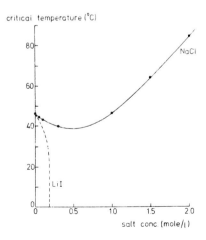

Fig.8. Critical temperatures of the phase separations as a function of the salt concentrations.

Micellar weight determinations were performed at room temperature after suppressing the phase separation with 0.2 M LiI. The micellar solutions, however, still show strong thermodynamic non-ideality, which hampers detailed analyses. The light scattering diagram (Zimm-plot) is shown in Fig.9.

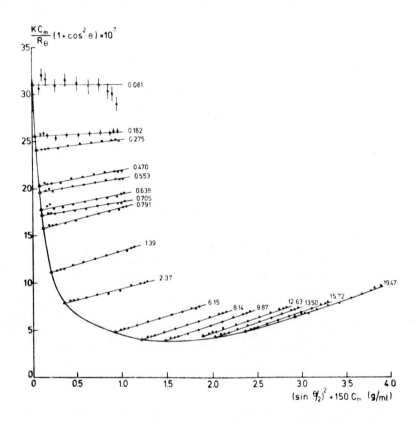

Fig.9. "Zimm plot" of dioctanoyllecithin at 25.0 °C, 0.2 M LiI and λ_v = 546 nm. The numbers to the right of the lines refer to micellar concentrations in mg/ml.

Since the CMC is low and association numbers are high, no appreciable errors are introduced by subtracting a constant monomer contribution instead of a variable concentration derived from association equilibrium equations. We attempted to interpret these data with the help of eqs.[12] and [13] (2c,11).

$$\frac{K\,c_m\,(1+\cos^2\theta)}{R_\theta} = \frac{1}{\langle M\rangle_{w.app.}} + \frac{16\,\pi^2\,n_o^2}{\lambda_o^2} \cdot \frac{\langle r_g^2\rangle_z}{\langle M\rangle_w} \cdot \sin^2(\theta/2) \quad [12]$$

$$\frac{1}{\langle M\rangle_{w.app.}} = \frac{1}{\langle M\rangle_w} + A_2\,c_m \quad [13]$$

K equals K^1 from eq.[8] and multiplied by $(dn/dc)^2_{micelle}$. $\langle r_g^2\rangle_z$ stands for the z-average radius of gyration squared and is defined by:

$$\langle r_g^2\rangle_z = \langle M\rangle_w^{-1}\int_0^\infty M\,f(M)r_g^2\,dM \quad [14]$$

where $f(M)$ is the weight distribution function.

The results, extrapolated to $\theta = 0$, are shown in Fig.10.

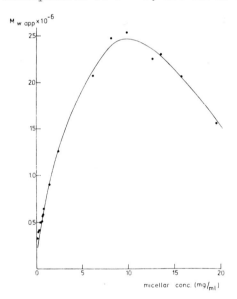

Fig.10.   Apparent weight average micellar weights as a function of micellar concentrations.

The molecules are so large that the application of the theory of the excluded volume of rigid particles cannot be used for the evaluation of A_2. At low concentrations we assume the contribution of the virial coefficient to be negligible and according to our association model we assume $\langle M\rangle_w$ to increase linearely with $\sqrt{c_m}$.

The angular dependence of the scattering contains information on the micellar shape. We found $[\langle r_g^2 \rangle_z / \langle M \rangle_w]^{\frac{1}{2}}$ to be independent of concentration and equal 3.1×10^{-9} cm.mol$^{\frac{1}{2}}$.g$^{-\frac{1}{2}}$. Adoption of the open association model then implies that $\langle r_g^2 \rangle_z / \langle M \rangle_w$ is also independent of the micellar weight. This phenomenon is compatible with disc- or random coil-like micelles, but not with straight rigid rods. However, on assuming a disc shape, we would obtain a disc thickness of 3.6 Å, and a molecular area of 440 Å^2, which clearly is not realistic. More sensible results can be obtained with a coil-like structure. At c = 10 g/l and $\langle M \rangle_w = 2.38 \times 10^6$ we for instance obtain an end to end distance of 1170 Å $(= \sqrt{6}\langle r_g^2 \rangle_z)$. This is not unreasonable comparing it with a flexible rod with the same molecular weight and a radius of 20 Å and contour length of 4260 Å. On the basis of these observations we assumed that the micelles of di-C$_6$ and di-C$_7$ also have spherocylindrical and not disc-like shapes.

V. COMPARISON OF EXPERIMENTS WITH ASSOCIATION MODEL

In Fig.11 we have plotted experimental results from di-C$_6$, di-C$_7$ and di-C$_8$ in one graph.

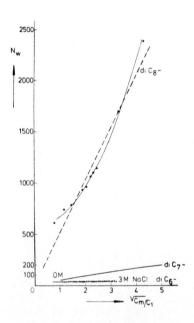

Fig.11. The weight average association number for three lecithin homologues as a function of the square root of the ratio of the micellar and monomer concentrations. The broken line for di-C$_8$ was calculated from $N_w = 2.38 \times 10^7 \times (526.5)^{-1} \times (c_m)^{\frac{1}{2}}$. The full line for di-C$_6$ refers to NaCl-free solutions while the dotted line refers to solutions containing 3 M NaCl.

We see that the association numbers strongly increase with chain length. The approximate values for the parameters K, K/K_1, n and $1-KC_1$ are given in Table 2, from which we see that the general trends are in agreement with the model discussed in chapter II.

TABLE 2
Parameters of the Association Model for three Lecithin Homologues

	K in $l.mol^{-1}$	K/K_1	n	$1-KC_1$
di-C_6	60	1.10	27	0.17
di-C_7	700	1.24	30	0.006
di-C_8	4050	(1.33)	(40)	0.002

In the case of di-C_8 it was impossible to calculate K/K_1 and n independently. Many different combinations can be used as long as $(K/K_1)^{\frac{n-1}{2}}$ is around 255. Separate estimations of n and K_1 can in this case, of course, only be obtained at concentrations low compared to the CMC where deviations from eq.[7] occur and eq.[6] has to be used.

The agreement with the theory concerning the effect of NaCl on micelle formation of di-C_7 has already been shown in Fig.6 where the rather straight line is virtually unaffected by the addition of NaCl. Taking the results for di-C_6 from Fig.4, we arrive at the same conclusions, although the effects are much smaller in that case.

VI.   CONCLUSIONS

In this paper we have presented a simple theory for the dependence of the average micellar weight on the concentration of non-ionic surfactants. The theory predicts an increase in micellar polydispersity with increasing micellar concentrations, hydrocarbon chain length and concentration of a salting-out electrolyte.

Micellar weights of three short-chain lecithin homologues containing two equal fatty acid esters of 6, 7 and 8 carbon atoms were determined by light scattering and ultracentrifugation. The results were interpreted with the help of equations for an association equilibrium between monomers and micelles and thermodynamic non-ideality based on the excluded volume of rigid particles.

The micelles of di-C_6 are rather monodisperse with micellar weights around 16 000 to 18 000. NaCl addition leads to a small increase in the average micellar weights. Di-C_7 gives polydisperse micelles ($\langle M \rangle_w / \langle M \rangle_n \approx 2$) and weight average weights ranging from 20 000 to 120 000. Addition of NaCl in this case leads to a very strong increase in the micellar weights at a fixed micellar concentration. The average micellar weight of di-C_8-lecithin increases strongly with lipid concentration and average weights between 250 000 at low concentrations upto several millions at concentrations around 1 % have been obtained.

In general, the experimental results are in good agreement with the association theory.

VII. REFERENCES

1. Pieterson, W.A., Vidal, J.C., Volwerk, J.J., and de Haas, G.H., <u>Biochem.</u> 13, 1455 (1974).
2a. Tausk, R.J.M., Karmiggelt, J., Oudshoorn, C.A.M., and Overbeek, J.Th.G., <u>Biophys. Chem.</u> 1, 175 (1974).
 b. Tausk, R.J.M., van Esch, J., Karmiggelt, J., Voordouw, G., and Overbeek, J.Th.G., <u>Biophys. Chem.</u> 1, 184 (1974).
 c. Tausk, R.J.M., Oudshoorn, C.A.M., and Overbeek, J.Th.G., <u>Biophys. Chem.</u> 2, 53 (1974).
 d. Tausk, R.J.M., and Overbeek, J.Th.G., <u>Biophys. Chem.</u> 2, 175 (1974).
3. Mukerjee, P., <u>J. Phys. Chem.</u> 76, 565 (1972).
4. Mukerjee, P., <u>J. Phys. Chem.</u> 69, 4038 (1965).
5. Yamakawa, H., "Modern Theory of Polymer Solutions", ch.5. Harper & Row, New York, 1971.
6. Cassasa, E.F., and Eisenberg, H., <u>Advan. Protein Chem.</u> 19, 287 (1964).
7. Fujita, H., "Mathematical Theory of Sedimentation Analysis", (E. Hutchinson and P. van Rysselberghe, Eds.), Academic Press, New York, 1962.
8. Adams, E.T., and Filmer, D.L., <u>Biochem.</u> 5, 2971 (1966).
9. Isihara, A., and Hayashida, T., <u>J. Phys, Soc. Japan.</u> 6, 40 (1951).
10. Tanford, C., <u>J. Phys. Chem.</u> 76, 3020 (1972).
11. Vrij, A., and van den Esker, M.W.J., <u>J. Chem. Soc. Faraday Trans.II</u> 68, 513 (1972).

ENTHALPIES OF LYOTROPIC LIQUID CRYSTALLINE PHASES
1. LAMELLAR AND HEXAGONAL PHASES IN THE SYSTEM *N*-PENTANOL - SODIUM *N*-OCTANOATE - WATER

Per J. Stenius, Jarl B. Rosenholm and Marja-Riitta Hakala

Department of Physical Chemistry, Åbo Akademi

I. SUMMARY

The relative molar enthalpies of the three components in the lamellar and the hexagonal liquid crystalline phases formed at 25°C in the system sodium *N*-octanoate - *N*-pentanol - water have been measured calorimetrically. The principal factor determining the enthalpies is the composition of the system. The enthalpies of transfer of a component from an aqueous solution to a liquid crystalline phase are small. It appears that the factors determining the formation of lyotropic mesophases must be mainly entropic.

II. INTRODUCTION

Several reviews of the phase equilibria in binary systems of a surfactant and water as well as in ternary systems of a surfactant, water and a third amphiphilic or non-polar compound have been published (1-4). It has been shown that if the system consists of an ionic surfactant (e.g., a fatty acid salt), a normal alkanol (with 5-10 carbon atoms) and water the main types of lyotropic liquid crystalline phases (mesophases) formed are: (i) a structure consisting of hexagonally arranged, indefinitely long rods of surfactant and alcohol in an aqueous continuum (middle soap or E-phase)[1], (ii) a structure of lamellar bilayers of surfactant and alkanol with intervening water layers (neat soap or D-phase) and (iii) a structure consisting of hexagonally arranged rods of surfactant and alkanol in a continuous alkanolic medium (F-phase). A number of other mesophases, the structure of which is still to some extent under discussion may also occur. Fig. 1. shows the phase diagram of the three-component system

---

1. In this paper we use the alphabetic notation introduced by Ekwall (3,4); notations used by other authors are given in table I of ref. 4.

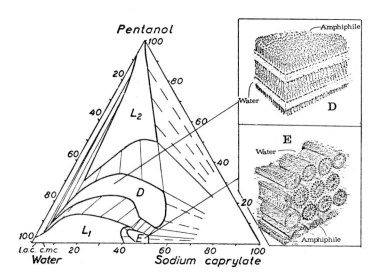

Fig. 1. Phase diagram for the system sodium octanoate – pentanol – water at 20 °C (4). The structure of the mesophases is shown schematically. L1=aqueous solution. L2=isotropic solution of water and octanoate in pentanol.

discussed in this paper, sodium N-octanoate (NaC_8) – N-pentanol (C_5OH) – water. This system was mainly chosen for its relative simplicity when developing the calorimetric methods.

The main instrument in determining the structure of these phases has been X-ray investigation (2,5). From studies of the phase diagrams for several systems, it may be inferred that the main factor determining the separation of these phases from isotropic solutions are (a) the charge density on the micellar surface and (b) the volume fraction of micellar substance together with the shape of the micellar aggregates. The regions of existence of these phases appear to be governed by the ability of the polar groups and the counter-ions to bind water (4). These conclusions are qualitatively supported by studies of the binding of counter ions and water in the systems, above all by nmr methods (6-9). It is, however, obvious that any attempt to give a quantitative model of the factors that stabilize the phases should be founded on experimental knowledge of the energy contents of the components as a function of the composition of the phases. In this paper, we present the first results of an attempt to determine the enthalpies of the components in lyotropic mesophases.

III. THERMODYNAMIC EQUATIONS

Notation: We give the component and its state within parantheses after the symbol for each quantity referring to this component. Thus, for example, $x(C_5OH, D)$ means the mole fraction of pentanol in D-phase. Different compositions in the same phase are denoted by roman numerals, for example, $x(C_5OH, D, I)$. A mesophase is generally denoted by M, otherwise the phase notations given in fig. 1 will be used.

Standard state: We choose as standard states pure H_2O, pure C_5OH and infinitely dilute NaC_8 in H_2O, respectively. We wish to determine the relative molar enthalpies of these components as functions of the composition of the three-component system formed by them. These are defined by

$$l(A) = h(A) - h^{\ominus}(A) \qquad (1)$$

where $h(A)$ is the partial molal enthalpy of component A and $\ominus$ denotes the standard state. The $l(A)$ may be calculated for each component from the total enthalpy content L of the system relative to the reference states:

$$L = n(H_2O) \, l(H_2O) + n(C_5OH) \, l(C_5OH) +$$
$$+ n(NaC_8) \, l(NaC_8) \qquad (2)$$

Experimentally, we determine L for a mesophase M in three steps:

(i) we dissolve phase M in an aqueous solution (L1,I) of NaC_8 and C_5OH to form another solution (L1,II):

$$n(M) + n(L1,I) \rightarrow n(L2,II) \qquad (3)$$

where $n(M) = n(C_5OH,M) + n(H_2O,M) + n(NaC_8,M)$; $n(L1,I) = n(H_2O,L1,I) + n(C_5OH),L1,I) + n(NaC_8,L1,I)$ and $n(L2,II) = n(H_2O, L1, I) + n(H_2O,M) + n(NaC_8,L1,I) + n(NaC_8,M) + n(C_5OH,L1,I) + n(C_5OH,M)$. The enthalpy of this process is

$$\Delta H_3 = n(L2,II) \, L_m(L2,II) - n(M) \, L_m(M) -$$
$$- n(L1,I) \, L_m(L1,I) \qquad (4)$$

The L_m are the molar enthalpy contents.

(ii) We dissolve pure C_5OH in aqueous solutions of NaC_8(L1,III), (L1,IV) to form the two aqueous solutions in eq. (3):

$$n(C_5OH,l) + n(L1,III) \rightarrow n(L1,I) \qquad (5)$$

$$n'(C_5OH,l) + n(L1,IV) \rightarrow n(L1,II) \qquad (6)$$

where $n(C_5OH,1) = n(C_5OH,L1,I)$ and $n'(C_5OH,1) = n(C_5OH,L1,II)$

The enthalpies of these processes are

$$\Delta H_5 = n(L1,I) \; L_m(L1,I) - n(L1,III) \; L_m(L1,III) \qquad (7)$$

$$\Delta H_6 = n(L1,II) \; L_m(L1,II) - n(L1,IV) \; L_m(L1,IV) \qquad (8)$$

since, by definition, $L_m(C_5OH,1) = 0$.

(iii) We dilute the binary solutions of NaC_8 in H_2O to infinite dilution:

$$n(L1,III) + aq \rightarrow n(NaC_8,aq) \qquad (9)$$

$$n(L1,IV) + aq \rightarrow n(NaC_8,aq) \qquad (10)$$

The enthalpies of these processes are

$$\Delta H_9 = - \; n(L1,IV) \; L_m(L1,III) \qquad (11)$$

$$\Delta H_{10} = - \; n(L1,IV) \; L_m(L1,IV) \qquad (12)$$

We find from equations (4), (7), (8), (11), (12):

$$L(M) = n(M) \; L_m(M) = - \; \Delta H_3 - \Delta H_5 + \Delta H_6 - \Delta H_9 + \Delta H_{10}$$

where $\qquad\qquad\qquad\qquad\qquad\qquad\qquad\qquad\qquad\qquad (13)$

$$L_m(M) = \frac{L(M)}{n(M)} = x(H_2O,M) \, l(H_2O,M) + x(C_5OH,M) \, l(C_5OH,M)$$
$$+ \; x(NaC_8,M) \, l(NaC_8,M) \qquad (14)$$

The partial molar enthalpies can be determined by partial differentiation of $L_m(M)$; for example

$$l(C_5OH,M) = L_m(M) - x(H_2O,M) \, \frac{\partial L_m(M)}{\partial(H_2O,M)} -$$
$$- \; x(NaC_8,M) \, \frac{\partial L_m(M)}{\partial x(NaC_8,M)} \qquad (15)$$

IV. EXPERIMENTAL METHODS AND MATERIALS

The heats of dissolution of mesophases in aqueous solution (ΔH_2, eq (4)) were determined with an LKB 10700 Batch Microcalorimeter system. About 300 mg of the mesophase was weighed into one compartment of the calorimeter cell and about 5 cm^3 of aqueous solution was pipetted into the other. The cell was then turned over several times in the calorimeter. The mesophase generally dissolved within 30 min, requiring 10-20 turnings of the cell. The hetas evolved were of the magnitude 1-2 J. Two precautions were important: (a) the mesophases had to be weighed into the cells as rapidly as possible to avoid drying out; (b) especially in the case of the D-phase it was necessary to dissolve the phase into an aqueous solution containing about 10 % pentanol in order to avoid distillation of pentanol from the D-phase to the solution during the equlibration period preceding the turning over of the cell. Fig. 2 show L_m for the E-phase in the binary system NaC_8/H_2O and is typical of the reproducibility of experimental results.

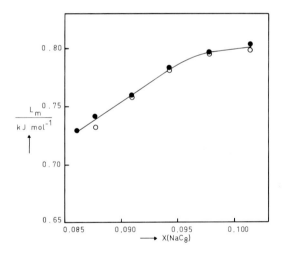

Fig. 2. $L_m(E)$ for the binary system sodium octanoate/ water as a function of the mole fraction of NaC_8 in the E-phase at 25°C.

The heats of dissolution of pentanol (ΔH_5, ΔH_6, eqs (7) and (8)) and the heats of dilution of sodium octanoate (ΔH_9, ΔH_{10}, eqs (11), (12)) were determined with a LKB 8700 precision calorimetry system, using a titration calorimeter unit built at our department. The heats evolved per titration step were typically of the magnitude 1-12 J, with an estimated error of

401

about 1 %. The heats of dissolution of solid NaC_8 and liquid pentanol in water were determined with the ampoule calorimeter of the LKB 8700 calorimetry system.

NaC_8 was prepared by neutralization of the corresponding acid (Fluka <u>puriss</u>. grade) with NaOH, as described elsewhere (9). The pentanol was Merck Ag <u>pro analysi</u>. The water was ordinary distilled water. All solutions were prepared volu-metrically. The mesophase samples were prepared and their homogeneity checked as described in ref. (5) and stored in sealed ampoules at 25°C. All measurements were made at $25 \pm 0.1^\circ$C.

V. RESULTS AND DISCUSSION

Fig. 3 shows L_m for the binary system NaC_8/H_2O, from infinitely dilute solution to the upper concentration limit of phase E. It is striking that the enthalpy content seems to

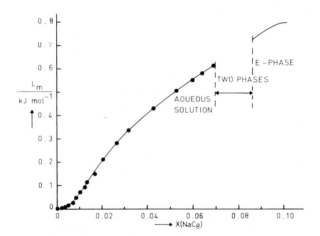

Fig. 3. The relative molar enthalpy of aqueous solutions of NaC_8 and mesophase E as a function of the mole fraction of NaC_8 at 25°C.

be dependent on the composition of the system only: the curve for the aqueous solution can be directly extended over the two-phase region to the E-phase. There is a very sligth difference between the slope of the curve in concentrated aqueous solution and in E-phase, as is seen in fig. 4 which shows the partial molar enthalpy for NaC_8. This is also seen from table 1, which gives some enthalpy changes that may be roughly estimated from this curve. The enthalpy content is strongly dependent on the

TABLE 1

Enthalpies of Transfer of Sodium N-Octanoate in Aqueous Systems at 25°C.

===

Process	ΔH, $kJ.\,mol^{-1}$
$NaC_8(aq) \rightarrow NaC_8(at\ the\ c.m.c.)$	$\simeq\ \ 10.5$
$NaC_8(at\ the\ c.m.c.) \rightarrow NaC_8(satd.soln.)$	$\simeq\ -3.3$
$NaC_8(satd.\ soln.) \rightarrow NaC_8(E\text{-}phase,\ in\ equil.\ with\ satd.soln.)$	$\simeq\ -0.3$
$NaC_8(aq) \rightarrow NaC_8(s)$	$\simeq\ -1.7$
$NaC_8(s) \rightarrow NaC_8(satd.\ E\text{-}phase)$	$\simeq\ \ 3.4$

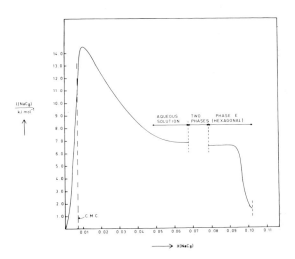

Fig. 4. The partial molar enthalpy of sodium octanoate in aqueous solution and in mesophase E as a function of its mole fraction at 25°C

concentration in the narrow concentration region below and just above the c.m.c. (0.35 mol/dm^{-3} (10)); once micelles are formed it changes much slower and the enthalpy change on transition of NaC_8 from saturated aqueous solution to E-phase is only about -0.3 kJ/mol. Obviously, there can be no drastic changes in either the binding of the water or in the hydrocarbon moieties when E-phase is formed. At higher concentrations the partial molar enthalpy decreases. This very probably

reflects changes in the hydration of the polar groups: at the
upper concentration limit of phase E the amount of water
appears to be just sufficient to hydrate the carboxylate groups
and the sodium ions (about 9 molecules of water/molecule of
surfactant.

Fig. 5 gives a summary of the measurements made of the
enthalpies of dissolution of D- and E-phase into the aqueous
solution. The points have been systematically chosen to fall
on lines with the same ratio $n(NaC_8)/n(C_5OH)$, $n(H_2O)/n(NaC_8)$
and $n(H_2O)/n(C_5OH)$, respectively. This considerably simplifies
the calculation of partial molar enthalpies.Details of the
method by which these are calculated will be published elsewhere[1].

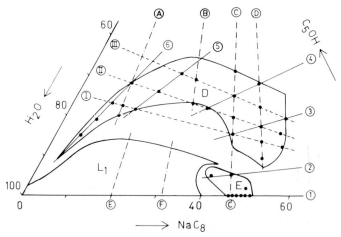

*Fig. 5. The composition of the samples for which the
relative molar ebthalpies have been determined plotted in part
of the phase diagram in fig. 1. The letters and numbers refer
to lines for which L_m is given in figures 6-8. Heats of solubilization of C_5OH have been determined along lines E and F.*

Figures 6-8 present the enthalpies of formation plotted
as functions of the concentration of each of the components
at constant molar ratio of the other components. In fig. 6 we
note that the enthalpy content of the D-phase increases as we
pass towards lines corresponding to higher contents of sodium
octanoate. There is no dramatic difference between the
enthalpy contents of the E-phase and the D-phase. The main
factor determining the enthalpies is still the composition of
the system as a whole. This is even more evident in fig. 7, in
which the data have been re-plotted to show the enthalpies as

1. Bedö, Zs., Blomqvist, K. and Rosenholm, J.B., To be
published.

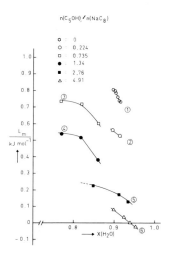

Fig. 6. L_m (D) and L_m (E) given as a function of $x(H_2O)$ for different constant molar ratios of C_5OH to NaC_8 at $25°C$. The numbers refer to the lines given in fig. 5.

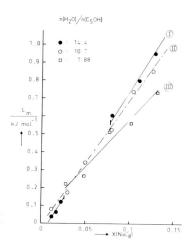

Fig. 7. L_m (D) as a function of $x(NaC_8)$ for different constant molar ratios of H_2O to C_5OH at $25°C$.

functions of the sodium octanoate content of the D-phase for different constant molar ratios of water to pentanol. For line I, we pass through a two-phase region, but the line drawn through the points corresponding to low NaC_8 contents passes directly through the points at high NaC_8 contents.

Finally, fig. 8 shows the effect of increasing the
pentanol content at constant molar ratio of water to NaC_8. The
enthalpies decrease rapidly as the amount of pentanol increases,
especially at high concentrations of NaC_8. The enthalpies of
D- and E-phase samples that are almost in equilibrium with
each other (lines C and C´) are almost equal.

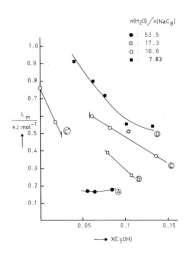

*Fig. 8. L_m (E) and L_m (D) as functions of $x(C_5OH)$ for
different constant molar ratios of H_2O to NaC_8 at $25°C$.*

The first conclusion to be drawn from these data is that the
principal factor determining the enthalpies of these phases is
their composition. This is, of course, not too surprising. The
environment of the components does not change very much once
micelles have been formed: it is a location in borders between
lipophilic and hydrophilic moieties. One would, however,
expect a clear difference in the enthalpies of transfer to a
lamellar or to a hexagonal phase. No such difference can be
found. To illustrate this, fig. 9 shows the partial molar
enthalpy of C_5OH as a function of the overall composition of
the system. The enthalpies for a few processes of interest may
be estimated from these curves and are given in table 2. We do
not have sufficient data yet to plot enthalpies for the
different components in aqueous solution and in mesophases in
equilibrium with these solutions. However, it is evident from
fig. 9 that when the first amounts of pentanol are solubilized
in an aqueous solution, the enthalpy content of the pentanol
increases strongly. Once a certain minimum amount has been dissolved, however, the partial molar enthalpy remains almost
constant up to (at least) the upper border of the D-phase. In
more concentrated solutions the steep part of the curve is

TABLE 2

Enthalpies of Transfer of N-Pentanol in Sodium N-Octanoate –
Water Systems at 25 ^{o}C

Process	ΔH, kJ mol^{-1}
$C_5OH(1) \rightarrow C_5OH(aq)$	$- 6.5 \pm 0.1$[a]
$C_5OH(1) \rightarrow C_5OH(inf.dil.$ in 1 M $NaC_8)$	$\simeq - 7$
$C_5OH(1) \rightarrow C_5OH(solubilized$ in 1 M $NaC_8)$	$\simeq - 1$
$C_5OH(1) \rightarrow C_5OH(solubilized$ in 2.3 M $NaC_8)$	$\simeq - 3$
$C_5OH(1) \rightarrow C_5OH(5$ %, in E-phase)	$\simeq - 8.5$
$C_5OH(1) \rightarrow C_5OH(D$-phase, 20-30 % NaC_8, in equil. with aq.soln)	$- 2.5 - - 3$

a. Values given in the literature range between -6.36
and -8.08 kJ mole^{-1} (11).

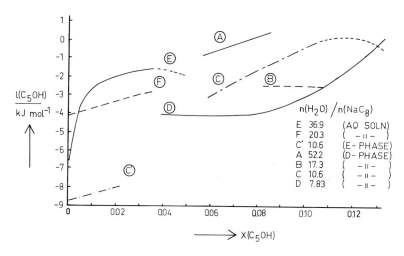

Fig. 9. The partial molar enthalpy for C_5OH in different
states in the system H_2O-C_5OH-NaC_8 as a function of its mole
fraction at 25°C.

absent (possibly because pentanol does not have any effect on
micellization in these solutions), the relative molar enthalpy
being almost constant (about -3 to -4 kJ mol^{-1}) throughout.
Although the lines in aqueous solution and in the D-phase do
not represent the same H_2O/NaC_8 ratios, fig. 9 shows that there
cannot exist any large differences in enthalpy between pentanol
in a saturated aqueous solution and in D-phase. On passing from

E-phase to D-phase, however, the change in enthalpy is much
larger. In this case the pentanol passes from a clearly defined
cylindrical to a lamellar aggregate. This difference between
the enthalpies of transfer from aqueous solution and from the
E-phase gives support to the notation that the effect of
adding pentanol to aqueous micelles is to change their shape
towards a more oblate one, such that the change in hydration
when transferring pentanol from a micelle to D-phase is very
small. We would like to stress, however, that all changes in
enthalpies within the aqueous phase, once micelles have been
formed, are relatively small. If there is a change in the mean
shape of the micelles within the aqueous solution region as
pentanol is added, it is not, in any case, clearly reflected
in the relative molar enthalpies.

Fig. 10 shows the partial molar enthalpies of water in
D- and E-phase. There is a marked difference in the curves for

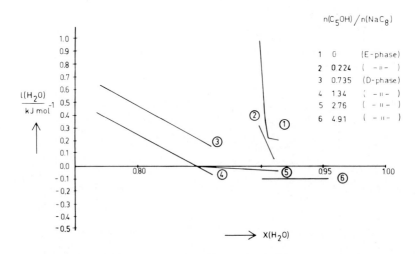

*Fig. 10. The partial molar enthalpy for H_2O in different
states in the system $H_2/-C_5OH-NaC_8$ as a function of its mole
fraction at 25°C.*

these two phases. In D-phase, the enthalpies change linearly;
in the water-rich part they are almost independent of the
concentration. The dependence on the composition is much more
marked in the E-phase. Obviously there is a clear difference
between the binding of water in these two phases.

We summarize our results as follows:

(1) We have shown that it is possible to measure the
enthalpies of lyotropic liquid crystalline phases with an
accuracy sufficient to calculate the relative molar enthalpy

of each component.

(ii) The principal factor determining the enthalpy of each component is the composition of the system, once micelles have been formed.

(iii) The formation of mesophase is accompanied by very small enthalpy changes, irrespectively of whether the phase is a lamellar or a hexagonal one.

(iv) Any theory of the formation of lyotropic mesophases probably should be based on entropic rather than enthalpic effects. However, our results have been obtained at one single temperature, and we would not like to speculate more about this question until these enthalpies are known at several temperatures. So far, of course, our results are also limited to one single system. Preliminary experiments on the system NaC_8-decanol-water indicate, however, that the effects are quite similar to those reported here.

VI. ACKNOWLEDGEMENTS

We wish to thank Professor Ingvar Danielsson for discussions and Jarl-Erik Jansson for invaluable help with the experiments. Finacial support from the Finnish Research Council for Natural Sciences is gratefully acknowledged.

VII. REFERENCES

1. Winsor, P.A., Chem. Rev., 68, 1 (1968).

2. Luzzati, V., in "Biological Membranes" (D. Chapman, Ed.) p. 71, Academic Press, New York, 1968.

3. Ekwall, P., Danielsson, I. and Stenius, P., in "Surface Chemistry and Colloids", (M. Kerker, Ed.), MTP Intern. Rev. Sci., Phys. Chem. Ser. 1, Vol. 7, p. 97, Butterworths, London, 1972.

4. Ekwall, P., Advan. Liquid Crystals, 1, 1 (1975).

5. Ekwall, P., Fontell, K. and Mandell, L., Acta Polytech. Scand., Chem. Ind. Met. Ser., 74, I (1968).

6. Lindman, B. and Ekwall, P., Molec. Cryst., 5, 79 (1968).

7. Tiddy, G.J.T., J. Chem. Soc., Fl, 68, 369 (1972).

8. Persson, N.-O. and Lindman, B., J. Phys. Chem., 79, 1410 (1975).

9. Rosenholm, J.B. and Lindman, B., <u>J. Colloid Interface Sci.</u>, In press.

10. Mukerjee, P. and Mysels, K.J., in "Critical Micelle Concentrations of Aqueous Surfactant Systems", NSRDS-Nat. Bur. Stand. (U.S.) 36, p. 165, Washington, D.C., 1971.

11. Krishnan, C.V. and Friedman, H.L., <u>J. Phys. Chem.</u>, 73, 1572 (1969).

THE CRITICAL MICELLE CONCENTRATIONS OF MIXTURES OF LITHIUM PERFLUOROOCTANE SULFONATE AND LITHIUM DODECANE SULFATE IN AQUEOUS SOLUTIONS

Minoru Ueno,Keiichi Shioya,Tetsutaro Nakamura
and Kenjiro Meguro
Science University of Tokyo,Department of Chemistry
Faculty of Science

I. INTRODUCTION

Micellar properties in mixtures of two anionic hydrocarbon surfactants with comparable hydrophobic groups have so far been extensively studied(1 - 6).

However, relatively few studies for the mixtures of perfluorocarbon and hydrocarbon surfactants have so far been reported(7,8).

Lange and Beck(6) have reported that critical micelle concentration values(CMC) of mixtures of two sodium alkyl sulfates are smaller than those in the ideal mixing and then the variation of the CMC values with the mole fraction is depicted by a concave curve. The negative deviation is very well known to be attributed to the mixed micelle formation in the mixed solution of the two hydrocarbon surfactants.

However, micellar properties in the mixed aqueous solutions of a perfluorocarbon and a hydrocarbon surfactants would be expected to show some behaviors different from those in the mixture of two homologous alkyl surfactants, because chain-chain interactions between perfluorocarbon and hydrocarbon in both surfactant molecules are different from those between two hydrocarbon chains.

In this paper, properties in the mixed aqueous solutions of lithium perfluorooctane sulfonate(LiFOS) and lithium dodecane sulfate(LiDS) have been studied by changing the mole fraction of these two surfactants and are discussed on the basis of the experimental data obtained by surface tension,electroconductivity and dye methods, respectively.

II. EXPERIMENTALS

A. Materials

Lithium perfluorooctane sulfonate(LiFOS was obtained neutralizing the respective sulfonic acid with lithium hydroxide,which was obtained by distillation of the potassium salt(supplied by Dainippon Ink Chem. Ind. Co. Ltd.) in the presence of above 95% sulfuric acid(b.p. 150,14 Torr)(9), and was purified by repeated crystallization from dioxane till it showed no minimum in the surface tension curve.

Lithium dodecane sulfate(LiDS) was prepared from dodecanol of a high purity by sulfation with chlorosulfonic acid according to the method of Dreger et.al. (10). The clude product was purified by washing several times with a quantity of ether and then by repeated crystallization from the mixed solvents consisting of n-hexane and ethanol. LiDS showed no minimum in the surface tension curve.

The water used in preparing the solutions was prepared by distillation of alkaline permanganate solution made up from ion-exchanged water and had a specific electroconductivities of $1.0-1.5 \times 10^{-6}$ohm^{-1} cm^{-1}

B. Measurements

The surface tensions of aqueous solutions of single and mixed surfactants were measured with a Wilhelmy-type surface tensiometer SIMAZU ST-1 at 25 C.

The electroconductivities of single and mixed surfactants were determined using a seal cell and a conductance bridge, Toa Denpa Model CM-1DB, at 25 C.

The obtained values were checked by measuring the specific conductivities of standard potassium chloride solutions.

The absorption spectral change of pinacyanol chloride in the aqueous solutions of single and mixed surfactants were measured using a Hitachi spectrophotometer(Model ESP-3T) at 25 C. The sample solutions were permitted to stand in the dark place for at least 15 hrs to attain the equilibrium.

III.RESULTS AND DISCUSSION

Shinoda and co-workers(9) have reported that Krafft points of fluorocarbon surfactants are raised with an increase of chain length as in the case of ordinary surfactants and the increment of the Krafft points with an increase of chain length of fluorocarbon surfactants is much larger than that of cor-

responding ordinary alkane sulfonate(11).   Thus it
is also important to select a suitable counter ion
of fluorocarbon surfactant whose Krafft point is low
in order to dissolve the surfactant at lower temperature.   Accordingly LiFOS whose Krafft point was
below 0 C was employed in this work,since both Krafft
points of potassium and sodium corresponding salts
were very high(9).

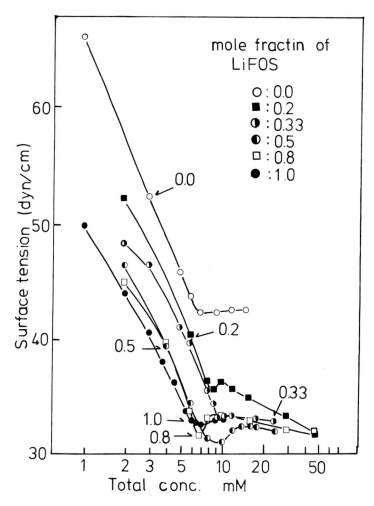

Fig. 1.   Surface tension v.s. total concentration curves of single and mixed surfactants
at various constant mole fractions of LiFOS.

LiDS was also selected as ordinary hydrocarbon surfactant with the common counterion.

Fig. 1 shows the plots of surface tension values against the logarithm of total concentrations of the single and mixed surfactants at various constant mole fractions of LiFOS. LiFOS shows lower surface tension than that of LiDS, and these CMC values give 8.3 mM for LiDS and 6.3 mM for LiFOS, respectively.

Their surface tension curves for the mixed system exhibit a trend to shift extremely downwards with the increasing mole fraction of LiFOS and again upwards at 0.8 mole fraction of LiFOS as shown Fig. 1.

This suggests that LiFOS in the mixed solutions makes relatively larger contribution to the lowering surface tensions than that of LiDS. On the other hand, the break points corresponding to the mixed

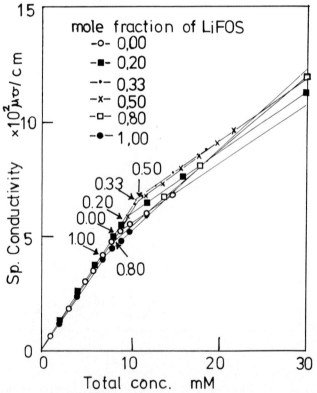

Fig. 2.   *Specific electroconductivity v.s. total concentration curves of single and mixed surfactants at various constant mole fraction of LiFOS.*

micelle concentration shift toward the higher concentration, compared to those of single surfactants, and again toward the lower concentration at 0.8 mole fraction of LiFOS. Such a phenomenon has not been so far observed in the mixed system of ordinary surfactants

Fig. 2 shows the plots of specific conductivity against the total concenrarions of single and mixed solutions of two surfactants at various constant mole fractions of LiFOS. As evident from this figure, the break points corresponding to the mixed CMC shift toward the higher concentration than that of single surfactants with increasing mole fraction of LiFOS, and again approach a value between those of single surfactants at 0.8 mole fraction.

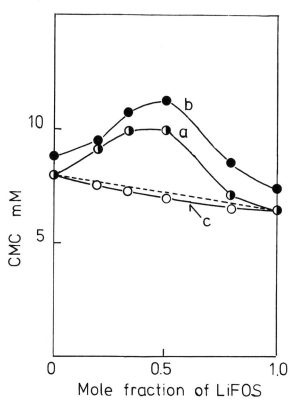

Fig. 3. Mixed CMC v.s. mole fraction of LiFOS curve : (a) obtained from surface tension and (b) obtained from conductivity. (c) ; calculated by Lange'equation.

Total concentration values of mixed CMC corresponding to these break points in Figs. 1 and 2 obtained by two methods are plotted in Fig. 3(a) and
(b) as a function of mole fraction of LiFOS together
with curve (c) of mixed CMC for the two surfactants
obtained using Lange's equation for comparison(6).

Both curves (a) and (b) show a great positive
deviation from ideal mixing and give a maximum at
about 0.5 mole fraction of LiFOS.   On the other
hand, curve (c), which is obtained provisionally regarding the CMC of LiFOS as that of a hydrocarbon
surfactant by using Lange's equation, shows evidently
a negative deviation from ideal mixing.   In general, the mixed CMC values plotted as a function of

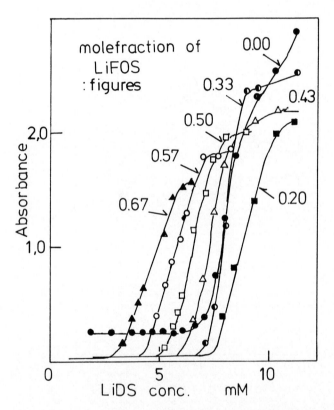

*Fig. 4.   Absorbance of pinacyanol chlo-
ride in aqueous solutions of single and
mixed surfactants v.s. LiDS concentra-
tion(spectral changes at α - band).*

mole fraction of one component in mixed two hydrocarbon surfactants are found to be deviated negatively
from ideal mixing(6).   As evident from a comparison
of curve (a) with curve (c), this suggests that the
mixed micelle formation in the mixtures of perfluorocarbon and hydrocarbon surfactants become more difficult than that of single surfactant or those for
mixtures of two hydrocarbon surfactants, and this
tendency is more remarkable at about 0.5 mole fraction of LiFOS.
      In order to confirm this positive deviation of
the CMC from ideal mixing in further detail, absorption spectra of pinacyanol chloride were measured in
mixed aqueous solutions of LiFOS and LiDS at various

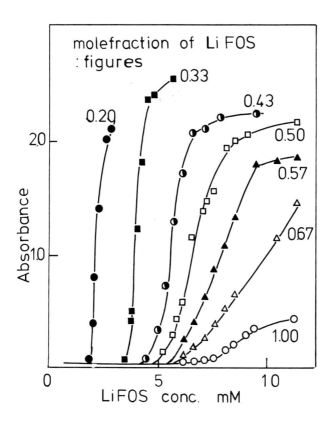

*Fig. 5.   Absorbance of pinacyanol chlo-
ride in aqueous solutions of single and
mixed surfactants v.s. LiFOS concentration (spectral changes at α - band).*

constant mole fractions of LiFOS.   The spectral
changes at α - band of 612 nm are plotted in Fig. 4
against LiDS cncentrations.   As with Fig. 4, the
spectral changes are also plotted in Figs. 5 and 6
against LiFOS and total concentrations,respectively.

As seen in these figures, the absorbances for
each constant mole fraction of LiFOS increase abruptly
at characteristic concentrations of LiDS, LiFOS or
total ones of two surfactants, and after this become
constant at some higher concentrations.

Accordingly, CMC values of single and mixed surfactants contributing to mixed micelle formation can
be taken as a midpoint in the range of the concentrations from a steep increase to a plateau of the absorbance according to Corrin and Harkins(12).

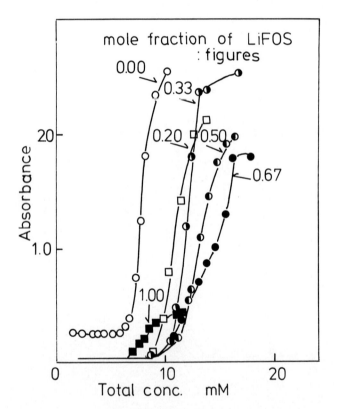

Fig. 6.   *Absorbance of pinacyanol chloride in aqueous solutions of single and
mixed surfactants v.s. total concentration
(spectral changes at α - band).*

418

Their CMC values obtained here are plotted in Fig. 7 (a), (b) and (c), respectively as a function of mole fraction of LiFOS, in which curves (a),(b) and (c) are obtained from the CMC values of LiDS, LiFOS and total concentrations, in Figs.4,5 and 6, respectively.

The CMC values of single LiDS, LiFOS and the mixed CMC values taken as total concentrations contributing to mixed micelle formation, as shown in Figs 7 (a), (b) and (c), were found to be deviated positively from ideal mixing, and the mixed CMC values taken as total concentrations yielded a maximum at 0.5 mole fraction as well as the results obtained by the surface tension and conductivity measurements.

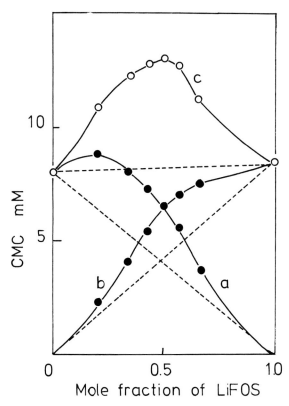

Fig. 7.   CMC v.s. mole fraction of LiFOS curves : (a) obtained from CMC values of LiDS in Fig. 4,(b) obtained from CMC values of LiFOS in Fig. 5, and (c) obtained from mixed CMC values as total concentrations in Fig. 6.

Furthermore, this curve (c) was consistent with the solid line obtained by summation of the CMC values of LiDS and LiFOS at constant mole fractions of LiFOS.

The results obtained by this dye method also provide a strong evidence that the mixed micelle formation in the mixed aqueous solutions become more difficult than that of each single surfactants or two hydrocarbon surfactants.

Consequently, in such a mixed system of LiDS and LiFOS, the positive deviation of the mixed CMC from ideal mixing may result in the characteristic properties that both surfactant molecules tend to exclude each other in aqueous mixed solutions.

IV. REFERENCES

1. Shinoda K., J. Phys. Chem., 58,541(1954).
2. Shinoda K., ibid, 58, 1136(1954).
3. Mysels J. K. and Otter R. J., J. Colloid Sci., 16, 462(1961).
4. Mysels J. K. and Otter R. J., ibid,16,474(1961).
5. Inoue H. and Nakagawa T.,J. Phys. Chem.,70,1108 (1966).
6. Lange H. and Beck K.-H.,Kolloid-Z.U.Z-Polym.,251, 424(1973).
7. Klevens H. B. and Raison M., J. Chimie, 51,1 (1954).
8. Mukerjee P. and Mysels K.J..,"Colloidal Dispersion and Micellar Behavior",The American Chemical Society Symposium, part 17,239(1975).
9. Shinoda K., Hato M. and Hayashi T., J.Phys.Chem., 76, 909(1972).
10. Dreger E. E., Keim I., Miles G. D., Shedlovsky L. and Ross J., Ind. Eng. Chem., 36, 610(1944).
11. Tartar N. V. and Wright K.A., J. Am.Chem. Soc., 61, 539(1939).
12. Corrin M. L. and Harkins W.D., J. Am. Chem.Soc., 69, 679(1949).

EFFECTS OF POLYOXYETHYLENE CHAIN LENGTH UPON MIXED SURFACTANT SOLUTIONS

Kenjiro Meguro,Hiroyuki Akasu,Minoru Ueno
and Tomoo Satake
Science University of Tokyo,Department of Chemistry
Faculty of Science

I. INTRODUCTION

Whereas a number of investigations have been carried out on bulk properties of the mixture of an anionic surfactant and nonionic surfactant (1 - 6), there has been little work reported on surface properties of the mixed surfactant solutions (7). In the most previous works, evidently the distribution of polyoxyethylene chain length in nonionic surfactants used prevented surface studies because of the selective adsorption of more hydrophobic portion in the nonionic surfactants. Recently we studied the surface tensions of the mixed solutions of sodium dodecyl sulfate and a pentaethyleneglycol n-dodecyl ether or hepta oxyethyleneglycol dodecyl ether, and confirmed that the surface tension curve for sodium dodecyl sulfate(SDS) had only one break point in the absence of heptaethyleneglycol dodecyl ether(5ED),whereas the curves had two breaks in the presence of 5ED, and the surface tension remained constant between the two breaks, in spite of the large change of concentration of 5 ED(8).

These results have been interpreted in terms of mixed surface layer and mixed micelles consisting of 5ED and SDS(8).

In this reports, we extend this study to the mixed solutions consisting of SDS and a series of homogeneous polyoxyethyleneglycol dodecyl ethers whose ethylene oxide chains varied from 0 to 8, respectively.

II. EXPERIMENTALS

A. Materials

A series of nonionic surfactants used were homogeneous ploethyleneglycol n-dodecyl ethers, i.e., $C_{12}H_{25}O(CH_2CH_2O)_mH$ (m = 0 to 8). All these surfactants except lauryl alcohol (m = 0) were supplied from Nikko Chemicals Co. Ltd., Tokyo.

These samples were confirmed having both homogeneous poly-oxyethylene chain length and sufficient purity by ga chromatography, IR spectrum, cloud point and surface tension.

Lauryl alcohol(m = 0) was a commercial product of the extra pure grade.   Since its purity was guaranteed by a result of gas chromatography, the reagent was used without further purification.

Sodium dodecyl sulfate(SDS) was synthesized according to a modified method of Dreger, et al(9), i.e., instead of acetic acid, chloroform was used as a solvent for dodecanol.

Commercial sodium chloride of the extra pure grade was roasted before use, and distilled water was prepared by the distillation from alkaline permanganate solution, after being passed through an ion exchange column.

7,7,8,8,-tetracyano quinodimethane(TCNQ) was purified repeated recrystallization from acetonitrile solution.   The melting point of this sample determined by DTA coincided well with the result published(10).

B.   MEASUREMENTS

The surface tension was determined by a modified Wilhelmy type self-recording surface tensiometer, SIMAZU ST-1.   The measurements were carried out at 25 0.2 C at 5 min intervals until the values agreed with each other within 0.1 dyne/cm for 1/2 hr, and the closely agreeing values were accepted.   the time needed for attaining this steady state was usually about 2 hrs.

The solubilization procedures of TCNQ were as follows ;

At first, a definite volume of chloroform solution containing about 1×10^{-4} mole of TCNQ was poured to L-shaped glass tubes.   After TCNQ was free of solvent in vacuo, 15 ml of mixed surfactant solutions of various concentrations was added.   Then the tubes were sealed and shaken for 72 hrs in a thermostat at 25 C.   The absorption spectra of these solutions were measured in 1.0 cm quartz cells using a recording spectrophotometer, HItachi EPS-3T at 25 C.

III. RESULTS AND DISCUSSION

Surface tension   curve for single 7 ED is shown in Fig.1 together with ones in addition of SDS.

This curve for 7 ED yields a break point at 7.2×10^{-5} M, which coicides well with the published CMC of 7 ED(11).

Addition of SDS resulted in the decrease of the break corresponding to the CMC of the mixed micelle(12) and the slope of the curves below the break.   When the SDS concentration reached 6.3 mM, the gradient was extremely small and sharp

break could not be found.   However, the slope increased
again by the further addition of SDS and the surface tension
at a lower concentration of 7 ED approached a value for micellar solution of SDS.   This approach is consistent with the
view that the mixed micelle has an anionic character above CMC
of SDS(13).

For the elucidation of the mechanism on the appearance of
two break points in surface tension curve of mixed surfactant
solutions in this case, we classified the curve into three parts
as shown in Fig. 2.   In region I, the surface tension decreased
gradually, and this decrease may be owing to the adsorption of
nonmicellar surfactant ions or molecules of SDS and 7 ED to
the surface.

By increasing the concentration of 7 ED, the value of surface tension was approaching to the region II, and finally
arrived to the region II.

Schick(7), Lange(14) and Clint(15) have reported that the
CMC of mixed micelle would be lowered more than that of each
surfactant, and they had estimated the CMC of mixed micelle
from the break point of surface tension.

If these results were taken into consideration for the

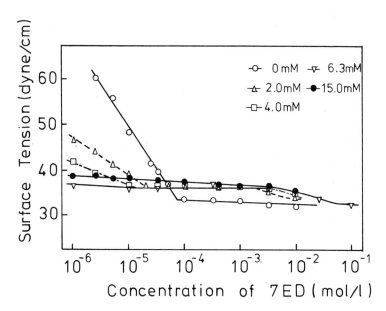

Fig. 1.   The surface tension of homogeneous
heptaethylene glycol n-dodecyl ether in the
presence of sodium dodecyl sulfate (SDS).

origin of first break which located in diluted concentration region, the changing point from region I to II would be the CMC of mixed micelle of SDS and 7 ED.

Apart from the explanation of the detail on region II, it is important to give some meaning for the changing point from region II to III. For this purpose, it needs to clarify whether this change is caused by the change in bulk phase or caused by the change in surface only. To give the answer for this question, the charge transfer solubilization of TCNQ into the solution of mixed surfactant system were applied.

According to our earlier report, TCNQ solubilized in surfactant solution of the concentration above CMC, exhibits the characteristic absorption spectra specified to the kind of surfactant. For instance, the absorption of TCNQ solubilized in 10 mM of SDS solution has a broad absorption maximum around 600 nm which is regarded as charge transfer absorption between TCNQ and the donor part of SDS in micelle.

On the other hand, the spectrum of TCNQ in 7 ED aqueous solution is completely different from that of SDS, and the spectra have absorption maxima at 480, 715 and 860nm, respectively (16, 17).

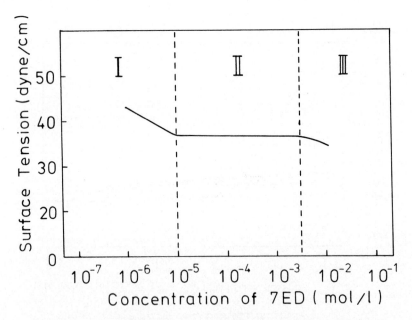

Fig. 2. *Schematical diagram of surface tension curve for 7 ED in the presence of SDS. This curve is classified into three parts of each region I, II and III.*

When TCNQ was solubilized in a mixed solution of 7 ED and SDS, the spectrum shown in Fig. 3 can be obtained.   It is noticiable that the three characteristic bands such as 480, 715 and 860 nm can be found in the absence of SDS, but these bands disappeared by the presence of SDS.   Considering that the appearance of the characteristic band of TCNQ in 7 ED so-lution depends on the charge transfer interaction between TCNQ and donor part such as oxygen part in ethyleneoxide chain, in micelles the disappearance of this bands must be caused by the disturbance of SDS penetrated into the micelle of 7 ED to the charge transfer interaction between TCNQ and 7 ED.

This fact proved same time the existence of mixed micelle between 7 ED and SDS.

By increasing the concentration of 7 ED in the mixed sur-factant system, the band at 480, 715 and 860 nm become to be appear again.

The appearance of these bands is remarkable at the con-centration range from 10 mM to 100 mM of 7 ED.   If the ab-sorption intensity of these bands appeared by the addition of 7 ED against the concentration of 7 ED,Fig. 4 could be obtained.

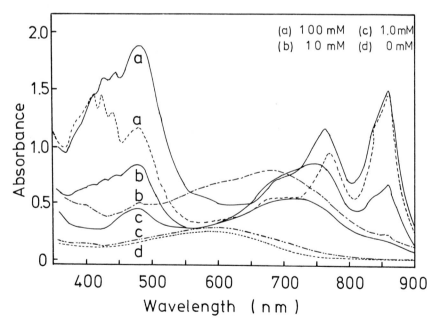

Fig. 3.   The absorption spectra of TCNQ solubi-lized in 7 ED solutions with 10 mM of SDS(broken lines) and without SDS(full lines).

Fig. 4 shows the difference between two absorption data at 860 nm in the presence and absence of SDS versus 7 ED concentration. The increase of intensity of 860 band by the addition of 7 ED is more remarkable in this figure, especially at the concentration of 10^{-3} to 10^{-2} mM.

Considering the fact that the concentration of 7 ED at changing point from region II to III is situated in 10^{-3} to 10^{-2} mM of 7 ED, and the appearance of 860 nm band of TCNQ is also situated in same concentration range, the changing point from region II to III must be depended on the change in bulk phase.

Next problem is to solve the question, that is, what kind of change in mixed micelle system does proceed at the changing point from II to III.

In order to clarify this problem, the effects of chain length of polyoxyethylene group on the surface tension of mixed surfactant system were investigated. In Fig. 5, the surface versus the concentration of nonionic surfactant were shown for the system consisting of a series of homogeneous polyoxyethyleneglycol dodecyl ethers and SDS. The concentration of SDS is maintained at 6.3 mM, and the number of ethyleneglycol group are varied as 0, 1, 2, 3, 4, 5, 6, 7 and 8, respectively.

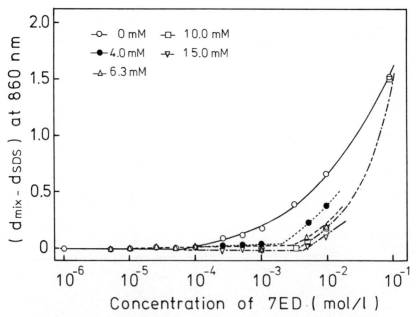

Fig. 4. *Difference between two absorbance data in the presence and absence of SDS at 860 nm v.s. 7 ED concentration.*

For the system consisting of dodecyl alcohol(m = O) and SDS, the solubilization of dodecyl alcohol into SDS solution is small comparatively, because the concentration of SDS is maintained at 6.3 mM which is just below its CMC.

Therefore, the increase of dodecyl alcohol added to the SDS solution causes the separation of dodecyl alcohol phase.

The similar behavior can be found in the system of mono-oxyethyleneglycol ether(m = 1).   Both dodecyl alcohol and monooxyethyleneglycol dodecyl ether must be more surface active than that of major component SDS in this case, and then they are adsorbed more strongly at the water-air interface.

As the concentration of SDS is 6.3 mM below CMC, and so-lubilization is essentially small, the surface tension versus concentration curves for these compounds show a common type of adsorbed film.

When the chain length of polyoxyethylene group reached to 2 ED, saome changes appear in the surface tension-concentra-tion curve.   It is in 2 ED that the appearance of region II is first noted.   As the concentration of SDS is kept 6.3 mM also in this case, the break point between region I and II could be considered as the CMC of the mixed micelle, but, con-sidering the narrowness of region II, in which the mixed micelle

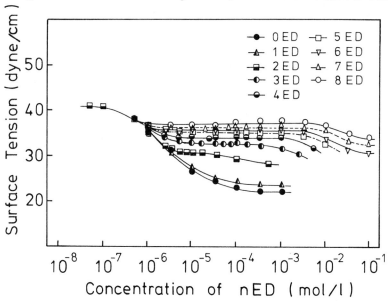

Fig. 5.   Effects of polyoxyethylene chain length on the surface tension of the mixture containing of a polyoxyethyleneglycol n-dodecyl ether and sodium dodecyl sulfate(SDS).   Concentration of SDS is maintained as 6.3 mM.

427

can exists, the mixed micelle in this case can not hold the large amount of 2 ED in side of its micelles.

As shown in Fig. 5, the range of region II becomes wider and wider with the increase of chain length of polyoxyethylene chain.   If the concentration at the changing point of region II to III was assumed to be the barrier concentration between mixed micelle phase and a phase containing the mixed micelle and single sonionic surfactant micelle or nonionic surfactant excess mixed micelle, the result of TCNQ solubilization data can be explained reasonably.

The appearance of single nonionic surfactant micelle or nonionic surfactant excess mixed micelle by the one kind of phase separation has the deep relation of the appearance of 480, 715 and 860 nm bands.

Now, the adsorption of nonionic surfactant onto the adsorbed film of SDS at air-water interface was assumed to obey the szyszkowski equation.   Then we can derive the following equation by a combination of szyszkowski and Gibbs equations;

$$(1/\Gamma) = (1/A) + (1/ABC) ------------(1).$$

where, Γ, A, B and C are surface excess, $b\gamma_0/RT$, $1/a$ and con-

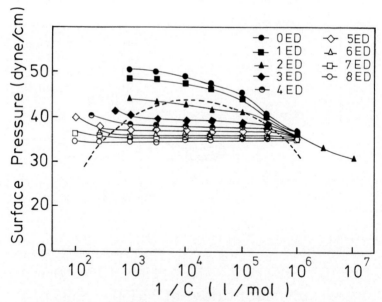

Fig. 6.   Surface pressure v.s. the reciprocal of concentration of nonionic surfactants.

centration of nonionic surfactants, respectively.   Here,as
the concentration of SDS remains constant and A and B are constant, $1/\Gamma$ corresponding to surface area depends on $1/C$.

Fig. 6 shows the relation between the surface pressure
and the reciprocal of the concentrations of nonionic surfactants.   It is very noticiable that the relation similar to
van der waals equation in pressure-volume of real gas can be
obtained in this case.   Here, the chain length has a role
as the temperature in van der waals equation.   As shown by
the dotted line in Fig. 6, transition points exist depending
on the polyoxyethylene chain length in these curves.

As the result, it is concluded that the so called critical
chain length of polyoxyethylene group in nonionic surfactant
molecule exists for the mixed micelle formation.

The effect of temperatures on the surface tension of homogeneous 5 ED in the presence of SDS is shown in Fig. 7.

As with other measurements,the concentration of SDS was
maintained at 6.3 mM in this case.   As shown in Fig.7, each
curve has two break points.   The break at lower concentration is almost independent of the temperature,while the break
at higher concentration shifts toward the lower concentration

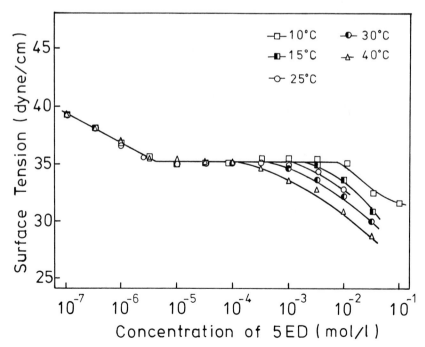

Fig. 7.   Temperature dependence of the surface
tension of 7 ED in the presence of 6.3 mM of SDS.

of 5 ED with the increase of temperature, and then the plateau becomes shorter.

In general, polyoxyethyleneglycol dodecyl ether has a tendency to loose a packing of ethylene oxide chains in micelle surface by increasing the temperature, because of the lowering of hydration at ethylene oxide chains. Therefore, the temperature rising has a effect corresponding to the shortening the ethylene oxide chain length. Accordingly, the critical chain length of polyoxyethylene group for the mixed micelle formation can be related to the temperature whose increase gives an effect similar to hydrophobicity of nonionic surfactant with shorter ethylene oxide chain length for surfactants with longer ethylene oxide chain length.

IV. REFERENCES

1. Nakagawa T. and Inoue H., J. Chem. Soc. Japan, 78, 104 (1956).
2. Yoda O., Meguro K., Kondo T. and Ino T., ibid.,77, 904 (1956).
3. Meguro K., Kondo T. and Ino T., Bull. Chem. Soc. Japan, 31, 472 (1958).
4. Corkill J. M., Goodman J. F. and Tate J. R., Trans.Faraday Soc., 60, 986 (1964).
5. Tokiwa F. and Aigami K., Kolloid-Z.Z. Polym., 239, 687 (1970).
6. Tokiwa F. and Moriyama N., J. Colloid Interface Sci.,30, 338 (1969).
7. Schick M. J. and Manning D. J., J. Am. Oil Chem. Soc., 43, 133 (1966).
8. Akasu H., Ueno M. and Meguro K., J. Am. Oil Chem. Soc., 51, 519 (1974).
9. Dreger E. E., Keim G. I., Miles G. A., Shedlovsky L. and Ross J., Ind. Eng. Chem., 36, 610 (1944).
10. Acker D. S. and Hertler W. R., J. Am. Chem. Soc., 84, 3370 (1969).
11. Deguchi K. and Meguro K., J. Colloid Interface Sci., 38, 596 (1972).
12. Same to number 8.
13. Akasu H., Nishii A., Ueno M. and Meguro K., ibid, 54,278 (1976).
14. Lange H. and Beck K.-H., Kolloid-Z.Z. Polym., 251, 424 (1973).
15. Clint J. H., J. Chem. Soc. Faraday Trans. I., 71, 1327 (1975).
16 Deguchi K. and Meguro K., J. Colloid Interface Sci.,49, (1974).
17. Deguchi K., Mizuno T. and Meguro K., ibid, 48, 474(1974).

HYDROPHILIC-HYDROPHOBIC PROPERTIES OF LONG-CHAIN
IONIC SURFACTANTS

Israel J. Lin

Associate Professor, Dept. of Mineral Engineering,
Technion - Israel Institute of Technology, Haifa, Israel.

I. A B S T R A C T

The physico-chemical properties of ionic surfactants is intimately associated with their amphipathicity, i.e., the balance between the polarity of the head group and nonpolarity of the hydrocarbon chain. These properties are closely affected by such factors as chain length, nature of functional groups, unsaturation in the hydrophobic group, branching, polyoxyethylene - and polyoxypropylene grouping, the nature of the hydrophilic group and its position in the molecule, etc., the changes can be described and explained in terms of the effective chain length (n_{eff}), hydrophile-lipophile balance (HLB), and hydrophobicity index (H.I.).

Detailed knowledge of the relationship between surfactant structure and properties will facilitate choice of the best surfactant for a particular purpose in processes such as flotation, flocculation, foam separation, emulsification, detergency, corrosion inhibition, activated comminution etc., and related interfacial phenomena.

The effect of various structural factors - such as chain length, structure of hydrophobic group and position of hydrophilic group - on the critical micelle concentration (CMC) and the hydrophile-lipophile balance (HLB) of certain ionic surfactants was described on earlier occasions by Lin and others[1,2]. The aim of this paper is correlation of these physico-chemical parameters with the effective chain length (n_{eff}) of the hydrocarbon chain.

II. GENERAL BACKGROUND

Classification of surfactants according to such empirical parameters as HLB, CMC, n_{eff} and H.I. is a handy technique in choosing the best reagent in such processes as detergency, emulsification, foam separation, flotation, flocculation, activated comminution, zone refining, or partition chromatography. It is also useful in interpretation of such interfacial phenomena as reduction in surface tension, adsorption, micellization, hemi-micellization, zeta-potential change, wetting, dispersibility, etc. Their meaning was also dealt with in detail on earlier occasions, and they can be briefly defined as follows;

1. The term "effective chain length"[1,3,4] refers to the equivalent straight-chain compound which yields the same physical results as the one under investigation. It is reduced by coiling, branching, and hydrophilic substitution (double or triple bonds, halogens etc.) and increased by hydrophobic substitution (e.g. addition of a benzene ring). Although essentially an empirical concept, it was also estimated mathematically[3,5,6].

2. The HLB represents the relative degree of hydrophilicity or hydrophobicity of the component groups of the surfactant, and is obtainable in a variety of ways, perhaps from Davies' equation:

$$HLB - 7 = \Sigma(hydrophilic\ group\ numbers) - \Sigma(hydrophobic\ group\ numbers) \qquad [1]$$

The group numbers - empirical values reflecting the relative hydrophilicity or hydrophobicity of the functional groups of a surfactant molecule - are found in tabulated form in literature[7].

For homologous series of straight-chain hydrocarbon surfactants with $nx(CH_2)$ groups in the hydrocarbon radical, eq. [1] may be written as

$$HLB - 7 = \Sigma(hydrophilic\ group\ numbers) - n(group\ number\ per\ CH_2) \qquad [2]$$

It was shown[8] that the hydrophobic group number for CH_2 derives from the term $\phi'/(2,303\ kT)$, where ϕ' is the free-energy change involved in transferring a methylene group from an aqueous to a hydrocarbon phase and equals $1,09\ kT$; k - the Boltzmann constant, and T - absolute temperature. The same term was shown to represent hydrophobic-energy change in the micellization process.

ϕ' characterizes the alkyl chain and remains constant irrespective of the nature of the ionic head, provided each successive methylene group added to the chain is sufficiently removed from the functional group and free of any steric or electronic interaction.

More recently this treatment was extended to fluorocarbon surfactants, classified numerically on the HLB scale[6]. The analogous term ϕ'_{CF_2} was found to have an average value of 2,0 kT, from which the group number for CF_2 was calculated as 0,870.

For a given homologous surfactant series, the HLB decreases as the number of carbon atoms in the hydrocarbon chain increases. Eq. 2 has a fundamental significance in terms of the free energy of micellization and its validity was checked for both anionic and cationic surfactants[1]. Given as CH_2 or CF_2 groups, n_{eff} is that value of n which the analogous straight-chain compound must have to yield the same degree of surface activity as its structurally-modified counterpart. Eq.[2] becomes accordingly,

$$HLB - 7 = \Sigma (hydrophilic\ group\ numbers) - [\phi'/(2,303\ kT)]n_{eff} \qquad [3]$$

3. Micelle formation can be expedited, i.e., made to occur at a lower concentration, by several means: (i) lowering the temperature, (ii) increasing the chain length of the surfactant molecule, (iii) admixture of organic compounds or inorganic salts. The resulting micellization can bring about marked changes in the properties of the aqueous system.

It is well known that in homologous series of straight-chain surfactants, the CMC decreases logarithmically with increasing number of methylene units in the alkyl group. Thus,

$$log\ CMC = A - Bxn \qquad [4]$$

This equation is valid only over a limited range of chain length. Eq.[4] may be written in a general form as

$$log\ CMC = A - B\ x\ n_{eff} \qquad [5]$$

where A and B are experimentally determinable empirical constants, for a given temperature.

Lin et al.[3,6] derived expressions for the CMC of aqueous solutions of long-chain hydrocarbons as function of the n_{eff}. With no inorganic salt added, the relationship reads:

$$log\ CMC = - \phi'n_{eff}/[2,303\ kT\ (1+K_g)] + const. \qquad [6]$$

where K_g is the number ratio of counter ions to long-chain ions in the micelle.

4. The effect of n on solubility is represented by a well-known equation analogous to [4] and [5], namely

$$log\ S_n = log\ S_o - B'\ x\ n \qquad [7]$$

where S_n is the solubility of homolog n, S_o that of a homolog containing no CH_2 groups (i.e. the intercept of the log S - vs. - n plot on the ordinate), and the constant B' represents the contribution of the methylene group, equal to ϕ'_{CH_2} / (2,303 kT).

5.   Combination of equations [3] and [6] yields, for a given homologous series[9],

$$log\ CMC = a + b\ (HLB) = a + HLB/(1+K_g)$$ [8]

where a and b are again empirical constants for a given temperature. Log CMC plotted against HLB shows the linear pattern presented in Fig. 1.

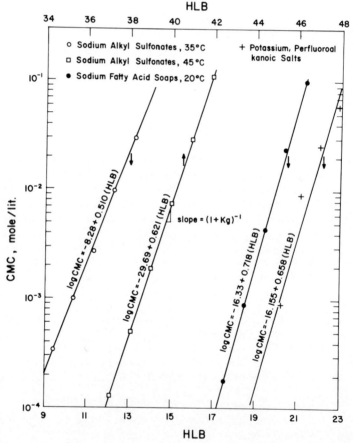

Fig. 1. *Plot of HLB values against CMC for different surfactants.*

6.   The hydrophobicity index (H.I.), related to n_{eff} by the expression

$$H.I. = n_{eff} / n$$ [9]

incorporates the contributions of other parameters such as structural variations, e.g. the position of the polar head, and the presence of double or triple bonds, of aromatic groups, or of cis - and trans configurations, etc. Values of H.I. greater than 1 indicate increased hydrophobicity of the molecule, while values less than 1 indicate the opposite.
Substitution of the above difinition in eq. [5] yields:

$$log\ CMC = A - B\ (H.I.\ x\ n) \qquad\qquad [10]$$

The CMC is a function of the structure and make-up of the organic compound. Thus, it increases with decreasing chain length, and also varies with the functional groups present and with their relative position (isomerism, etc.).

III. RELATIONSHIP OF PROPERTIES AND STRUCTURE IN SURFACTANTS: LITERATURE DATA.

1)   Unsaturation:
    a) Aliphatics. Replacement of $(-CH_2\ -CH_2-)$ in a surfactant by $(-CH=CH-)$ or $(-C\equiv C-)$ reduces the CMC by a factor of 2 to 4. On the basis of data for straight-chain ionic surfactants[10], this was shown to be equivalent to shortening the chain by 1 to 1,5 (CH_2) groups (using the n_{eff} concept), and implies lower hydrophobicity of the molecule, with a parallel increase of the HLB (see Eq. [3]). Partition-coefficient[11] and spreading-pressure[12] measurements also showed that in the series alkane $(-CH_2\ -CH_2-)$ < alkene $(-CH=\ CH-)$ < alkyne $(-C\equiv C-)$ hydrophilicity increases in that order.
    b) Aromatics. It is well known that the $(-CH=\!=\!CH-)$ bond in a benzene ring is considerably less hydrophobic than its $(-CH_2\ -CH_2-)$ counterpart. These is also good evidence[12] that it is less hydrophilic than the ordinary $(-CH=\ CH-)$ bond. Thus its presence in alkylbenzene sulfonates results in a lower CMC compared with alkyl sulfonates with the same number of carbon atoms but lacking the benzene ring[1]. The hydrophobic equivalent of a benzene ring for n_{eff} calculations in straight-chain surfactants is taken as 3,5 (CH_2) units, and its hydrophobic group number is given as 1,662[1].

2)   Position of Head-Group; Chain Structure:

    a) The first example is sodium dodecylbenzene sulfonate, for which the n_{eff} was calculated through the effect of variation of the benzene-ring position on the hemi-micelle concentration[13]. These n_{eff} values, substituted in eq. [3], in turn yield the HLB.
    b) Branching in the sodium salts of nonylbenzene sulfonic, decylbenzene sulfonic and propyl α- sulfostearic acids increases the CMC by a factor of 2 to 4[10], associated with lower hydrophobicity.

c) Lin et al.[1] studied the effect of the head-group position on the basis of Evans's results[14] for seven isomers of sodium n-tetradecyl sulfate with the sulfate group ranging from the terminal to the median position. The closer the head group to the middle of the chain, the higher the CMC and the lower the n_{eff}.

d) Carbon atoms on short branches on a hydrophobic alkyl group, or on the shorter portion of a hydrophobic group when the hydrophilic group is not terminal, have about one-half the effective length of a carbon atom in a straight alkyl chain with hydrophilic terminal groups[15].

3) Position of Ester Group:

a) Studies of ester-type surfactants were reported by Hikota[16]. The effects of the position of the ester group on the CMC and n_{eff} are listed in Table I and plotted in Fig. 2.

TABLE 1
CMC and n_{eff} of sodium sulfoalkyl alkanoates[16-17].

Compound	CMC, mol/lit	CMC vs. n	K_g (Ref.3)	ϕ'_{CH2}
$C_8H_{17}COO(CH_2)_2SO_3Na$	0,046			$\phi'/[(1+K_g)2,303kT] = -B; \quad \phi' = -1,07kT$
$C_9H_{19}COO(CH_2)_2SO_3Na$	0,025			
$C_{10}H_{21}COO(CH_2)_2SO_3Na$	0,0105	logCMC=-0,293(n+2)+ 1,603	0,59	
$C_{11}H_{23}COO(CH_2)_2SO_3Na$	0,0062			
$C_{12}H_{25}COO(CH_2)_2SO_3Na$	0,0022			
$C_9H_{19}COO(CH_2)_3SO_3Na$	0,019			
$C_{10}H_{21}COO(CH_2)_3SO_3Na$	0,009	logCMC=-0,293(n+3) + 1,778	0,59	
$C_{11}H_{23}COO(CH_2)_3SO_3Na$	0,0043			
$C_9H_{19}COO(CH_2)_4SO_3Na$	0,013	logCMC=-0,358(n+4)+ 2,768	0,59	
$C_{11}H_{23}COO(CH_2)_4SO_3Na$	0,0025			

As the graphs for the different classes of compounds are practically parallel, their spacing can be used as measure of the
hydrophilicity difference of the $- COO(CH_2)_j SO_3 Na$ and
$-OCO(CH_2)_j SO_3 Na$ groups. For a given number of carbon atoms,
shifting of the ester group from the ionic head to a more

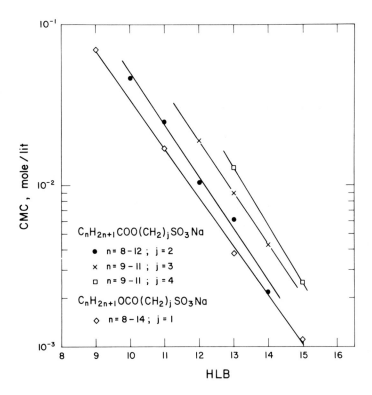

*Fig. 2. CMC vs. total number of $-CH_2-$ groups for sodium
sulfoalkyl alkanoates and sodium sulfoalkyl esters.*

central position in the hydrocarbon chain resulted in a higher
CMC value. The relation between the latter and the total number of CH_2 groups is logCMC$=-0,293(n+3) + 1,778$ for
$C_n H_{2n+1} COO(CH_2)_3 SO_3 Na$ and logCMC$=-0,147(9+j) + 0,011$ for
$C_9 H_{19} COO(CH_2)_j SO_3 Na$. It is thus seen that the CMC-reducing
power of the CH_2 unit situated between the ester and sulfonate
groups is approximately half that in the hydrocarbon chain in
its effect on the log CMC $(0,293/0,147 \cong 0,5)$. Using the concept of n_{eff}, we have:

$$log\ CMC = -0,293\ (n + j/2) + 1,314 \qquad [11]$$

for n=8 to 12, and j=2 to 4, n in this case being the number of CH_2 groups in the fatty acid portion and j the number of those situated between the sulfonate and ester groups.

Table I shows also that for a given total number of methylene groups, shifting of the ester group from the terminal to a more central position in the alkyl chain again results in a higher CMC value. This means that the hydrophilicity of the surfactant increases as the ester group approaches the middle of the chain .

b) In another paper [18], Hikota reported CMC values for sodium alkyl sulfoacetates and sodium alkyl sulfopropionates. These results (Table II) show that the latter - in which the ester linkage (hydrophilic group) is closer to the middle of the alkyl chain - have higher CMC values than the corresponding sulfoacetate salts.

TABLE 2
CMC and n_{eff} of sodium sulfoalkyl esters [18].

===

Compound	CMC, mol/lit	CMC vs. n
$C_8 H_{17} OCO\ CH_2 SO_3 Na$	0,070	
$C_{10}H_{21} OCO\ CH_2 SO_3 Na$	0,017	
$C_{12}H_{25} OCO\ CH_2 SO_3 Na$	0,0038	$\log CMC = -0,301\,(n+1)+1,540$
$C_{14}H_{29} OCO\ CH_2 SO_3 Na$	0,0011	
$C_8 H_{17} OCO\,(CH_2)_2 SO_3 Na$	0,0487	
$C_{10}H_{21} OCO\,(CH_2)_2 SO_3 Na$	0,0131	
$C_{12}H_{25} OCO\,(CH_2)_2 SO_3 Na$	0,0028	$\log CMC = -0,294\,(n+2)+1,630$
$C_{14}H_{29} OCO\,(CH_2)_2 SO_3 Na$	0,0007	

The same conclusions as in § 3a above may be reached for change of the length of the CH_2 group in the hydrophilic portions - $OCO(CH_2)_j SO_3 Na$ and - $\overset{..}{C}OO(CH_2)_j SO_3 Na$.

4) Polyethylene Chain (E.O. groups):

a) A class of surfactants of particular interest is that represented by compounds in which polyethylene (or polypropylene) groups are present together with the hydrocarbon radical, and which have special surface-active properties dictated by the E.O. grouping.

For non-ionic compounds of this class, the CMC* was found to increase with the number of E.O. groups[19], in keeping with the attendant increase in hydrophilicity.

By contrast, for compounds with an ionizable group at the end of the polyoxyethylene chain, it was found (Tables III and IV) that the CMC decreases initially with increasing number of

TABLE 3

CMC for homologous series of anionic surfactants.

Compound		CMC, mol x lit^{-1}	Ref.
$C_nH_{2n+1}(OCH_2CH_2)_mOSO_3Na$			
n = 12	m = 0	0,008	a, b, f
12	1	0,0047	a, b, f
12	2	0,003	a, b, f
12	3	0,0019	a, b,
12	4	0,0012	a, b
12	17,5	0,000042	a, b
m = 14	m = 0	0,0027	a, b
14	1	0,0014	a, b
14	2	0,001	a, b, f
14	3	0,00069	a, b
n = 16	m = 0	0,00051	c
16	1	0,00021	c
16	2	0,00013	c
16	3	0,00007	c
16	4	0,00008	c
n = 18	m = 0	0,00013	c
18	1	0,00011	c
18	2	0,00008	c
18	3	0,00005	c
18	4	0,00007	c
$CH_3(CH_2)_7CH=CH(CH_2)_8(OCH_2CH_2)_mOSO_3Na$			
	m = 0	0,0036	a, b
	m = 1	0,0020	a, b
	m = 2	0,0017	a, b
	m = 3	0,0012	a, b
$[C_{12}H_{25}(OCH_2CH_2)_mSO_4]_2Ca$			
	m = 0	0,0012	e
	m = 1	0,00046	e
	m = 2	0,00037	e
	m = 3	0,00031	e
$[C_{10}H_{21}(OCH_2CH_2)_mO]_2POOK$			
	m = 0	0,00047	d
	m = 1	0,00021	d
	m = 2,6	0,0001	d
	m = 5,5	0,00007	d
	m = 8,6	0,00014	d
	m = 13,0	0,00022	d

a. Götte, V.E., Proc. Int. Congr. Surface Activity, Vol. III, Butterworths, Lond., 45 (1960).
b. Schick, M.J., J. Phys. Chem., 68, 3585 (1964).
c. Weil, J.K., et al., Ibid, 62, 1083, (1958); Weil, J.K., et al., J. Am. Oil Chems' Soc., 36, 241 (1959).
d. Petrov, A., and Y.S., Smirnov, Chem., Phys. Chem. Anwendrungstech Grenzjlaechenaktive Stoffe, Ber. Int. Kong. 6th, 11 Sept. (1972) Pub (1973) Band 3, pp. 615-622.
e. Hato, M. and K. Shinoda, J. Phys. Chem., 77, 378 (1973).
f. Barry, B.W. & R. Wilson, J. Pharm. Pharmac., 26, 125 (1974).

* The values of n_{eff} together with the corresponding values of the H.I. and HLB are given elsewhere[2].

TABLE 4
CMC, for homologous series of cationic surfactants, at 25°C.

===

Compound	CMC, mol x lit^{-1}	Ref.
$C_{10}H_{21}N(CH_3)_3Cl$	0,071	a
	0,0611	b
$C_{10}H_{21}(OCH_2CH_2)N(CH_3)_3Cl$	0,030	a
	0,030	b
$C_{10}H_{21}(OCH_2CH_2)_2N(CH_3)_3Cl$	0,023	a
$C_{10}H_{21}NC_5H_5Cl$	0,065	a
$C_{10}H_{21}(OCH_2CH_2)NC_5H_5Cl$	0,0256	a
$C_{10}H_{21}(OCH_2CH_2)_2NC_5H_5Cl$	0,021	a
$C_{12}H_{25}N(CH_3)_3Cl$	0,0201	a
$C_{12}H_{25}(OCH_2CH_2)N(CH_3)_3Cl$	0,0071	a
$C_{12}H_{25}(OCH_2CH_2)_2N(CH_3)_3Cl$	0,0052	a
$C_{12}H_{25}NC_5H_5Cl$	0,0156	a
$C_{12}H_{25}(OCH_2CH_2)NC_5H_5Cl$	0,0069	a
$C_{12}H_{25}(OCH_2CH_2)_2NC_5H_5Cl$	0,0049	a
$C_{16}H_{33}NC_5H_5Cl$	0,0009	a
$C_{16}H_{33}(OCH_2CH_2)NC_5H_5Cl$	0,00029	
$C_{16}H_{33}(OCH_2CH_2)_2NC_5H_5Cl$	0,00018	a

a) *Mandru I., J. Colloid Interface Sci., 41, 430 (1972).*
b) *Hoyer H.W. et al., J. Phys. Chem., 65, 1807 (1961).*

E.O. groups up to a critical number somewhere between 3 and 5, beyond which it increases as in the nonionic compounds. The above critical number is justified by the findings of Ford and Furmidge[20] who reported that 1 to 3 E.O. groups in cationic surfactant do not suffice for emulsification but a minimum of 3 or 4 are needed; similarly, according to Schick[21], "At least 4 to 6 E.O. units per molecule are required, depending on the nature of the hydrophobic group, in order to produce a surfactant with distinctly hydrophilic characteristics".

In the range of m covered by Tables III and IV, lengthenning of the polyoxyethylene chain in the molecule is equivalent to extension of the hydrocarbon radical.

b) Surfactants will a small number of E.O. groups in the molecule exhibit a similarly anomalous behavior when used as collectors in mineral flotation. Weehuizen et al.[22] described a class of collectors of the ether-carboxylic acid type (general formula $R_n(OC_2H_4)_mOCH_2COOH$) which have marked collecting power with respect to salt-type minerals (e.g. fluorspar, barite and calcite), sulfides (e.g. sphalerite, pyrite and chalcopyrite), oxides (e.g. ilmenite), and silicates (e.g. spodumene and feldspar). Of interest here is the effect of m in recovery of fluorite by acids with $R_n = 12$. Results are presented in Fig.3 which shows an ascending order of m=1>o>2>3, again indicating the unusual effect of single E.O. groups. In this case the critical number of E.O. units is 2, owing to the fact that the E.O. group is not directly attached to the polar head.

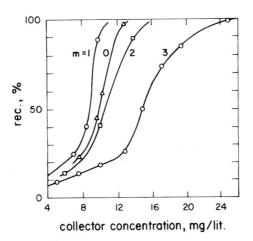

Fig. 3. *Collecting properties of ether-carboxylic acids of the dodecyl series on fluorite.*

IV. CONCLUSIONS

(i) The effective chain length is a common derived parameter for the different typs of substitutions and stractural modifications; it represents the degree of hydrophobicity of any substituted entity.

(ii) The effect of variation of the molecular structure on the
HLB is predicted from the observed effects on CMC.
(iii) The oxyethylene groups in long-chain ionic surfactants
have a dual reinforcing effect, associated either with the
nonpolar or the polar part of the molecule. For a given hydrocarbon radical and polar head with increasing length of
the E.O. chain, the CMC, micelle formation, and surface activity pass through extremed values. These surfactants exhibit
unusual properties dissimilar to those of non-ionic compounds.
(iv) Given sufficient variety of substituted molecules with
known H.I. values, design and synthesis of molecules with pre-
determined degrees of hydrophobicity should be easier, in turn
permitting more efficient choice of surfactants for a specific
purpose.

ACKNOWLEDGMENT

This study was supported by the Mineral Engineering
Research Center, Technion, Haifa, Israel.

V.   REFERENCES

1. Lin, I.J., et al., J. Colloid Interface Sci. 45, 378 (1973).
2. Lin, I.J., and Marszall L., Ibid. (in Press).
3. Lin, I.J., et al., Colloid Polymer Sci., 252, 407 (1974).
4. Jorne J., and Rubin E., J. Colloid Interface Sci., 38,
      639 (1972).
5. Lin, I.J. and Metzer A., J. Phys. Chem., 75, 3000 (1971).
6. Lin, I.J., Ibid., 76, 2019 (1972).
7. Davies J.T., and Rideal E.K., "Interfacial Phenomena",
      2nd ed., Academic Press, N.Y. (1963).
8. Lin, I.J., and Somasundaran P., J. Colloid Interface Sci.,
      37, 731 (1971).
9. Lin, I.J., AIME Trans., 250, 225 (1971).
10. Mukerjee, P., and Mysels K.J., "Critical Micelle Concentra-
      tions of Aqueous Surfactant Systems", U.S. Government
      Printing Office (1971).
11. Leo, A., et al., Chem. Revs., 71, 525 (1971).
12. Pomerantz P., et al., J. Colloid Interface Sci., 24, 16
      (1967).
13. Dick S.G., et al., Ibid., 37, 595 (1971).
14. Evans, H.C., J. Chem. Soc. 579 (1956).
15. Rosen, M.J., J. Am. Oil Chems. Soc., 52, 431 (1975).
16. Hikota, T., Bull. Chem. Soc. Jap., 43, 2236 (1970); Ibid,
      43, 3913 (1970).
17. Hikota, T., and Meguro K., Yukagaku (J. Jap. Oil Chem.
      Soc.,) 23 No. 3, 31, (1974): J. Am. Oil Chems' Soc.,
      52, 419 (1975).

18. Hikota, T., and Meguro, K., J. Am. Oil Chems' Soc., 46, 579 (1969): Ibid, 47, No.5, 158 (1970).

19. Carless, J.E. et al., J. Colloid Sci., 19, 201 (1964), ; Schick, M.J., Ibid, 17, 801 (1962).

20. Ford, R.E., and Furmidge C.G.L., J. Colloid Interface Sci., 22, 331 (1966).

21. Schick, M.J., J. Am. Oil Chems' Soc., 40, 680 (1963).

22. Weehuizen, J.M., et al., Proceedings 11th Int. Mineral Process. Congr., Cagliari, 121 (1975).

INTERACTIONS OF SURFACTANTS WITH PROTEINS

I. MICROCALORIMETRY AND FOAMING

Fouad Z. Saleeb and Timothy W. Schenz
General Foods Corporation

The interactions between an ionic surfactant, dioctyl sodium sulfosuccinate, and sodium caseinate have been investigated by microcalorimetry, foam studies and zeta potential measurements. The binding has been categorized into two major types. At low surfactant/protein ratios, the binding was predominantly hydrophobic in nature, while at higher concentrations, ionic binding occured and was dependent upon pH. The foaming properties of protein surfactant solutions could be interpreted in view of the hydrophobic and ionic regions of binding.

I. INTRODUCTION

Interactions between surfactants and proteins have primarily been studied in systems of biological importance (1,2). The binding of a surfactant to a protein has been reported to occur at discrete protein binding sites. (3). In other systems the binding appears to be less specific and more heterogeneous in nature (4,5). In either case, the binding of surfactant molecules to proteins has generally been found to be dependent on the concentration of the surfactant, pH and ionic strength of the system (2).

The present study examined the interaction of two ionic surfactants, dioctyl sodium sulfosuccinate and cetyltrimethylammonium bromide, with sodium caseinate. Surfactant/caseinate systems were studied by microcalorimetry, surface tension measurements, electrophoresis and foamability. These results have been interpreted in terms of hydrophobic and ionic binding sites.

II. EXPERIMENTAL

A. Materials

The sodium caseinate isolate was obtained from New Zealand Milk Products, Inc., Rosemont, Illinois, and used without further purification. The anionic surfactant was dioctyl sodium sulfosuccinate (DSS) from American Cyanamid Company (Pearl River, New York). The cationic surfactant was cetyl trimethyl ammonium bromide (CTAB) from K&K Laboratories, Inc., Plainview, New York. Both were used as received.

The protein solution contained 10 g/l sodium caseinate. The surfactant stock solutions were 0.025 M. Both protein and surfactant solutions were adjusted to the desired pH with NaOH or HCl and to an ionic strength of 0.01 with NaCl. The surfactant concentration was varied from 6x10-4 M to 1x10-2 M.

B. Methods
1. Foaming

A protein-surfactant mixture was placed in a glass column (4cm x 30cm) fitted with a medium sintered glass disk at the bottom through which water-saturated air was passed at 200 ml/min. The solution was foamed for 120 seconds at which time the initial foam height and liquid level were recorded. The foamability was related to foam volume expansion expressed as percentage increase above the liquid level (6,7).

2. Titration Microcalorimetry

A titration microcalorimeter (model 1050/ 1150, Tronac, Inc., Orem, Utah) was used in the isothermal mode for the titration studies. The surfactant solution was added automatically and continuously into the reaction vessel containing the protein solution. The amount of heat (millicalories) evolved or absorbed for the specified time interval was then plotted against the total surfactant concentration after corrections were made for chemical (dilution) and non-chemical effects (baseline drift). The rate of addition of surfactant was adjusted so as not to exceed the time response of the calorimeter.

3. Zeta Potential

A Zeta Meter Inc. apparatus (New York, New York) was used to measure the electrophoretic mobility of protein particles in surfactant solutions. Zeta potentials were calculated using the Smoluchowski equation (8).

4. Surface Tension

Surface tension measurements were made in the usual way using a Rosano surface tensiometer (Biolar Corporation, N. Grafton, Massachusetts) and a sand-blasted platinum blade. Differences in the surface tension of surfactant solutions with and without protein were attributed to the binding of the surfactant to the protein. A binding isotherm was constructed from these differences.

III. RESULTS AND DISCUSSION

The initial foam volumes for a series of surfactant/protein solutions (DSS/sodium caseinate) are shown in Figure 1. Foaming was measured at pH 6.0 and at pH 3.6, 1.2 pH units above and below the protein's isoelectric point.

At pH 3.6, the initial foam volume reached an optimum at a surfactant/protein ratio of 0.1 millimoles DSS/g protein followed by defoaming. At 0.5 millimoles DSS/g protein, the initial foam volume was essentially zero. Further addition of surfactant re-established the foam-ability of the system. At pH 6.0, a minimum in foamability was observed at around 0.05 millimoles DSS/g protein followed by a general increase in foaming.

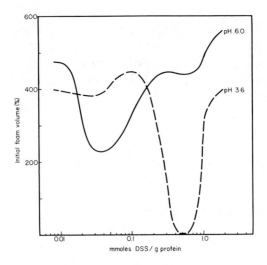

Fig. 1. Initial foam volumes for DSS/protein solutions; total protein concentration = 10 g/l.

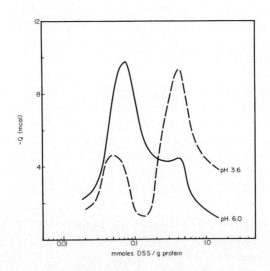

Fig. 2. Thermometric titration of sodium caseinate by DSS; total protein concentration = 10 g/l.

The calorimetry data are shown in Figure 2. At pH 3.6, two distinct exothermic peaks were observed at 0.05 and 0.4 millimoles DSS/g protein; at pH 6.0 the peaks were at 0.07 and 0.4 millimoles DSS/g protein. The first peaks at the low surfactant/protein ratio, being invariant at pH's above and below the isoelectric point of the protein, indicated binding that was predominantly hydrophobic in nature (1,9). The difference in the heights of the two peaks may be explained by the increased number of exposed hydrophobic sites on the protein at pH's above the isoelectric point (10). The binding at 0.4 millimoles DDS/g protein was dependent on pH (second peaks) and pointed to some type of ionic interaction (11).

Further evidence for this designation of hydrophobic and ionic binding is presented in Figure 3, in which is shown the calorimetry data for the interaction of a cationic surfactant (CTAB) with sodium caseinate. The ionic portion of the curves was essentially reversed when compared to the anionic DSS surfactant (Figure 2). However, the initial hydrophobic binding was invariant under a pH change. The relative magnitudes of the first peaks also were the same as with the anionic surfactant data (Figure 2).

The amount of DSS adsorbed on the caseinate, shown in Figure 4, was calculated from surface tension measurements on surfactant solutions with and without protein. Up to a concentration of 0.6 millimoles/l of DSS, the amount of surfactant adsorbed on the protein was less than 0.02 millimoles/g protein and was nearly independent of pH. This corresponded to a surfactant/protein ratio of 0.1 millimoles/g protein (first peak of the calorimetry data). Above this concentration, the binding increased dramatically and was a function of pH, showing the ionic nature of the binding.

The zeta potentials of the DSS surfactant/ protein solutions were calculated and are shown in Figure 5. At pH 3.6, the protein possessed a

449

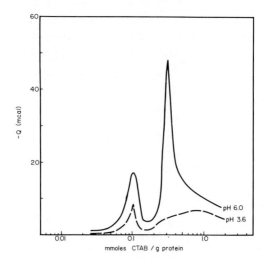

Fig. 3. Thermometric titration
of sodium caseinate by CTAB;
total protein concentration =
10 g/1.

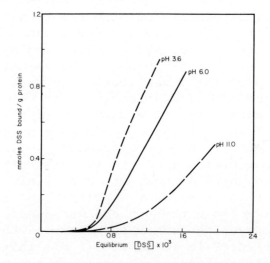

Fig. 4. Binding isotherm of DSS
onto sodium caseinate; total
protein concentration = 10 g/1.

net positive charge. As negatively charged
surfactant molecules progressively adsorbed on
the protein, the zeta potential of the protein
passed from a positive to a negative value. At
pH 6.0, both protein and surfactant were
negatively charged and very little change in the
value of the zeta potential was observed with
increasing DSS concentration. The slight
decrease in zeta potential in the region below
0.1 millimoles DSS/g protein can be explained
by the increased size of the surfactant/protein
complex, which results in an apparent effective
charge (charge/area) decrease.

Comparison of the foaming data in Figure 1
with the calorimetry data in Figure 2 revealed
that the enhancement and inhibition of foaming
corresponded with calorimetry peaks at those
concentrations that represent the binding of the
surfactant to the protein.

At pH 6.0, where both the protein and DSS
were negatively charged, the initial binding
of surfactant to protein produced a complex that
had foaming characteristics that were relatively
poor. After this predominantly hydrophobic
binding, ionic interactions became appreciable
(Figure 2), resulting in a structure which
could support a foam.

A pH 3.6, the slight initial decrease in
foam volume could be ascribed to binding that
was predominantly hydrophobic in nature. Since
hydrophobic binding of surfactant molecule onto
a protein results in an ionic group of the
surfactant being exposed to the aqueous phase,
the surfactant/protein complex becomes more
hydrophilic, more soluble and thus has less
tendency to adhere to an air bubble. The optimum
foam volume at 0.1 millimoles DSS/g protein
coincided with zero zeta potential (Figure 5),
This was in agreement with the observation that,
provided there was no precipitation of the
protein, maximum foaming occured at the iso-

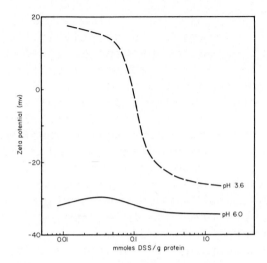

Fig. 5. Zeta potential of DSS surfactant/protein solutions; total protein concentration = 10 g/l.

electric point of the protein (7). In the present system, a pseudo-isoelectric point was achieved by addition of an anionic surfactant to a positively charged protein. In order to reach a pseudo-isoelectric point for any given system, it becomes necessary to balance surfactant and protein concentrations against the pH of the system.

Upon further addition of surfactant beyond 0.1 millimoles DSS/g protein (Figure 1, pH 3.6), the decrease in foam volume was due to the formation of a highly soluble complex. This complex resulted from the combined effect of an ionic interaction, and a second layer adsorption of surfactant molecules in a tail to tail manner (1,11). The increase in foam volume above 0.5 millimoles DSS/g protein was due to higher concentrations of free surfactant in the system.

IV.   ACKNOWLEDGEMENTS

We wish to acknowledge the work of Steven M. Schechter and Laurie A. Brennan and thank Henri L. Rosano for his helpful discussions. We also are grateful to General Foods Corporation for permission to publish this research.

V.   REFERENCES
1.   Tanford, C., The Hydrophobic Effect, Chapter 14, John Wiley & Sons, New York, 1973.
2.   Helenius, A. and Simons, I., Bioch. Biophys. Acta 415, 29 (1975).
3.   Scatchard, G., Ann. N.Y. Acad. Sci. 51, 660 (1949).
4.   Karush, R. and Sonnenberg, M., J. Amer. Chem. Soc. 71, 1369 (1949).
5.   Karush, R., J. Amer Chem. Soc. 72, 2705 (1950).
6.   Cante, C.J., et al., J. Colloid Interface Sci., 45, 242 (1973)
7.   Graham, D.E. and Phillips, M.C., S.C.I. Int. Symp., Brunnel University, 1975. Academic Press, London and New York. In press.
8.   Smoluchowski, M., Z. physik. Chem. 92, 129 (1918).
9.   Greene, F.C. and Kasarda, D.D., Cereal Chem. 48, 601 (1971).
10. Spector, Arthur A., J. Lipid Res. 16, 165 (1975).
11. Milik, W.U. and Saxena, V.P., in "Colloidal Dispersions and Micellar Behavior" (K.L. Mittal, Ed.), American Chemical Society, Washington, D.C., 1975.

STABILITY AND INSTABILITY OF SURFACTANT SYSTEMS

R. Despotović, Lj. A. Despotović, V. Horvat,
N. Filipović-Vinceković, D. Mayer and B. Subotić
"Ruđer Bošković" Institute

*Colloids prepared and aged in aqueous media in the pre-
sence of cationic, nonionic and anionic surface active agents
or surfactants show a dependence on the chemical properties
and the concentration of the present surfactant. The main aim
of the present investigation is to establish in which manner
the surface active agents are incorporated at the solid/
/liquid interface and in which way the surface active agents
cause surface reactions. The obtained results show complex
mutual phenomena, indicating, as a reasonable possibility,
the formation of submicellar species with distinguished colloid properties.*

I.   INTRODUCTION

Surfactant systems possess valuable and scientifical-
ly very interesting properties. Investigation results of
" surfactant + inorganic sol " systems in aqueous media are
directing to the conclusion that there is a mutual colloid
interaction. Similar conclusion can be also drawn for the
surfactant systems in nonaqueous media (1) . It is well
known that the surfactants exert very strong effects on the
colloid stability of various inorganic suspension. In comparison with coagulation by inorganic electrolytes surface

active agents or surfactants can cause the flocculation in
the same inorganic sol at concentrations lower 1000 to 10000
times   (2, 3) . In the majority of results the critical
changes of colloid stability and instability are reached below the critical micellar concentration of applied surfactant,
indicating, as a reasonable possibility, the formation of sub·
micellar surfactant species with properties characteristic
for colloid. Behavior of mixed surfactant/surfactant solutions confirms such a conclusion   (4) . Many of the literature data direct us toward the conclusion that the colloid
stability and instability of an inorganic sol in the presence
of surfactant is due to colloid properties of both inorganic
sol and submicellar surfactant species. In an attempt to
understand a mechanism by which the stability of an organic
sol, i.e. surfactant system depends on the bulk properties,
a series of experiments were performed. The results discussed
here indicate the mutual dependence of stabilities for both
components in the observed system.

II.   EXPERIMENTAL

A. Materials

        All used inorganic salts were of BDH Analar quality.
Sodium n-dodecyl sulphate (T^-) was of a special purity
grade supplied by BDH . n-Dodecylammonium nitrate (T^+) was
prepared as previously described (4) . TRITON X 305 (T^o)
is a standard product of Rhom & Haas . All used water was
twice distilled in an all-Duran 50 apparatus.

B. Preparation of Systems

        All the prepared systems were kept at constant tempe-
rature of 293 K in an Haake ultrathermostate. The systems
were prepared by direct mixing of equal volume of precipita-

tion components. The following designations have been used
for the prepared systems:

- Sol I ,   prepared by mixing of 0.002M AgNO$_3$ and 0.004
                M NaI containing various concentrations of T^+:
- Sol II-A ,   prepared by mixing of 0.002 M AgNO$_3$ and
                0.004 M NaI containing various concentrations
                of T^- :
- Sol II-B ,   prepared by mixing of 0.002 M AgNO$_3$ + var.
                concns. T^- with 0.004 M NaI :
- Sol II-C ,   prepared by mixing of 0.002 M AgNO$_3$ and T^-
                of various concentrations :
- Sol II-D , T^- of various concentrations in aqueous media:
- Sol II-E ,   prepared by mixing of T^- of various concns.
                and 0.002 M NaI :
- Sol II-F ,   prepared by mixing of 0.002 M AgNO$_3$ and
                0.004 M NaI :
- Sol III ,   prepared by mixing of 0.002 M AgNO$_3$ with a
                mixture of 0.004 M NaI + var. concns. of T^o :
- Sol IV ,   prepared by adding T^+ into T^- solutions.
                Total surfactant concentration is
                $$c_{T^+} + c_{T^-} = 0.0002 \text{ M} .$$

C.   Diffusion Experiments

Diffusion process in AgI - I^- systems (sols I,
II-B, III) was measured by the radiometry. A radioactive
indicator 131-I had high specific radioactivity. The applied technique was the same as previously described (5) .
The systems to be analysed were aged for 100 minutes before
labeling, and the changes in radioactivity were determined
100 minutes after labeling.

D.   Tyndallometry

Tyndallometric values τ were registered by Pulfrich
photometer with a turbidimetric extension. All the systems

were prepared (sols I - IV) in the same way; the precipitating components were poured 7 times from one tube to another (*in statu nascendi* method) . Sodium iodide soluiton or silver nitrate solution were in contact with a surfactant solution for t_d minutes before mixing with another precipitating component. Prepared sols were aged for t_A minutes before turbidimetry.

E. Sedimentation

The silver iodide sols were prepared from labelled sodium iodide containing various concentrations of surfactant T^- by the standard procedure (sol II-B) . The contact time was t_d = 10 minutes, mixing with magnetic stirrer for 15 seconds and the aging of sols t_A = 10000 minutes. Using homogenized suspension the radioactivity A_0 was determined. The radioactivity A_t was determined by measuring the radioactivity of clear supernatant above the precipitated AgI . The fraction f_s of silver iodide which sedimented spontaneously was computed from $f_s = (A_0 - A_t) /$ $/ (A_t)$. The conditions under which radiosedimentation analysis was performed are described in detail in an earlier paper (6) .

F. Microelectrophoresis

The sols I - IV , aged for t_A = 100 minutes were used for microelectrophoretic measurements. The electrophoretic mobility and particle charge were determined by the aid of an ultramicroscope with a double capillary electrophoretic cell. The mean of seven samples was presented as the results.

G.   Surface Tension

        The surface tension σ (dynes / cm) determinations
were conducted (sols I, II-A,B,C,E, IV) on a Lecomte du
Nouy semiautomatic torsion balance   (Krüss, Hamburg). 11 samples composed the system; at constsnt total surfactant concentration of $c_{T+} + c_{T-} = 0.0002$ M   the samples differed
10 percent in a combination of the T^+ and T^- components.
The results are presented as the average value of 7 or 10
measurements.

H.   Conductometry

        Conductivity of the sols II-C and IV were detremi-
ned by means of a Cambridge high frequency conductometer using a conductivity cell with a normal platinum electrodes all
in a thermostated body.

III.   RESULTS AND DISCUSSION

        Silver iodide sols containing various surface active
substances are the most frequently examined colloid system of
the " inorganic sol + surfactant " type. Many data report
interesting facts and give conclusions about interaction and
interrelation between inorganic sol and the surfactant present in the systems (2, 3, 5 - 8) . Contrary to the relative simple interactions between sols particles and inorganic
ions, surfactant solutions depending on its chemical nature
and ionic or nonionic form cause more complex and substantially different phenomena especially in respect to the stability of sols (3 - 8) . Nonionic, cationic and anionic
surfactants interact by various mechanisms with silver iodide
sols since the radiometry data (Fig. 1. F $vs.$ log c) are
different for the sols prepared by the "same" procedure:

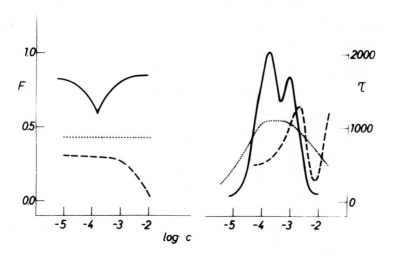

Fig. 1. Systems: AgI + NaI + *surfactant* T . *Left;*
AgI - I⁻ *diffusion schematicaly presented as a fraction ex-*
change F *versus* log c T ; *full line, sols I, II-B ; dot-*
ted line, sol III ; dashed line, cationic dye . Right; Tur-
bidity τ *plotted versus* log c T *for the following sols;*
full line, sol I ; dashed line, sol II B and dotted line
sol III.

nonionic surfactant (sol III) does not influence AgI - I⁻
exchange; in the presence of cationic or anionic dodecyl de-
rivatives (sols I and II-B) the rate of the exchange process
is decreasing by the increase of surfactant concentration and
after passing minimum it starts rising as the surfactant concentration is increasing. F minimum corresponds to the flocculation concentration f_c of DDANO$_3$ or to the "critical"
flocculation concentration c_f of NaDDSO$_4$ (Fig. 1. τ *vs.*
log c). Contrary to the broad τ maximum for systems with
nonionic surfactant, cationic DDANO$_3$ induces complex interactions and reaches twice τ maxima. First maximum

corresponds to the increasing adsorption of surfactant at the AgI surface causing flocculation of inorganic sol with equilibrium amount of adsorbed surfactant. At the same time sol particles reach zero point of electrophoretic mobility and lowest rate of AgI $-$ I$^-$ exchange. Based on stated facts it can be supposed that the selected or active points at the crystal plane occupied by the surfactant are preferentially responsible for the rate of AgI $-$ I$^-$ exchange or diffusion. Nonionic species are not accumulated at the active points and the AgI $-$ I$^-$ diffusion is not hindered. By the decrease of τ value over the f_c , particle charge is changing from zero to positive sign and the exchange rate is subsequently increasing up to the critical stabilization concentration c_s (second τ maximum) . According to Tamamushi model (9,10) and Karaoglanov induction crystallization the crystal planes are overcharged and they act as a nucleus for the agglomeration or the accumulation of surfactant at the crystal plane / / solution interface . Such approach lead to the conclusion of submicellar associates with distinguished colloid properties. Then the f_c and c_f values correspond to the critical concentrations for associates formation. If such model is accepted the flocculation process must be considered as the process of mutual coagulation and the process is substantially different as compared with coagulation by inorganic electrolytes. Collected experimental data do confirm our approach. Cationic dye as for example rhodamine 6G (11) are completely drawn from the solution by silver iodide sol particles. Surroundedby dye crystallites, overcharged sol particles have lower exchange rate at higher dye concentrations (Fig. 1. F vs. log c , dashed line) . Increase of the surfactant concentration over c_s and increasing of F values resulting in the decrease of τ value appears to be a consequence of the formation of primary silver iodide particles in the presence of the surfactant associates (4) . The

461

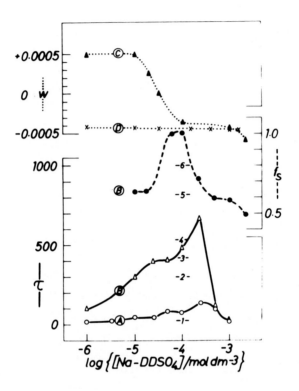

Fig. 2. Turbidity τ *plotted versus log c NaDDSO₄ for the following systems; curve A , sol II-A at t_d = 10 and t_A = 10 minutes; curve B , Sol II-B at t_d = 10 and t_A = 10 minutes; labelled* τ *values for 100 minutes aged sols at 0.0001 M NaDDSO₄ are numbered as follows: -1- sol II-C , -2- sol II-A .at t_d = 10 minutes, -3- sol II-A at t_d = 1000 minutes, -4- sol II-F , -5- sol II-B at t_d = 10 minutes and -6- sol II-B at t_d = 1000 minutes.*

Fraction f_s of sedimented AgI (sol II-B at t_d = = 10 and t_A = 10000 minutes) is expressed as a function of NaDDSO₄ concentration. Electrophoretic mobility w of II-C and II-D sols is plotted versus log c NaDDSO₄ .

role of surfactant associates is demonstrated by the very strong influences of NaDDSO$_4$ (sol II-B) as one anionic surfactant on negative silver iodide particles (12) . For all
described experiments the critical micellar concentrations of
applied surfactants are at markedly higher surfactant concentrations. Concept of interaction of colloid associates with
inorganic sol is examined at the surfactant concentration below c_M (= 0.0015 M) . Electrophoretic data for NaDDSO$_4$
in aqueous media show the same values at broad surfactant
concentration and below c_M (Fig. 2. w vs. log c). Conductivity data for the sol II-C correspond to the sum of the
NaDDSO$_4$ conductivity + conductivity of 0.001 M AgNO$_3$ indicating that there are no chemical interaction at the induced
conditions (0.000001 M to 0.001 M NaDDSO$_4$) . For the
II-C sol (aged for 100 minutes before electrophoresis)
electrophoretic data show that silver ions are adsorbed at
the surfactant associates most probably according to the
Sepulveda palisade adsorption (13) . The palisade interaction between NaDDSO$_4$ and AgNO$_3$ depends on contact time
t_d , so that the processes of formation of primary silver iodide particles must also depend on t_d value. Tyndallometric
value τ measured as a function of contact time t_d and aging time t_A show higher value for the systems formed for
longer contact times (Fig. 2., τ vs. log c , values -3-
and -5-). τ values are also higher in the systems in
which AgNO$_3$ was mixed with NaDDSO$_4$ and lower for the
systems in which NaI was mixed with NaDDSO$_4$ before the
precipitation (curves A and B , -2- and -3- τ data).
Resulting curve for pure AgI + NaI (sol II-F) system lies
between values obtained for the II-A and II-B sols (Fig. 2.,
-4- τ value). Phenomenon is nonconvinient for clasical approach because turbidity of the negative sol depends on the
anionic surfactant concentrations. Concept of the associate
formation with colloid properties below c_M can explain

463

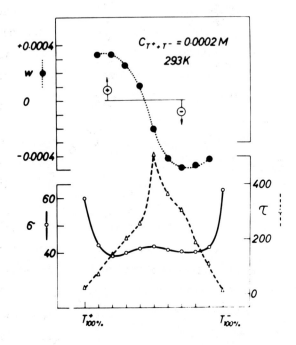

Fig. 3. Systems: DDNO$_3$ (T$^+$) + NaDDSO$_4$ (T$^-$) in aqueous media at 293 K (sol IV) . Turbidity τ , surface tension σ and electrophoretic mobility w plotted versus various T$^+$ + T$^-$ combinations at constant total surfactants concentration c$_{T^+}$ + c$_{T^-}$ = 0.0002 M .

described phenomenon. Radiosedimentation data confirm the supposed role of surfactant associates since the factor f_s of sedimentation follows the turbidity curve (Fig. 2.), and complete sedimentation (f_s = 1) is reached at the same surfactant concentrations as tyndall maximum. Surface tension measurements indicate that the micelles are not formed at the same experimental conditions.

All presented results definitely lead to the conclusion, that the observed phenomena are obtained with associates of distinguished colloid character. Colloid associates of cationic and anionic surfactant must interact similarly to two opposite charged colloid particles if are put into contact. Conductivity data show that there is no measurable chemical interaction in systems with mixed cationic and anionic surfactants (sol IV) . From electrophoretic data (Fig. 3. w versus log c) electrophoretic mobility w confirm the hypothesis of the formation of submicellar colloid associates. The colloid characteristics of both, the cationic and anionic species employed at 0.0002 M total concentration were verified by tyndallometry. The typical τ maximum is reached. The resulting curve of surface tension without maximum corresponding to the chemical reaction and different from average value corresponding to physical mixture for each T^+ / T^- ratio supports our hypothesis of mutual interactions between surfactant and inorganic sol particles. At the same time it confirms the role of each present component as relevant for resulting stability of all system.

IV.   REFERENCES

1. Kertes, A. S. and Gutmann, H., "Surfactants in Organic
    Solvents", The Institute of Chemistry, The Hebrew University, Jerusalem, Israel.
2. Ottewill, R. H. and Watanabe, A., Kolloid Z. 170, 38
    (1960);   170, 132 (1960);   173, 122 (1960); Despotović,
    R., Discussion Faraday Soc. 42, 208 (1966).
3. Despotović, R., La Rivista Italiana delle Sostance Grasse
    52, 377 (1975);   Despotović, R., Filipović-Vinceković,
    N., Mayer, D., Popović, S. and Szvoboda, M., Proc. International Conference on Colloid and Surface Science, Ed.

E. Wolfram, Akadémiai Kiadó Budapest (1975) Vol. I. p. 265.

4. Despotović, R., Despotović, Lj. A., Filipović-Vinceković, Horvat, V. and Mayer, D., Tenside-Detergents 12, 323 (1975).

5. Despotović, R., Filipović-Vinceković, N. and Mayer, D., Croat. Chem. Acta 47, 549 (1975); Despotović, R. and Leskovar, V., Ibid. 48, 1 (1976).

6. Despotović, R., Filipović-Vinceković, N. and Peček, N., Colloid & Polymer Sci. in press.

7. Parfitt, G. D., Ann. Reports A 64, 125 (1967).

8. Ottewill, R. H., Ibid. A 66, 183 (1969).

9. Tamamushi, B. and Tamaki, K., Kolloid Z. 163, 122 (1959).

10. Karaoglanow, Z., Z. analit. Chem. 106, 129; 309 (1937).

11. Despotović, R., Katanec, J. and Filipović, N., Colloid & Polymer Sci. 253, 306 (1975).

12. Despotović, R., Tenside-Detergents 10, 297 (1973).

13. Sepulveda, L., J. Colloid Interface Sci. 46, 372 (1974).

KINETICS AND MECHANISM OF PROTON EXCHANGE OF DODECYLAMMONIUM CHLORIDE MICELLES

E. K. Ralph and M. MacNeil
Memorial University of Newfoundland

ABSTRACT

Exchange rates for the amino protons of dodecylammonium chloride micelles with water protons have been measured in the pH range (1-6) using n.m.r. pulse techniques. The most important feature of the exchange is a reaction between NH_2 and NH_3^+ groups on the micelle surface, Eq. (1).

$$\underset{R}{\overset{H^+}{HNH}} . \underset{H}{\overset{H}{O-H}} . \underset{R}{\overset{H}{NH}} \quad \xrightarrow{\quad k_i \quad} \quad \underset{R}{\overset{H}{HN}} . \underset{R}{\overset{}{H-O}} . \underset{R}{\overset{H^+}{HNH}} \qquad (1)$$

Reaction (1) greatly facilitates proton exchange because the formation of a single NH_2 group can result in the rapid exchange of <u>all</u> of the micelle's remaining proton. Assuming that reaction (1) may be represented as a random walk on an array of hexagonally packed nitrogen centers, k_i is calculated as 5.4×10^8 sec^{-1} to 8.5×10^8 sec^{-1}, depending on the total surfactant concentration.

Rate constants for micellar acid dissociation and reaction with OH^- are discussed in relation to theoretical predictions for diffusion controlled reactions.

A STUDY OF MICELLES AND POLYELECTROLYTES AS
INTERACTING SPECIES WITH BISULFITE

Oh-Kil Kim and James R. Griffith
Naval Research Laboratory
Washington, D. C. 20375

I. ABSTRACT

The presence of an interaction of cationic micelles with sodium bisulfite (SB) as catalyst was previously explored in acrylamide polymerization. Further evidence for the interaction is discussed in this report based on observations of the UV absorption spectra and the rate of autoxidation of bisulfite in the presence of cationic micelles. The absorbance of bisulfite solution at 259 nm was greatly increased and a new absorption at 314 nm appeared in the presence of cetyltrimethyl ammonium bromide (CTAB), while negligible or no increase in the absorbance was observed at the respective wavelengths with the system of non-micellar analog. Such changes in the absorption spectra in SB/CTAB system are interpreted as being due to the equilibrium change of bisulfite to isomeric metabisulfite resulting from a complexation favored by high local concentration of bisulfite along the micellar surface.

Another evidence for the complexation is presented by the incomparably fast rate of oxidation of bisulfite in the presence of CTAB. The stoichiometry of the bisulfite complex was determined to be 1:1 and 2:1 with respect to the molar ratio of SB/CTAB from the oxidation rate plot. Non-micellar polycations showed somewhat a similar effect on the bisulfite oxidation. At any rate, it is certain that bisulfite becomes far more susceptible to oxidation by forming a complex.

THE EFFECT OF ALCOHOLS ON THE LAMELLAR TO SPHERICAL
STRUCTURAL TRANSITION IN AQUEOUS AEROSOL OT SYSTEM

W. C. Hsieh and D. O. Shah
University of Florida

ABSTRACT

The objective of the proposed research was to investigate the effect of various alcohols on the structural transition from lamellar to spherical micelles in the aerosol OT
System. The birefringent solutions of Aerosol OT in 1% NaCl
were titrated with various alcohols. Upon addition of a
specific amount of alcohol, the birefringent solution of
Aerosol OT became isotropic and clear. Upon further addition
of alcohol, it became turbid and subsequently phase separation took place. These transitions were studied also by the
band-width measurements of the high resolution NMR spectra
of the Aerosol OT System. In general, it was observed that
the higher the solubility of alcohol in water, the less effective they were in converting the lamellar structure to
spherical micelles. The relative ability of spherical
micelles of Aerosol OT to solubilize various alcohols was
determined from the second transition (isotropic to turbid).
The free energy of adsorption of various alcohols at the
Aerosol OT lammellae/water interface was calculated from the
titration data.

ELECTROLYTE EFFECTS ON NONIONIC SURFACTANTS

Dr. Steven A. Shaya
The Procter and Gamble Company

Electrolyte effects for nonionic surfactants are examined for critical micelle concentration, surface tension, partitioning, and cloud point phenomena. The effect of salts is shown to be consistent with salting theories in which a salting coefficient $k = (\partial \log \text{ surfactant activity}/\partial c_s)_{T,C}$ describes the result of electrolyte solvent interactions. Here C_s is the electrolyte concentration and c is the concentration of nonionic surfactant. The expected linearity of log cmc, surface tension, log partition coefficient, and

$$(\partial \log \text{cmc}/\partial C_s) = -k_{cmc} \qquad (1)$$

$$(\partial \gamma /\partial C_s) = -2.3\,\Gamma\,RTk \qquad (2)$$

$$(\partial \log P/\partial C_s) = k_p \qquad (3)$$

$$\lim c \to 0 \; (\partial T_{cp}/\partial C_s) = -RT_{cp}k_{cp}/\Delta S \qquad (4)$$

temperature of the lower consolute boundary, were found for $C_{12}H_{25}(OCH_2CH_2)_6OH$ with NaCl, Na_2SO_4, and $Na_3C_6H_5O_7$. The magnitude of k depends on the balance of salt effects on all species in equilibrium and so differs for each process. The effect of electrolyte on the activity of micellar surfactant can be determined from partitioning or surface tension measurements just above the cmc, and k_{cmc} is found to be $k_{monomer} -k_{micelles}$ as predicted by Mukerjee. k_{cmc} is consistent with salting out of the $C_{12}H_{25}$ hydrophobe.

The effect of cadmium or magnesium nitrate is anomalous in that it salts out the hydrophobic tail but salts in the ethoxylate chain; tetraethylammonium bromide presents the reverse anomaly, salting in the carbon chain while salting out the hydrophilic group. The cloud temperature dependence with salt is found to correlate with the effect on the polyethoxylate group as is predicted using a model in which salts influence the equilibria between surfactant in the hemispherical ends or cylindrical portion of cylindrical micelles. Micellar molecular weight is predicted to be most sensitive to electrolyte near the consolute boundary, and an exponential dependence on C_s is expected.

473

Solubilization of Vitamin K and Folic Acid by Mice=
lle Forming Surfactants - Felícita García, Bernice
Irizarry and Gabriel A. Infante, Catholic University of Puerto Rico, Ponce, Puerto Rico, 00731.

ABSTRACT:
 2-methyl-3hydroxy-1, 4 naphthoquinone (vitamin K),
 2-methyl-1, 4-naphthoquinone (vitamin K_3), and folic acid are solubilized by micellar cationic he-
 xadecyltrimethyl ammonium bromide (CTAB), anionic
 sodium lauryl sulfate (NaLS) and nonionic polyoxyethylene (15) nonylphenol (Igepal). The solubilization by the different site and the nature
 of this environment depends upon the solubilizate
 and the surfactant. Some differences in solubilization by the different surfactants suggest some dependence on the character of the micelles.
 Absorption spectroscopy and nuclear magnetic resonance spectroscopy have been utilized to determine the solubilities and the position of the
 subtrates within the micelles.
 Cationic CTAB solubilized vitamins K as high as
 250 times more than in pure water, while anionic
 NaLS has the lesser effect on the solubilities of
 these vitamins.
 It is hoped that the results of this work will
 lead to more chemical research studies on these
 biological important compounds.

Support for this work from Research Corporation
is gratefully acknowledged.

How Petroleum Oils Become Solubilized in Water -
Gloria Jové, Julián López, Juan Negrón and Gabriel
A. Infante, Catholic University of Puerto Rico,
Ponce, Puerto Rico 00731.

ABSTRACT:
One of the most common pollutants in water are
oils and grease. A relatively high concentration of grease and oils in water (30 ppm) decreases the dissolved oxygen and increases the
biochemical oxygen demand. The development of
a strong petrochemical industry in Puerto Rico
comes together with the possibilities of oil
spills in our coastal areas with detrimental consequences to our environment. In the present,
work, we studied different micelle forming surfactants that could recover the petroleum and
its products from sea and fresh water and others
that could solubilize petroleum and its products
in sea and fresh water.
Surprisingly, one of the micelles studied in
concentrations as high as 0.1M solubilized completely a relative high ammount of petroleum
(5,000 ppm) and its derivates in water. The significance and implications that this discovery
could have in the energy and pollution studies
will be discussed.

Support for this work from Research Corporation
is gratefully acknowledged.

PHASE BEHAVIOR OF THE $C_{10}E_4$-WATER AND $C_{10}E_4$-HEXADECANE-WATER SYSTEMS

Donn N. Rubingh
The Procter & Gamble Company

The phase behavior of pure tetraoxyethylene glycol monodecyl ether ($C_{10}E_4$) with water and water plus hexadecane has been determined between 0 and 90°C. The $C_{10}E_4$-H_2O system displays what has been termed a double cloud point within certain concentration limits. The phase diagram shows that this phenomenon is a consequence of discontinuities in the tie line length which results in sudden changes in volume and refractive index of the disperse phase droplets. We show that a consistent interpretation of our data requires the existence of four 3 phase lines with zero variance.

The ternary diagram gives both the temperature and concentration dependence of solubilization. The oil solubilization region L_1 has a critical temperature above which it does not exist. The solubilization capacity remains reasonably constant up to concentrations approaching those where neat phase begins to separate. An unusual feature of the $C_{10}E_4$-hexadecane-H_2O system is the existence of two isolated hexadecane solubilization regions. These regions are contiguous with the corresponding one-phase regions in the two component $C_{10}E_4$-H_2O system.

The L_1 region and the corresponding water solubilization region L_2 merge at 37°C. If one associates a normal micellar structure with the L_1 region and an inverted structure with L_2 then 37°C is the temperature at which micellar inversion takes place. Oil solubilization increases as one approaches this temperature from below and water solubilization increases as one approaches this temperature from above.

ELECTRIC FIELD JUMP INVESTIGATION

OF THE CETYLPYRIDINIUM IODIDE MICELLE SYSTEM

Hans-Heinrich Grünhagen
Max-Planck-Institut für Biophysikalische Chemie
Göttingen

Chemical and physical relaxation effects are observed when aqueous solutions of the amphiphilic electrolyte cetyl-pyridinium iodide (CPJ) above the critical micelle concentration (CMC) are subjected to electric fields of the order of 10^4 to 10^5 V/cm. For this investigation a cable discharge pulse generator for single shot square wave pulses has been developed with a high voltage range up to 50 kV and a pulse width from 1 μsec to about 1 msec (1). The field effects can be monitored by transmission optics (absorption, electrodichroism, electrobirefringence) and by emission optics (fluorescence, light scattering) (2). Above but close to the CMC (4.2×10^{-4} M at 40° C) the charge transfer band of iodide bound to micellar pyridinium reveals three chemical relaxation processes of the CPJ-micelle system, their concentration dependent relaxation times being of the order of 50 nsec, 5 μsec, and 50 μsec (3). The relaxations can be attributed to counterion dissociation (fast effect) and to monomer-micelle equilibration (slow effect). Only at higher concentrations of CPJ electrodichroitic effects are observed (4). This is in agreement with the existence of spherical micelles at concentrations close to the CMC and of anisometric aggregates at higher concentrations.

REFERENCES

1. Grünhagen, H.H., Messtechnik, 19 (1974).
2. Grünhagen, H.H., Biophysik 10, 347 (1973).
3. Grünhagen, H.H., J. Colloid Interface Sci. 53, 282 (1975).
4. Grünhagen, H.H., Chem. Phys. Lipids 14, 201 (1975).

A 6
B 7
C 8
D 9
E 0
F 1
G 2
H 3
I 4
J 5